Lecture Notes in Networks and Systems

Volume 175

The series "Lecture Notes in Networks and Systems" publishes the latest developments in Networks and Systems—quickly, informally and with high quality. Original research reported in proceedings and post-proceedings represents the core of LNNS.

Volumes published in LNNS embrace all aspects and subfields of, as well as new challenges in, Networks and Systems.

The series contains proceedings and edited volumes in systems and networks, spanning the areas of Cyber-Physical Systems, Autonomous Systems, Sensor Networks, Control Systems, Energy Systems, Automotive Systems, Biological Systems, Vehicular Networking and Connected Vehicles, Aerospace Systems, Automation, Manufacturing, Smart Grids, Nonlinear Systems, Power Systems, Robotics, Social Systems, Economic Systems and other. Of particular value to both the contributors and the readership are the short publication timeframe and the world-wide distribution and exposure which enable both a wide and rapid dissemination of research output.

The series covers the theory, applications, and perspectives on the state of the art and future developments relevant to systems and networks, decision making, control, complex processes and related areas, as embedded in the fields of interdisciplinary and applied sciences, engineering, computer science, physics, economics, social, and life sciences, as well as the paradigms and methodologies behind them.

Indexed by SCOPUS, INSPEC, WTI Frankfurt eG, zbMATH, SCImago.

All books published in the series are submitted for consideration in Web of Science.

More information about this series at http://www.springer.com/series/15179

Meenakshi Tripathi · Sushant Upadhyaya
Editors

Conference Proceedings of ICDLAIR2019

 Springer

Editors
Meenakshi Tripathi
Malaviya National Institute
of Technology (MNIT)
Jaipur, Rajasthan, India

Sushant Upadhyaya
Malaviya National Institute
of Technology (MNIT)
Jaipur, Rajasthan, India

ISSN 2367-3370 ISSN 2367-3389 (electronic)
Lecture Notes in Networks and Systems
ISBN 978-3-030-67186-0 ISBN 978-3-030-67187-7 (eBook)
https://doi.org/10.1007/978-3-030-67187-7

This Springer imprint is published by the registered company Springer Nature Switzerland AG
The registered company address is: Gewerbestrasse 11, 6330 Cham, Switzerland

Preface

The conference (ICDLAIR) provided a platform for researchers and professionals to share their research and reports of new technologies and applications in DL, artificial intelligence, and robotics like biometric recognition systems, medical diagnosis, industries, telecommunications, AI Petri nets model-based diagnosis, gaming, stock trading, intelligent aerospace systems, robot control, law, remote sensing and scientific discovery agents and multiagent systems; artificial vision and robotics; and natural language and Web intelligence.

The ICDLAIR aimed to bridge the gap between these non-coherent disciplines of knowledge and fosters unified development in next generation computational models for machine intelligence.

January 2020

Meenakshi Tripathi
Sushant Upadhyaya

Organization

Program Chairs

Holger Schäpe   Springer Nature

Meenakshi Tripathi   Malaviya National Institute of Technology (MNIT), Jaipur, India

Sushant Upadhyaya   Malaviya National Institute of Technology (MNIT), Jaipur, India

Contents

Indoor Navigation Techniques: A Comparative Analysis Between Radio Frequency and Computer Vision-Based Technologies

Steve Stillini and Lalit Garg(⊠)

Department of Computer Information Systems, Faculty of Information and Communication
Technology, University of Malta, Msida MSD2080, Malta
lalit.garg@um.edu.mt

1 Background

It has always been a human desire to see how far up into the sky their creation can reach. It is like humanity to push further than ever and create things better than anyone before they only dreamt of. This complexity of achievements achieved by human beings, however, brings forward other challenges that need to be faced by present scientists.

When human beings started exploring the enormity of the world, the need to navigate around it arose quickly. From early times we see sailors trying to navigate and try to find ways that can lead them to their destination. The early Pacific Polynesians were the first to use the motion of the stars, weather, and the position of certain wildlife species or the size of waves to find the path from one island to another. The use of celestial navigation brought forward a new era were sailors tracked the movement of the stars to find paths to where they need to go. Further, in time the first useful invention to help was the magnetic compass. Evolving in time, the inventions of instruments such as the chronometer, the sextant, and several instruments that map the world around us came by.

The most recent break-in navigation brought forward by the Global Positioning System is still being used. It is very beneficial for outdoor navigation where a person needs to travel from point A to B while keeping track of what is his current location relative to the Earth is.

Nowadays, as technology is advancing, architects are designing spaces that are far from buildings but looking more like indoor cities—ranging from the Boeing Everett Factory as the largest usable space by volume with a floor space of 4.3 million square feet, to the Burj Khalifa, putting its name in the Guinness Book of Records as the current tallest skyscraper in the world, with a height of 828.8 m. When considering these large buildings around us nowadays and larger structures being built as we speak, the need for indoor navigation has never been felt this high. Despite having the GPS, this cannot be used indoors and we have to come up with a new way of finding our way indoors.

Surely enough, the use of Augmented Reality has risen these days exponentially. The number of ways this technology can be applied is infinite. Indoor navigation is, therefore, a handy application of AR.

Using signs to guide people indoors is currently the way people can find their way in large indoor spaces. It was quite beneficial up to some years ago before augmented reality

© The Author(s), under exclusive license to Springer Nature Switzerland AG 2021
M. Tripathi and S. Upadhyaya (Eds.): ICDLAIR 2019, LNNS 175, pp. 1–10, 2021.
https://doi.org/10.1007/978-3-030-67187-7_1

came by. The more sophisticated approach that can be used nowadays gives directions to the person according to where his current location is. It is, therefore, more reliable as the person does not have to rely on a sign which can be found, maybe, every 200 m or so, but one can get live feedback regarding his location, and the direction the user needs to follow to reach the destination, in instant time. Apart from that signs could be written in a specific language that is not understood by the whole population. In Fig. 1 below, one can see a traditional way of navigating through a building using signs. Figure 2 in the other hand, shows how the user experience can be improved by using augmented reality to overlay the directions of the user.

Fig. 1. A traditional way of navigating through a building using signs.

Fig. 2. Augmented reality to overlay the directions of the user.

2 Technology

In order to make Indoor navigation work, one might decide what approach he wishes to follow. One can either use RF signals from BLE beacons in case of using Bluetooth or access points in case of using Wi-Fi. In this case, the hardware needs to be bought. BLE beacons do not cost much money when purchased in large amounts. In case of using access points, most environments already have routers installed so there is no need to invest in any other hardware. This approach, however, needs a lot of infrastructural updates to the environment because BLEs need to be in the line of sight.

On the other hand, if the visual-based approach (Fig. 2) will be taken tools such as the AR toolkit, we are needed to overlay the directions on the map of the environment. The tools are open source and can be used by everyone. Apart from that one might need a camera and a powerful laptop/computer.

3 Features

What an app of this type offers to the user is a sense of understanding relative to where he is in a particular building. Buildings nowadays are massive, and sometimes people feel lost inside them, especially if they have never been in them before. Apart from telling the user what his current location is, an app like this can also offer the user some direction to his destination. By using augmented reality, the user does not have to worry about his orientation as the application takes care of that, and quickly leads the user to the room, object, or place he wants to go to. This app can be used by shopping malls to map clothes stores, cafeterias, restaurants, bars or even public bathrooms. Apart from that, another feature can be implemented with the use of augmented reality. In places like museums, etc. augmented reality can be implemented to overlay specific information for individual artifacts. By scanning the object or barcode relative to the object, the information would be overlaid on the real artifact. Also, this can be used in cloth shops. A user can scan the barcode of a shirt and see all available colors from that shirt for example. A recent implementation for this was in a Ferrari showroom where the user could scan a car and overlay different features such as different colors, suspension, rims, etc.

4 Advantages and Applications

Inevitably, a system of this sort can save people much time. The importance of time nowadays is not denied by anyone. Imagine going to a huge building, and you do not have any idea where a shop you are looking for is. A system like this can direct you to that place instantly. Imagine you are a university student and are trying to find a class in a particular faculty. If a system like this is implemented in University students would not feel lost especially during the first few days. Imagine implementing a system like this at the hospital. People would not lose hours looking for a particular ward especially if they are older adults. Imagine implementing a system like this in an airport. People do not have to worry about finding the departure gate anymore or having to worry that they are going to miss the next flight. What if we apply a system like this in a history museum and overlay the environment with an interactive story timeline. Imagine how more information that would be compared to the traditional approach used in museums. The amount of time a user can save by using indoor navigation apps is unlimited.

5 Example

A quite good example of this app is by showing how it may look. Figure 3 shows an indoor navigational system in an airport using augmented reality. In this case, a visual base approach is being taken where the user scans the environment around him. The sequential images from the live feed are then compared with those stored in a database which also have the location assigned to them. When an image from the user

Fig. 3. An indoor navigational system in an airport using augmented reality.

matches one in the database, the app would know the location of the user. It then overlaid navigational information on the real environment to lead the user to the destination.

6   Limitations

A system that is based on image recognition to give the user direction is highly dependent on the way one positions the device. Apart from that, if something is obscuring the pattern that needs to be recognized that the app will not work correctly. For this reason, it cannot be used in case of fire emergency evacuations as smoke might reduce the clarity of the environment around the user and will not be able to recognize a pattern.

7   Literature Review

The first thing that needs to be considered is how to localize the device used indoors without the use of GPS. To start with, one needs to find a way to map an indoor area, so when opening the application, it needs to figure out the location relative to that map.

7.1   Radio Frequency Based Indoor Navigation

According to [1, 4], Bluetooth beacons are one method of how this application can be implemented. Bluetooth is a wireless technology that allows smartphones, smartwatches, tablets, game controllers, mice, etc. to communicate with one another within a small range. Despite Bluetooth works on a full band, it is regulated by the 'Bluetooth Special Interest Group' (SIG), and IEEE has standardized it. Bluetooth utilizes seventy-nine channels to send the data over the network starting with the primary channel at a frequency of 2404 MHz up to 2480 MHz increasing with 1 MHz on each iteration. In order not to interfere with other RF signals, Frequency Hopping [13] is used, FH is a technique where information is transmitted on either one of the channels available for a small amount of time and then sent to another channel after a while in case interference is used [1].

To transmit data, Bluetooth devices should first establish a connection. A device is able to connecting with seven devices simultaneously. It is possible by connecting using a model called 'masterslave', where the device that starts the connection would be considered as the master of the other devices. When a connection is established between the devices, this allows them to receive and transmit data between them [1].

Bluetooth is designed as technology that utilise low power using a battery. It can be used over a relatively small distance which are dependent on the output power, as well as other interferences such as reflection that can be caused by obstacles. Bluetooth is known for being used as an indoor navigation purpose [14] utilizing less power consumption than Wi-Fi and serve a broader range of devices. Even though the primary purpose of Bluetooth is to transfer data between two devices or more when considering indoor navigation, a connection might not always be needed. It is because there is no need for data to be exchanged between devices since they are only serving as a reference that broadcasts information regarding their position, which is enough for this purpose. Over the past decade, Bluetooth sensors are being used to only broadcast data and therefore without the need to establish a connection. This technique has increased in popularity especially in sports and healthcare devices. An example of these is Fitbit, which is a wristband used to monitor heart rate amongst other features [1].

Apart from that, the boost of what is known as the Internet of Things (IoT) gave rise to the use of Bluetooth, encouraging the use of small sensors implemented in everyday objects, allowing them to communicate together. It is today an essential part of the World Wide Web [15]. The need for low power consumption sensors started to emerge. In reply to this need a new Bluetooth standard called Bluetooth Low Energy (BLE) was developed and introduced to the market.

BLE is a subsystem of the traditional Bluetooth system. It is capable of broadcasting data with the use of minimal power consumption. It makes it perfect for applications running on small batteries which need to run for a prolonged period. It is, therefore, more suitable to use BLE instead of the traditional Bluetooth beacons for the aim of indoor navigation. Beacons are a tiny piece of hardware that can broadcast packets of data quickly. The packets received contains information regarding the beacon, as well as telemetry reading usually utilized in distances calculations. Devices such as tablets and mobiles will then be used to gather BLE signals and decipher the data broadcasted, which will then be used in the process of indoor navigation.

One of the purposes BLE was invented was because no connection is needed to broadcast data. They are commonly used as a one-way communication where BLEs serves as the transmitter and receivers pick the data and are not able to establish connections with the transmitter.

Beacons have increased in popularity in the last few years as well as the indoor navigation systems that rely on them. Apart from the fact that these sensors are inexpensive, they are straightforward to configure in the first place. Not only, but also, Google and Apple have created dedicated beacon protocols which make the management and communication with these beacons very easy. Taking into considerations all the features BLE can offer, they are the best for the job when one is considering Bluetooth for indoor navigation.

RF signals broadcasted by a wireless device are used interchangeably according to the localization method being used. In the past few years, some of these methods have been improved [16], with the most popular being Triangulation and Fingerprinting [2, 3]. RF signals can be calculated to be used as a reference. The strength of the signal can be measured using a signal parameter known as the Received Signal Strength Indicator (RSSI). It is the calculation of the power on an incoming radio signal expressed in dBm.

Fingerprinting utilizes the RSSI measurements from a cluster of devices to generate a fingerprint for a particular spot in a building. It can be achieved by saving the measurements, and the address of their corresponding device in a database. When different Fingerprints of several locations are generated, scans are conducted, and a runtime fingerprint is created every time. The final scan is then compared to every stored fingerprint to get the most similar match representing the user's location.

Fingerprinting is very accurate and easy to develop. In the case of Wi-Fi [17], existing AP in buildings is used. A disadvantage that this method has is that it cannot be assured that the APs used to create the Fingerprint would be available. A workaround to this problem is to fill missing RSSI values with arbitrary measurements although this could lead to the possibility of losing the accuracy significantly.

Fingerprinting can also be implemented using BLE beacons [18]. They need to be positioned correctly to get as much reading as necessary. However, a significant amount of these beacons is required in order to cover a vast area, which might be very costly in the beginning.

Both for Bluetooth and Wi-Fi signals, fingerprinting has its disadvantages. To collect fingerprints from an environment around a building is very time-consuming. Also, to take maximum potential from a fingerprinting system the environment which is going to implement navigation need to be studied thoroughly in order to place the beacons in the best place possible. On the other hand, when using WiFi, it may be time-consuming to collect the sample of the fingerprints.

Another technique known as triangulation can also be used (Fig. 4). In this technique, RSSI is also used but differently from fingerprinting. For this technique to work three or more devices must be positioned in shape so they cover a specific area. After distances to the reference devices are calculated roughly by the receiving device and an interception point between these is found. This interception is usually the place where the person is at [19].

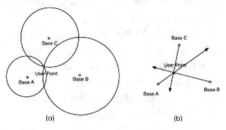

Fig. 4. Triangulation.

7.2 Computer Vision-Based Indoor Navigation

Another option is using computer vision by way of augmented reality SDK to map the environment. According to [5], AR supports precise localization, but the performance is highly dependent on the way the device is held by the user, especially when they get fatigued after a long time of holding it. Despite all, this computer vision-based technology tends to be more accurate [5] than beacon sensors which cannot find an approximation of the users' orientation.

A common researched approach of vision-based indoor navigation uses image recognition of a place through a live feed of a device's camera in the shape of ego-motion so the pictures would be compared with a sequential database of ortho-graphic pictures of that same environment, collected before. The collected pictures are assigned with their location and the device's inertial sensors aid in delivering orientation. This approach can be used to build a working Augmented Real-

Fig. 5. An augmented reality (AR) marker at an airport.

ity based direction instruction (Fig. 5). Also, this offers user localization. However, this technique faces an issue. This approach needs a lot of computational power and resources since a huge database map of pictures is used and that can lead to delays while comparing the images [6].

An alternative computer vision-based method that can be used and is widely studied [6, 8–12] uses AR markers that optically track the direction that the user needs to go and overlay it on the markers to perform the navigational process. There can be different types of Augmented Reality trackers available for a large variety of markers. Physical markers such as QR Codes, ID markers and Bar Codes use fiducial tracking [20] to detect patterns. These markers can be recognized because of their high contrast and geometric shape. Other types of physical markers, such as image markers, have proper visual content to be distinctly recognized amongst other patterns. Physical markers are strategically placed to cover the indoor space entirely.

In some cases, unique shapes within the building such as signs, furniture, etc. could be utilized as image markers. The problem with many physical markers is that markers need to be strategically placed inside the building, so they do not get hidden while navigating. It can internally impact the interior of the building. For picture markers to work enough picture markers need to be available, as few buildings have enough different markers that could be used to navigate. Also, there is always the possibility of not locating a trackable while navigating which might be a risk [7].

However, a 3-Dimensional tracking method exists, which is an enhanced form of optical tracking. This method is not widely used for navigating purposes. 3D maps are built by scanning a potential area of an environment. When enough visual knowledge of a trackable is gathered, then it could be utilized for Augmented Reality information overlay. This approach is not so computationally exhaustive for a device being used and the internal physical structure of the building does not get interrupted.

Locating different routes in a building is not as challenging as specifically finding a sign to be interpreted as an image marker since the whole building is possibly being considered as a trackable marker. Also, these types of markers are easily detected from a long distance. Information regarding direction is overlaid on these trackable markers with the use of Augmented Reality technology, which can produce exact results during the navigation process.

Even though the study of technology is significant for indoor navigation one cannot stop there. Human factors can give an insight into how a system of this sort should be built. According to [21], there exist two groups of Indoor navigation users. Convenience users are those people who are everyday navigators who prioritize ease of use and affordability over accuracy and reliability. On the other hand, High-End users like the blind/partially sighted, surveyors and emergency response are willing to make use of specialized tools and equipment which might be expensive in order to get a trustworthy and accurate position solution. It is therefore vital that before building an indoor navigation system, one needs to see who is going to use the system.

7.2.1 AR Toolkit

A convenient tool to perform Augmented Reality is the AR toolkit (Fig. 6). It is used to display a virtual object on the detected marker. AR toolkit has been utilized in this field to show directional information in a 3D format to deepen a user's ability to recognize the route for a destination [22].

AR toolkit is free to use the library for programmers to use while developing AR applications. It is generally used to overlay a predesigned 3D object on the detected marker. An excellent feature implemented in AR toolkit is that it is capable of accurately track in real-time the point of view of the user by using computer vision techniques to find the camera location and orientation relative to the marker's orientation so that the virtual object that rendered above the marker always seems aligned with the marker. The rendering from this tool provides smooth animation of the 3D object [22].

Firstly, a frame is selected from the video taken by the camera. This picture will be transformed into a black and white image according to the thresholding value, which is the technique of converting an image to binary. Following that, the tool will search for square shaped objects utilizing an image labeling approach where those components which are connected, and the size which nearly satisfies a fiduciary marker is found out. The process continues by recognizing contours' outlines that can fit the four-line segment as square shapes. Corners or points from the contour referenced previously will be then recognized. This square segment can have any orientation. Therefore, it needs to be normalized to the original orientation in the training phase of the marker. The interior shape of the normalized square regions will then be compared with the already trained markers, stored as binary data. If there is a match, the confidence value, which is the percentage of matching, will be computed [22].

When a marker is spotted, the orientation and position of that marker will be calculated relative to the camera being used. When the point of view is calculated, OpenGL API is utilized to overlay the Virtual Reality Mod-

ARTOOLKIT

Fig. 6. AR toolkit logo.

elling Language model on the marker according to the calculated point of view. The output shows the virtually created object on-screen perfectly aligned with real-world marker while repeating the whole process for real-time processing.

7.2.2   Accuracy of AR Toolkit

How the tool (AR toolkit) performs is of vital importance in ascertaining the performance of the whole system for indoor navigation. All navigational guidance and 3D arrows highly depend on how the tool detects the surrounding environment. Compared to the detection module of the AR toolkit, the result modules in the program tend to be more accurate. The accuracy of the AR toolkit library can be tested by checking its ability to discover a marker and tracking precision upon discovery of the marker. The percentage accuracy of the discovered marker with an already trained marker will be worked out by the AR toolkit upon detection of the same marker.

The precision of the AR toolkit to recognize a spot in a building, better known as a marker, highly depends on many conditions. Lighting conditions are one of them. Different variations of lighting can lead to AR toolkit not detecting the marker appropriately. If the lighting condition is close to the one in training, then the threshold value is ideal to recognise the marker easily. Nonetheless, the threshold value can be adjusted to fit several lighting conditions in the indoor space.

An experiment done in [9] is conducted using conditions where a 55 mm by 55-millimeter marker is positioned at x- and z- directions while alternating y-direction from the camera angle with the marker. The hardware used as part of the configuration of this experiment was a webcam with a resolution of 640px by 490px which had a frame rate of 15.

The output highlights the error of the AR toolkit to endlessly monitor a spot is small for the approximate range from 20 cm up to 70 cm from the marker to the camera. It is viewed as acceptable as the presumed working distance from the user to the marker is fixed within the range mentioned previously.

8 Future Perspectives

Without any doubt, new technology brings forward new applications, many improvements around us, and new challenges to those that embrace it. However, this challenge cannot be faced without having a solid base knowledge on how to implement the core aspects of these new technologies. I believe that the importance of indoor navigation is being taken very seriously and improvements are being made. The technology is there, while it is always being improved. The way forward now is how to augment it in our everyday lives to make it more realistic and usable. While keep improving the technology, we need to get a solid understanding of what the people want and keep them in the center of these new technologies. Applications like these are solely built to be used by a human being. It is, therefore imperative for having superior technology while also having the correct methods to implement this technology in everyday life. Whether it is using a glass to augment what is real with what is fake or using a mobile to give directions to a user is still not sure. Studies are still being conducted and many more need to be done until we come up with the solution. Maybe we can find even a better way to integrate indoor navigation in the everyday life of the person. Inevitably, new devices such as smart watches and the new whole era of IoT devices make this subject even more interesting to explore because there is no limit to where this technology can be applied.

In an application like this where accuracy and precision are paramount in the future, one can improve it by utilizing both augmented reality and RF signals to make it more reliable and accurate. This way if a pattern would not be recognizable by image recognition, there would still be another option as a fall back by using radio-based frequency navigation. In this way, we would be combining two different technologies which are both excellent on their own but can be improved by having the benefits of both joined together.

References

1. Herrera-Vargas, M.: Indoor navigation using Bluetooth low energy (BLE) beacons. Master's thesis, Turku University of Applied Sciences, June 2014
2. He, S., Chan, S.-H.G.: Wi-Fi fingerprint-based indoor positioning: Recent advances and comparisons. IEEE Commun. Surv. Tuts. **18**(1), 466–490 (2016)
3. Luo, J., Yin, X., Zheng, Y., Wang, C.: Secure indoor localization based on extracting trusted fingerprint. Sensors **18**, 469 (2018)
4. Subedi, S., Kwon, G.-R. Shin, S., Hwang, S.-S., Pyun, J.-Y.: Beacon based indoor positioning system using weighted centroid localization approach. Department of Information and Communication engineering, Chosun University, August 2016
5. Rehman, U., Cao, S.: Augmented reality-based indoor navigation using google glass as a wearable head-mounted display. In: Proceedings of IEEE International Conference on Systems, Man, and Cybernetics, pp. 1452–1457 (2015)
6. Kasprzak, S., Komninos, A., Barrie, P.: Feature-based indoor navigation using augmented reality. In: 2013 9th International Conference on Intelligent Environments (IE), pp. 100–107. ACM (2013
7. Koch, C., Neges, M., König, M., Abramovici, M.: Natural markers for augmented reality-based indoor navigation and facility maintenance. Autom. Constr. **48**, 18–30 (2014)

8. Kalkusch, M., Lidy, T., Knapp, M., Reitmayr, G., Kaufmann, H., Schmalstieg, D.: Structured visual markers for indoor pathfinding. In: The First IEEE International Workshop on Augmented Reality Toolkit, p. 8–pp (2002)

9. Kim, J., Jun, H.: Vision-based location positioning using augmented reality for indoor navigation. IEEE Trans. Consum. Electron. **54**(3), 954–962 (2008)

10. Huey, L.C., Sebastian, P., Drieberg, M.: Augmented reality based indoor positioning navigation tool. In: 2011 IEEE Conference on Open Systems (ICOS), pp. 256–260 (2011)

11. Chawathe, S.S.: Marker-based localizing for indoor navigation. In: Intelligent Transportation Systems Conference, ITSC 2007, pp. 885–890. IEEE (2007)

12. Delail, B.A., Weruaga, L., Zemerly, M.J.: CAViAR: context aware visual indoor augmented reality for a university campus, pp. 286–290 (2012)

13. Poole, I.: Bluetooth radio interface, modulation, & channels. radio-electronics.com. https://www.radio-electronics.com/info/wireless/bluetooth/radio-interfacemodulation.php. Accessed 12 Dec 2018

14. Yapeng, W., Xu, Y., Yutian, Z.: Bluetooth positioning using RSSI and triangulation methods. In: IEEE Consumer Communications and Networking Conf. (CCNC), pp. 837–842, 11–14 January 2013

15. Zanella, A., Bui, N., Castellani, A., Vangelista, L., Zorzi, M.: Internet of things for smart cities. IEEE Internet Things J. **1**(1), 22–32 (2014)

16. Sun, G., Chen, J., Guo, W., Liu, K.J.R.: Signal processing techniques in network-aided positioning: a survey of state-of-the-art positioning designs. IEEE Signal Process. Mag. **22**(4), 12–23 (2005)

17. Farshad, A., Li, J., Marina, M.K.: A Microscopic look at WiFi fingerprinting for indoor mobile phone localization in diverse environments. In: 2013 International Conference on Indoor Positioning and Indoor Navigation, pp. 1–10 (2013)

18. Faragher, R., Harle, R.: Location fingerprinting with bluetooth low energy beacons. IEEE J. Sel. Areas Commun. **33**(11), 2418–2428 (2015)

19. Wang, Y., Yang, X., Zhao, Y., Liu, Y., Cuthbert, L.: Bluetooth positioning using RSSI and triangulation methods. In: Consumer Communications and Networking Conference, pp. 837–842. Available from: IEEE Xplore Digital Library (2013)

20. Lakhani, M.A.: Indoor navigation based on fiducial markers of opportunity (2013)

21. Brown, M., Pinchin, J.: Exploring human factors in indoor navigation. In: Proceedings of the European Navigation Conference (2013)

22. Huey, L.C., Sebastian, P., Drieberg, M.: Augmented reality based indoor positioning navigation tool. In: Proceedings of IEEE Conference on Open Systems, pp. 256–260 (2011)

Sentiment Analysis of Hinglish Text and Sarcasm Detection

Abhishek Gupta[1], Abinash Mishra[2], and U. Srinivasulu Reddy[2](✉) [iD]

[1] Department of Computer Science and Engineering, Indian Institute of Information Technology, Tiruchirappalli, National Institute of Technology, Tiruchirappalli Campus, Tiruchirappalli, India
abhi.mittal021@gmail.com
[2] Machine Learning and Data Analytics Lab, Department of Computer Applications, National Institute of Technology, Tiruchirappalli, Tiruchirappalli 620015, India
{405117002,usreddy}@nitt.edu

Abstract. Today the term "Sentiment Analysis" is no newer to the world. It falls under the umbrella of Natural Language Processing which is a very interesting and creative field of artificial intelligence. One important aspect which needs to take in consideration before going for Sentiment Analysis is the kind of the language of the data which is supposed to be processed. People in urban areas of the northern part of India used to communicate in the mixed language of Hindi- English which is commonly termed as "Hinglish". While doing sentiment analysis one needs resources like Dictionary containing polarity for words, part of speech tagger for both of the languages. A lot of resources were developed for the English language, but this does not hold true for Hinglish. The aim of this research is not only to carry out sentiment analysis and sarcasm detection but also to contribute to the resource development for the Hinglish language. In this paper, Sentiment Analysis is done to classify sentences as positive, negative, sarcastic and non–sarcastic. This is done using extended sentiwordnet 3.0 and naïve Bayes classifier. From the current study it is analyzed that, sentiment analysis using SentiWordNet gives a better precision than the Naïve Bayes whereas the latter successfully classified the sentences into sarcastic and non-sarcastic.

Keywords: Hinglish · Hindi SentiWordNet · English SentiWordNet · Hinglish SentiWordNet · Code switch · Part of speech · Lexicon · Naïve Bayes Classifier

1 Introduction

1.1 Sentiment Analysis

Sentiment Analysis is very useful in today's era of advanced learning and technology. It has gain popularity due to its heavy potential to break into a person's inner world via extracting his/her emotions from the data which he/she continuously posting on social platforms like Facebook, twitter etc. If someone gets the set of emotions a person carrying with him, it becomes a cakewalk for that person to predict his/her opinion about something more accurately and that's what the Sentiment Analysis is all about. This

© The Author(s), under exclusive license to Springer Nature Switzerland AG 2021
M. Tripathi and S. Upadhyaya (Eds.): ICDLAIR 2019, LNNS 175, pp. 11–20, 2021.
https://doi.org/10.1007/978-3-030-67187-7_2

analysis of emotions has vast fields of applications ranging from enhancing selling on E-Commerce platforms to preparing influential political agenda for elections. While dealing with the analysis of emotions, one should not overlook the very aspect of the psychology of human being that [1] whenever a person stands at the peak of exhibiting his emotions, he expresses them in his mother tongue. This data is very rich in emotions. A lot of people on social media use other languages than English and there is a lot of transliteration involved due to easy typing. Hindi is such language which is widely being spoken in the northern part of the country [2]. Particularly in urban areas of the country, people consciously or unconsciously, frequently use English words while communicating in Hindi. This frequent switching between Hindi and English commonly known as "Hinglish". Hinglish example: "Aaj ka movie show houseful hai". It's Hindi containing English words written in Roman Script, which when translated in English means "Today's movie show is houseful". This Hinglish data, available on various social networking platforms, contains valuable information. One can exploit it using various text analysis means like sentiment analysis and sarcasm detection. This kind of diverseness in the language is another reason which prompts researchers to go for Hinglish sentiment analysis.

1.2 Sarcasm Detection

Sarcasm is something that diverts the sentiment and meaning of an utterance from its literal meaning. So, it is very important to detect sarcasm while doing sentiment analysis. Some of the main factors that constitute a sarcasm are a change in tone while speaking, facial expression, body movements, use of over intensified words etc. This is the reason that it is as easy to detect it in oral communication as tough in written form of communication. This is why sarcasm detection, in order to classify the data, involves a lot of groundwork on the text like extraction of lexical and syntactic features in the text. For example, leveraging certain lexical features like emoticons, interjections and N-grams in this regard. In the current work, classifier was trained on such features to filter sarcastic sentences from non-sarcastic ones.

2 Literature Survey

2.1 Sentiment Classification

Pandey et al. [3] used HindiSentiWordNet (HSWN) to find the overall sentiment associated with the document of Hindi movie reviews. They improved the existing HSWN by adding missing sentimental words related to Hindi movie domain. Subramaniam Seshadri et al. [4] added Hinglish words to knowledge base along with the English word in a bid to improve the result of sentiment analysis and higher accuracy. However, they limit their research work to Hinglish dictionary improvement and didn't cover the parts of speech tagging for Hinglish sentences. Mulatkar [5] used the Word Sense Disambiguation algorithm to found out the correct sense of words. Gupta et al. [6] used a pre- annotated corpus and additional inclusion of phrases, checking for the overall polarity of the review with negation handling. The authors successfully classified movie reviews, which are in Hindi language, as positive, negative and neutral. However, their approach didn't classify the movie reviews written in Hinglish language which amount a big chunk of movie reviews posted on social media regularly. Kaur et al. [7] did an extensive study of many machine learning methods such as Support Vector Machine (SVM), Naive Bayes, Decision Tree and showed that these methods are suitable while classifying literary artworks especially poetry. Yadav and Bhojane [8] proposed a system for sentiment analysis of Hindi health news, which used their own corpus to find the overall sentiment associated with the document. They used a Neural Network to train the polarity words stored in the database to make the processing faster.

From the above discussion, it is clear that accuracy of sentiment analysis depends upon the reliable resources such as Sent WordNet, pre-annotated corpus etc. In the current work, we developed a Hinglish SentiWordnet by mixing English and Hindi Sent Wordnet. Besides we also developed a part of speech tagger for Hinglish code-switch language.

2.2 Sarcasm Detection

Bouazizi et al. [9] performed sarcasm detection on Twitter data. First, they classified the features which are useful in sarcasm detection in four different sets and then, based on their presence in sentences and pattern of their occurring, classified sentences as sarcastic and non-sarcastic. Four sets of features which were proposed by them are pattern, sentiment-focused, syntactic & semantic and punctuation-focused features. The authors successfully filtered out the sarcastic sentences from dataset. However, their work is suitable for English language. For a code-switch language like Hinglish, there is a need to develop parts of speech tagger to identify the suitable features for sarcasm detection. Bindra et al. [10] used different Twitter tags as sentiment labels. A Twitter tag can be a reference to a specific user, hashtags or URLs. For example, "@" is used to tag a user like @Sachin. "#" is used for a hashtag like #sarcasm, #sad, #wonderful etc. These annotations used to develop the corpus for sarcasm and sentiment classification. They used the Twitter API to collect such sentences. Logistic regression (LogR) and support vector machine with sequential minimal optimization (SMO) were two classifiers they used in their experiment. Bharti et al. [11] proposed a Hadoop based framework that captured real-time sentences and processed them with a set of algorithms which identified sarcastic sentiment effectively. Apache Flume was used for capturing sentences in real time. For processing these sentences stored in the HDFS, they used Apache Hive. Further, Natural Language Processing (NLP) techniques like POS tagging, parsing, text mining and sentiment analysis were used to identify sarcasm in these processed sentences. However, such resources are neither well developed nor openly available for Hinglish language.

From the above discussion, it is clear that the presence of special symbols, exclamatory signs, hyperbolic utterances leads to sarcasm in a sentence. So, identification of such pieces of evidence is important which would be served as features to classifier for sarcasm detection.

In consideration to the above discussion, sentiment classification and sarcasm detection namely Dictionary approach and Machine Learning techniques were chosen for sentiment classification and sarcasm detection respectively. It is also evident from the above discussion that both the approaches have pros and cons. Lexical analysis can be used directly on data and does not require any pre-annotated data. For sarcasm detection, machine learning techniques are good. This required pre-annotated data as well as classified sets of features so that classifier can be trained to detect sarcasm.

3   Proposed Work

In this paper, authors aimed to carry out sentiment analysis and sarcasm detection for Hinglish sentences. To accomplish the same, authors suggested a hybrid approach which combines the idea of sentiment analysis with sarcasm detection. They first carried out sentiment analysis using dictionary-based approach and then the result of the same supplied as a test data to Naïve Bayes classifier for sarcasm detection. The final output is the combined result of the dictionary-based approach and the Naïve Bayes classifier. The authors developed resources like a Hinglish SentiWordNet for sentiment analysis and part of speech tagger for tagging Hinglish code-switch language sentences. The proposed solution is a hybrid of sentiwordnet based approach, for classifying sentiments of Hinglish text, and the Naïve Bayes classifier for sarcasm detection in Hinglish text.

3.1   Hinglish SentiWordNet Approach

The English sentiwordnet (ESWN) is extended by adding the Hindi sentiwordnet (HSWN) to it. To accomplish this, at first transformation of HSWN is done so that it becomes compatible with ESWN. After that ESWN is appended by adding transformed HSWN. Finally, more Hinglish words are added into extended SentiWordNet for better precision. Figure 1 shows the flow chart of the proposed method.

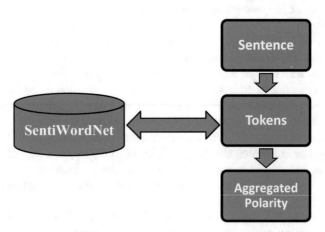

Fig. 1. Flow diagram for lexicon-based method.

In this approach when a sentence is fed to the proposed system, firstly the sentence is splitted into tokens. Then, each token is looked upon in the extended sentiwordnet. Sentiwordnet contains the sentiment score for each word. The Proposed system first checks for the category of the current token i.e. token is an adjective (a), noun (n), adverb (r) or a verb (v). After that, it looks for its sentiment score. Based on the score it returns the sentiment category of the token as either positive or negative. In this way proposed system performs for all tokens of a sentence. At last the proposed system takes the sum of all sentiment scores corresponding to all tokens of a sentence. If the total score is greater than zero, then classified the sentence as positive. If the total score is less than zero, then classified the sentence as negative.

3.2 Naïve Bayes Classifier Approach

The Naive Bayes classifier is one the important technique in machine learning. It is based on the Bayes theorem which calculates a conditional probability for an event to occur given the historical data of another event which already occurred. Thus, it predicts the occurrence of an event in light of another previously occurred event. In this way, it can be applied to predict whether a sentence is sarcastic or not.

Formula based on Bayes theorem is given below:

$$P(X \text{ and } Y) = P(X) * P(Y|X) \tag{1}$$

Here,

$P(X)$ = Probability of event X

$P(X \text{ and } Y)$ = Probability of event X and Y

$P(Y|X)$ = Probability of event Y given event X

Figure 2 shows the proposed approach for Sarcasm detection.

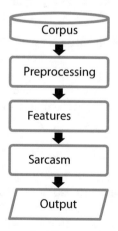

Fig. 2. Flow chart for sarcasm detection.

Preprocessing. The gathered data contains URLs and other non-useful data like a hashtag (#), annotation (@) which needs to be omitted. Some special characters are also present in the data, which are very useful for the detection of the emoticons. So before removing unwanted things first detect the emoticons present in dataset and replace them with one of the two generic words i.e. "zxp" and "zxn". Here "zxp" symbolize all emoticon with positive sentiment and "zxn" symbolize all emoticon with negative sentiment. Dataset also contains sentences with #sarcasm which is also useful in case of sarcasm detection. So that detect this special hashtag in all sentences and replace it with the generic word "zxs". Now remove all other special characters and unwanted data.

Feature Selection. Part of speech tags are used as the features for the classification model. To extract part of speech tags, first split sentences into tokens and then store them in a list. Now read the part of speech tags of required features and store them in another list in the same order in which they are present in the sentence. Then, return this list to the classification model for learning.

Part of Speech Tagging. Part of speech of a sentence are very useful in identifying the pattern on which sarcasm detection is based. These parts of speech served as features to the classifier. For code-switch languages, tagging part of speech is a multiple fold process. For Hinglish code-switch language, the author suggests a two-fold process for parts of speech tagging. In this process, the author first tagged the sentences with Stanford NLP POS Tagger. This tagger tagged the part of the English language of a sentence with appropriate tags and the parts of the Hindi language as FW (Foreign Word). After that author process the output from the first level tagging and replace the FW tag with appropriate language tag based on an algorithm. According to this algorithm, developed system looks at each and every token tagged as FW and replace FW tag with appropriate language tag based on Hindi grammar rules defined for verbs, nouns, adjectives etc. For example, if a token ends with "ta", "ti", or "te" than it would tag as verb. It tags all words which are in uppercase as CAPS, all occurrences of "zxs" as SAR, all occurrences of "zxp" as HBP, all occurrences of "zxn" as HBN, all Hindi noun word as NN and all Hindi verbs as VB.

Sarcasm Detection. The Pre-labelled data is used to train the classifier. The proposed system extract features from both of the files as discussed above and pass it to Naïve Bayes classifier during the learning phase. In this way, training of classifier is done. The classifier lookout for the presence of features which amount to sarcasm in test data as per learning and labelled each sentence accordingly as Sarcastic or Not- Sarcastic.

3.3 Performance Measure

To measure the performance of the above experiment, this paper uses confusion matrix and F- score. From the confusion matrix, accuracy and precision can be calculated. F-score is used to measure the performance of Naïve Bayes classifier. It equals to the harmonic mean of precision and recall. Formula to calculate F- score is given below:

$$F-\text{Score} = 2 * (P * R)/(P + R) \tag{2}$$

Here,

P = Precision, indicates the relevancy between correctly classified sentiments out of total classified sentiments. Mathematically,

$$P = (\text{True Positive})/(\text{True Positive}) + (\text{False Positive}) \tag{3}$$

R = Recall, indicates the relevancy between correctly classified sentiments out of total existing corresponding sentiments. Mathematically,

$$R = \text{(True Positive)}/\text{(True Positive)} + \text{(False Negative)} \qquad (4)$$

4 Results

4.1 Sentiment Classification

The test was carried out by giving dataset as input to the proposed system. The proposed system breaks the sentence into tokens. Then it looks for the respective sentiment score. Based on the score it returns the sentiment category of each token. Finally, the polarity of the sentence arrived at by adding all the corresponding sentiment scores of each token. The performance of Hinglish Sentiwordnet for the classification of Hinglish sentences as positive or negative is given in Table 1.

Table 1. Confusion table of sentiment analysis.

N = 1000	Actual positive	Actual negative
Predicted positive	534	27
Predicted negative	36	403

The performance of Hinglish SentiWordnet is evaluated and expressed in Table2.

Table 2. Performance of Hinglish SentiWordNet.

Data size	Precision	Recall
1000	95.18	93.68

Analyzing Tables 1 and 2, it is clear that the proposed Hinglish SentiWordnet performs well for classifying Hinglish Data.

4.2 Sarcasm Detection

The test was conducted using the Naive Bayes classifier and the following results were harvested. The test data was passed to the proposed system. Then said classifier lookout for the presence of features which amount to sarcasm as per learning and labelled each sentence accordingly as Sarcastic or Not- Sarcastic. The performance of Naïve Bayes classifier for sarcasm detection is given in Table 3.

Table 3. Confusion table for sarcasm detection.

N = 1000	Actual sarcastic	Actual non-sarcastic
Predicted sarcastic	199	16
Predicted non-sarcastic	4	781

The performance of Naïve Bayes Classifier is evaluated and expressed in Table 4.

Table 4. Performance of naïve bayes classifier.

Data size	F-score
1000	95.20

Analyzing Tables 3 and 4, it is clear that the proposed Naïve Bayes Classifier performs well for classifying Hinglish Data.

4.3 Combined Result of Sentiment Analysis and Sarcasm Detection

The author arrived at the hybrid result by combining the above said approaches in a sequential manner i.e. first performed the sentiment analysis using the Hinglish Senti Wordnet on test data and classified the sentences in two categories named positive and negative. Then this classified data is supplied as test data to train Naïve Bayes classifier

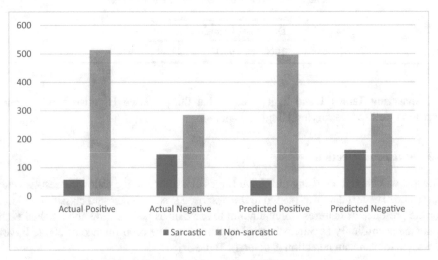

Fig. 3. Combined result of sentiment analysis and sarcasm detection.

for sarcasm detection. The trained Naïve Bayes classifier successfully labeled these sentences as sarcastic or non-sarcastic based on the features present in them. The final output is a combination of both approaches *i.e.,* Hinglish sentences get classified out of these four following categories: positive-sarcastic, negative-sarcastic, positive-non-sarcastic and negative-non-sarcastic. The combined result showed how many sentences are sarcastic and non-sarcastic in nature out of positive and negative sentences. The combined result is shown using a Bar Graph in Fig. 3. Also, the plot explains the actual and predicted class with respect to the sarcasm detection.

The above resulted graph reveals interesting conclusions. Originally, out of total sarcastic sentences, 28.08% sentences are in positive nature and remaining 71.92% sentences are in negative nature. The statistics of predicted data also establishes the same thing. In the resulted data, out of total predicted sarcastic sentences, 25.12% sentences are in positive nature and remaining 74.88% sentences are in negative nature. This result establishes the general inclination of sarcasm towards negative sense.

5 Conclusion

This work has analyzed the Hinglish language data. The author proposed a dictionary-based method to analyze sentiments. The author also proposed a machine learning based method to filter such sentences into two categories namely Sarcastic and Non-Sarcastic.

The author suggests that the English SentiWordNet can be extended by appending its content with the content of Hindi SentiWordNet. The author shows that such Extended SentiWordNet proved useful in sentiment analysis of Hinglish sentences. The system developed by the author breaks Hinglish sentences into tokens to check their polarity. The Hinglish SentiWordNet returns the polarity of each token which later aggregated to get the overall polarity of a sentence. In this way proposed system able to classify Hinglish sentences as positive or negative. The proposed study showed a better measure of F-score *95.20%,* precision *95.18%,* and a recall value of *93.6%.*

To categorize the sentence into sarcastic and non-sarcastic authors trained the naïve Bayes classifier with different sets of features which are responsible for the detection of sarcasm in such sentences. Once it analyzes the sentence, it annotates the sentences with a tag (sarcastic or non-sarcastic).

The limitation of the techniques discussed above guides the author towards future explorations. More concretely, it will be beneficial to incorporate the context, in which sentences are utter, in sentiment analysis and sarcasm detection. This kind of context-aware analysis significantly improves the performance of the proposed system. The context, in which sentences are utter, affects the polarity of sentences according to its sentimental nature. Sometimes it makes a normal comment a sarcasm if its sentimental nature is directly opposite to that of said comment.

The performance can further be improved by applying deep learning architecture and its variant towards the improvement in measure of F-score. Also, penalize algorithm can be implemented in order to reduce the miss-classification error, in turn the performance can further be improved from the existing.

References

1. Puntoni, S., de Langhe, B., Van Osselaer, S.M.J.: Bilingualism and the emotional intensity of advertising language. J. Consum. Res. **35**(6), 1012–1025 (2009)
2. Sailaja, P.: Hinglish: code-switching in Indian English. ELT J. **65**(4), 473–480 (2011)
3. Pandey, P., Govilkar, S.: A frameworkfor sentiment analysis in hindi using hswn. Int. J. Comput. Appl. **119**(19), 23–26 (2015)
4. Shehadri, S., Lohidasan, A., Lokhande, R., Save, A., Nagarhalli, T.P.: A new technique for opinion mining of hinglish words. Int. J. Innov. Res. Sci. Eng. Technol. **4**(8), 7184–7189 (2015)
5. Mulatkar, S., Bhojane, V.: Sentiment classification in Hindi. IOSR J. Comput. Eng. **17**(4), 100–102 (2015)
6. Gupta, A., Sonavane, D., Attarde, K., Shelar, N., Mate, P.: Sentiment analysis of movie reviews in Hindi. Int. J. Tech. Res. Appl. **41**, 31–33 (2016)
7. Kaur, J., Saini, J.R.: Emotion detection and sentiment analysis in text corpus: a differential study with informal and formal writing styles. Int. J. Comput. Appl. **101**(9), 1–9 (2014)
8. Yadav, M., Bhojane, V.: Design of sentiment analysis system for Hindi content. Int. J. Innov. Res. Sci. Eng. Technol. **4**, 12054–12063 (2015)
9. Bouazizi, M.: A pattern-based approach for sarcasm detection on Twitter. IEEE J. Mag. **4**, 5477–5488 (2016)
10. Bindra, K.K., Gupta, A.: Tweet sarcasm: mechanism of sarcasm detection in Twitter. Int. J. Comput. Sci. Inf. Technol. **7**(1), 215–217 (2016)
11. Bharti, S.K., Vachha, B., Pradhan, R.K., Babu, K.S., Jena, S.K.: Sarcastic sentiment detection in tweets streamed in real time: A big data approach. Digit. Commun. Netw. **2**(3), 108–121 (2016)

An Online Supervised Learning Framework for Crime Data

Shiv Vidhyut, Santosh Kumar Uppada$^{(\boxtimes)}$, and B. SivaSelvan

Computer Science and Engineering, IIITDM Kancheepuram, Kancheepuram, India
{coe15b017,coe18d005,sivaselvanb}@iiitdm.ac.in

Abstract. Crime analysis is a law enforcement function that involves systematic analysis for identifying and analyzing patterns and trends in crime and disorder. Crime analysis also plays a role in devising solutions to crime problems, and formulating crime prevention strategies. A lot of previous work has been done on analytic of crime demographics and geographic prior to this project. The unfortunate, frequent nature of crime, and the fact that it often follows a geographic and demographic pattern which requires constant updation makes this data set an excellent choice for online learning based models for analytics.

Keywords: Crime demographics · Online learning algorithm · Vowpal Wabbit · Exploratory data analytics

1 Introduction

Machine learning (ML) is a sub-field of artificial intelligence (AI) which brings in the capacity to learn and improve from experience without being explicitly programmed. Machine learning is a buzzword that involves training a specific model for a specific purpose which mimics the decisive ability of the humans on a particular scenario. It includes both learning and testing phase. The Learning phase majorly involves discovering patterns from input data to infer knowledge or learning from direct experience. The main focus is developing a program that learns and acts with very minimal to no human intervention. The first instance of what can be termed today as ML appeared in early 1958, which was a basic model of a perceptron, a neuron-like structure modelled based on our human brain. The next breakthrough in the field appeared in the late 70s, when Jeofrey Hinton introduced the world to the Back-propagation Algorithm, which is the foundation of modern neural networks, and deep learning. Additionally, the increase in computing power in the 90s shifted the focus back to traditional machine learning, and has been an active field of research ever since.

Machine Learning can be classified into two types based on different paradigms of classification: functionality and training methodology. Based on their functionality, they can be classified broadly into three types

- **Supervised Learning:** Supervised learning algorithms trains on data with known class labels and works on data having unknown class labels. Thus,

M. Tripathi and S. Upadhyaya (Eds.): ICDLAIR 2019, LNNS 175, pp. 21–36, 2021.
https://doi.org/10.1007/978-3-030-67187-7_3

a mathematical model is constructed that maps input with desired outputs. This model is called as supervised learning as the known class labels serve as supervisory signals. Every row is a training sample and represented as a vector/array in a mathematical model. The entire training data is represented in matrix form which serves as an input to a specific model. The function which learns the mapping of input and output labels is optimized by an iterative approach. A function reaches optimality only if it becomes capable to map an appropriate output label to the given input is which not a part of training data. The accuracy of the algorithm improves over time.

- **Unsupervised Learning:** Unlike supervised algorithms, unsupervised learning algorithms work on data with no defined class labels. These algorithms learn a pattern or similarity between the data items and cluster all similar data items. There are multiple functions that reflect the similarity measure between data items and the center of every cluster.

- **Reinforcement Learning:** Reinforcement Learning (RL) is another subfield of machine learning which works on punish and reward basis [2]. Based on the current environment, every correct move is rewarded and every wrong move is punished. The primary focus is to maximize the total rewards achieved [1].

Conventional machine learning paradigm majorly works in batches i.e., it trains once on the entire training dataset and the algorithm learns and acts on the test dataset accordingly. Every addition of new training sample to the training set will demand to start the training process from scratch. This retraining brings down the efficiency in terms of space and time. Scaling up the training set becomes a heavy task. Unlike batch learning algorithms, there is an alternative method called online learning which can adapt for data that arrives sequentially. At each step, the prediction accuracy gets increased and serves as a better predictor for upcoming data. Thus, online learning algorithms works best because in current scenario, building the entire data set for training set prior is difficult. As and when data is added, the predicting function is updated. Hence, online learning algorithms manage the data that comes with high velocity [18] (Figs. 1, 2, 3 and 4).

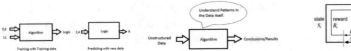

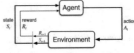

Fig. 1. Supervised learning

Fig. 2. Unsupervised learning

Fig. 3. Reinforcement learning

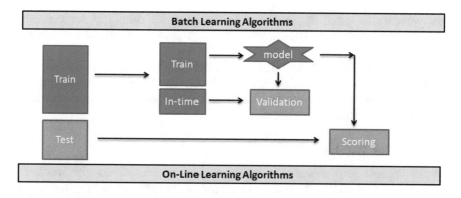

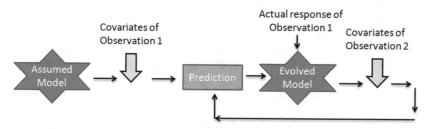

Fig. 4. Batch vs online learning

2 Literature Review

2.1 Online Learning

Perceptron [3] is the oldest algorithm for online learning. The perception takes an input, and multiplies it with the existing weight. The sign of the output scalar is used to determine which class the input belongs to. Subsequently, the true class label is received and is compared with the predicted value. The loss is calculated, and the weight is readjusted accordingly. The perception works inherently like an online model, where every training data point is presented to the model individually. Other first order algorithms appeared during the late 60s like the Winnow algorithm and the Passive-Aggressive Online Learning algorithm, but the breakthrough came when an Online version of the most famous of convex optimization algorithms- Gradient Descent was formalised [16].

The online gradient descent algorithm is one of the most frequently used convex optimization algorithms for a varity of tasks, primarily classification. In literature, different OGD variants have been proposed to improve either theoretical bounds or practical issues, such as adaptive Online Gradient Descent (OGD) (ogd), and mini-batch OGD [5], amongst others. Further, second order algorithms have been developed such as the Second Order Perceptron, Confidence Weighted Learning, Adaptive Regularization etc. The most recent development came in the Sketched Online Newton algorithm, which made significant improvements to speed up second order learning [6]. Similarly, for prediction tasks, Online Regression Algorithms have been developed with the concepts in parallel to linear regression [11,17]. The algorithm for online regression is implemented by using the following function to adjust the weights prior to a new input instance: The above function takes as input only the nth datapoint, and returns the modified set of weight vectors. Using this newly assigned weight vector, the value of the Test data can be predicted [10].

2.2   Vowpal Wabbit

Vowpal Wabbit is a machine learning system which pushes the frontier of machine learning with techniques such as online, hashing, allreduce, reductions, learning2search, active, and interactive learning. There is a specific focus on reinforcement learning with several contextual bandit algorithms implemented and the online nature lending to the problem well. Vowpal Wabbit is a destination for implementing and maturing state of the art algorithms with performance in mind. It is a command line-tool integrated with boosters to increase the speed of execution [19].

In a traditional learning setup, when it comes to categorical variables, label encoding or one hot encoding are popular of data-preprocessing. With online learning, we do not have the liberty of looking at the complete dataset to decide the number of categories prior, thus making such techniques invalid. Additionally, real data can be volatile and we cannot guarantee that new values of categorical features will not be added at some point. This issue hampers the use of the trained model when some new data is introduced. Vowpal Wabbit overcomes this issue by using **feature hashing**- a technique where the categorical data is hashed using an appropriate hash function. Thus, each categorical variable gets stored separately by default, and the hashed vector is used by VW while learning (Fig. 5).

Fig. 5. Vowpal Wabbit - a fast ML library

- Input Format: The input format for the learning algorithm is substantially more flexible than might be expected. Examples can have features consisting of free form text, which is interpreted in a bag-of-words way. There can even be multiple sets of free form text in different namespaces.

- Speed: The learning algorithm is fast – similar to the few other online algorithm implementations out there. There are several optimization algorithms available with the baseline being sparse gradient descent (GD) on a loss function [14].
- Scalability: This is not the same as fast. Instead, the important characteristic here is that the memory footprint of the program is bounded independent of data. This means the training set is not loaded into main memory before 2336 learning starts. In addition, the size of the set of features is bounded independent of the amount of training data using the hashing trick.
- Feature Interaction. Subsets of features can be internally paired so that the algorithm is linear in the cross-product of the subsets. This is useful for ranking problems. The alternative of explicitly expanding the features before feeding them into the learning algorithm can be both computation and space intensive, depending on how it's handled.

2.2.1 Features of Vowpal Wabbit

Some of the features that set vowpal wabbit apart, and allow it to scale to tera-feature (1012) data-sizes are:

- **The online weight vector:** VW maintains an in memory weight-vector which is essentially the vector of weights for the model that it is building. This allows fast access and computation. Unbounded data size: The size of the weight-vector is proportional to the number of features (independent input variables), not the number of examples (instances). This is what makes vowpal wabbit, unlike many other (non online) learners, scale in space. Since it doesn't need to load all the data into memory like a typical batch-learner does, it can still learn from data-sets that are too big to fit in memory (Fig. 6).
- **Cluster mode:** VW supports running on multiple hosts in a cluster, imposing a binary tree graph structure on the nodes and using the all-reduce reduction from leaves to root.
- **Hash trick:** VW employs what's called the hashing trick. All feature names get hashed into an integer using murmurhash-32. This has several advantages: it is very simple and time-efficient not having to deal with hashtable management and collisions, while allowing features to occasionally collide. It turns out (in practice) that a small number of feature collisions in a training set with thousands of distinct features is similar to adding an implicit regularization term. This counter-intuitively, often improves model accuracy rather than decrease it. It is also agnostic to sparseness (or density) of the feature space. Finally, it allows the input feature names to be arbitrary strings unlike most conventional learners which require the feature names/IDs to be both a) numeric and b) unique.
- **Parallelism:** VW exploits multi-core CPUs by running the parsing and learning in two separate threads, adding further to its speed. This is what makes vw be able to learn as fast as it reads data. It turns out that most supported algorithms in vw, counter-intuitively, are bottlenecked by IO speed, rather than by learning speed.

- **Checkpointing and incremental learning:** VW allows you to save your model to disk while you learn, and then to load the model and continue learning from where it was left off.
- **Test-like error estimate:** The average loss calculated by vowpal wabbit "as it goes" is always on unseen (out of sample) data (*). This eliminates the need to bother with pre-planned hold-outs or do cross validation. The error rate you see during training is 'test-like'.
- **Beyond linear models:** VW supports several algorithms, including matrix factorization (roughly sparse matrix SVD), Latent Dirichlet Allocation (LDA), and more. It also supports on-the-fly generation of term interactions (bi-linear, quadratic, cubic, and feed-forward sigmoid neural-net with user-specified number of units), multi-class classification (in addition to basic regression and binary classification), and more.
- **Handling categorical features:** More often than not learners accept only numeric features, which necessitates encoding of categorical variables. Vowpal Wabbit uses namespaces effectively to avoid encodings.

```
0 | price:.23 sqft:.25 age:.05 2006
1 2 'second_house | price:.18 sqft:.15 age:.35 1976
0 1 0.5 'third_house | price:.53 sqft:.32 age:.87 1924
```

Fig. 6. VW format example

2.2.2   Input Format

Vowpal Wabbit takes input as a .libsvm file- a text-file formatted in the following manner:

[Label] [Importance] [Base] [Tag]—Namespace Features —Namespace Features ... —Namespace Features

where

- Label is the real number that we are trying to predict for this example. If the label is omitted, then no training will be performed with the corresponding example, although VW will still compute a prediction.
- Importance (importance weight) is a non-negative real number indicating the relative importance of this example over the others. Omitting this gives a default importance of 1 to the example.
- Base is used for residual regression. It is added to the prediction before computing an update. The default value is 0.

- Tag is a string that serves as an identifier for the example. It is reported back when predictions are made. It doesn't have to be unique. The default value if it is not provided is the empty string. If you provide a tag without a weight you need to disambiguate: either make the tag touch the—(no trailing spaces) or mark it with a leading single-quote '. If you don't provide a tag, you need to have a space before the —.
- Namespace is an identifier of a source of information for the example optionally followed by a float (e.g., Metric Features:3.28), which acts as a global scaling of all the values of the features in this namespace. If value is omitted, the default is 1. It is important that the namespace not have a space between the separator—as otherwise it is interpreted as a feature.
- Features is a sequence of whitespace separated strings, each of which is optionally followed by a float (e.g., NumberOfLegs:4.0 Has Stripes). Each string is a feature and the value is the feature value for that example. Omitting a feature means that its value is zero. Including a feature but omitting its value means that its value is 1.

2.3   The BIGQuery API

The Chicago Dataset is a live dataset, maintained by the City of Chicago, USA. It is hosted on BigQuery, a Web service from Google that is used for handling or analyzing big data. It is part of the Google Cloud Platform. As a NoOps (no operations) data analytics service, BigQuery offers users the ability to manage data using fast SQL-like queries for real-time analysis. BigQuery has an API that can be used to integrate this process into the data analysis. BigQuery is compatible with SQL queries and can be used with Google Apps Script, Google Spreadsheets and other Google services [15] (Fig. 7).

```python
from bq_helper import BigQueryHelper
import pandas as pd

# https://www.kaggle.com/sohier/introduction-to-the-bq-helper-package
chicago_crime = bq_helper.BigQueryHelper(active_project="bigquery-public-data",
                                dataset_name="chicago_crime")

bq_assistant = BigQueryHelper("bigquery-public-data", "chicago_crime")
#List all availablt tables
bq_assistant.list_tables()
```

```
['crime']
```

Fig. 7. BigQuery API sample usage

Each dataset has its own workspace defined as a **project**, under which the data is stored. It can be accessed via BigQuery APIs that are available for a variety of platforms, such as REST APIs, the Python Toolbox, etc. The dataset itself is tagged using two parameters- its name, and the project repository its present in. Subsequently, data can be retrieved from the servers to run analytics on the same [12].

2.4   Crime Analytics

Crime analytics has been an active area of research in the US since the early 2000s, where the FBI has been trying to capitalize on the extensive data avaibale with the police departments and the body itself. Starting out as an experimental setup aimed to push the limits of technology in 2002, the first major breakthrough came ten years later, as a result of the National Institute of Justice's grant to Rutger's University on risk terrain modelling. Risk terrain modeling examines how the environment affects illegal activity [8]. In December 2012, Rutgers researchers applied their approach to the crime data displayed on old maps of Irvington, N.J., to demonstrate the method. They merged Esri's GIS software with police data and city maps to create new maps displaying the migration paths of neighborhood shootings in 2007 and 2008 [9].

Since then, crime analytics in general is an active field of study. Some of the major questions addressed by the community include the Total number of crimes committed per year per locaity, and its prediction; classification of crime by type), mapping predominant crime types by geographic location and identification of areal patterns [15], and statistical analysis of the same to provide better security and safety [7].

All previous attempts make use of historical data to demonstrate the significance of crime rate analysis. In this project, we attempt to exploit the power of the Vowpal Wabbit framework to demonstrate a live engine, which can be used real-time to predict and eliminate such crimes [13].

3   Implementation

3.1   The Problem Statement

Using the above data, we attempt to build models to answer the following questions:

1. Given the data - type of crime, date, day, geographic location and domesticity, can we predict whether the police arrests the individual?
2. Given the area, date, time and arrest data, can we predict what type of crime is most likely to happen/has happened in a certain location?
3. Given the historical data, can we predict the approximate number of crimes that may happen in the next month?

3.2   The Dataset

This dataset reflects reported incidents of crime (with the exception of murders where data exists for each victim) that occurred in the City of Chicago from 2001 to present, minus the most recent seven days. Data is extracted from the Chicago Police Department's CLEAR (Citizen Law Enforcement Analysis and Reporting) system. The data contains the following information:

- ID - Unique identifier for the record
- Case Number - The Chicago Police Department RD Number (Records Division Number), which is unique to the incident.
- Date - Estimated date when the incident occurred.
- Block - The partially redacted address where the incident occurred, placing it on the same block as the actual address.
- IUCR - The Illinois Unifrom Crime Reporting code. This is directly linked to the Primary Type and Description
- Primary Type - The primary description of the IUCR code.
- Description - The secondary description of the IUCR code, a subcategory of the primary description.
- Location Description - Description of the location where the incident occurred.
- Arrest - Indicates whether an arrest was made (Fig. 8).
- Domestic - Indicates whether the incident was domestic-related as defined by the Illinois Domestic Violence Act.
- Beat - Indicates the beat where the incident occurred. A beat is the smallest police geographic area – each beat has a dedicated police beat car. Three to five beats make up a police sector, and three sectors make up a police district.
- District - Indicates the police district where the incident occurred.
- Ward - The ward (City Council district) where the incident occurred.
- Community Area - Indicates the community area where the incident occurred. Chicago has 77 community areas.
- FBI Code - Indicates the crime classification as outlined in the FBI's National Incident-Based Reporting System (NIBRS).
- X Coordinate - The x coordinate of the location where the incident occurred in State Plane Illinois East NAD 1983 projection.
- Y Coordinate - The y coordinate of the location where the incident occurred in State Plane Illinois East NAD 1983 projection.
- Year - Year the incident occurred.
- Updated On - Date and time the record was last updated.
- Latitude - The latitude of the location where the incident occurred.
- Longitude - The longitude of the location where the incident occurred.
- Location - The location where the incident occurred in a format that allows for creation of maps and other geographic operations on this data portal.

The dataset is querried using the BigQuery API, and filtered acording to necessity. Post processing and clean up of the data, it is fed to the csvtolibsvm module, which converts the data into the format as required by Vowpal Wabbit.

```
bq_assistant.head("crime", num_rows=3)
```

unique_key	case_number	date	block	iucr	primary_type	description	location_description	arrest	domestic
2014	HJ193229	2003-02-19 22:35:00+00:00	105XX S STATE ST	0110	HOMICIDE	FIRST DEGREE MURDER	STREET	True	False
2315	HJ540270	2003-08-04 20:35:00+00:00	002XX W 105TH ST	0110	HOMICIDE	FIRST DEGREE MURDER	STREET	True	False
2312	HJ483963	2003-07-23 09:00:00+00:00	105XX S WENTWORTH ST	0110	HOMICIDE	FIRST DEGREE MURDER	STREET	False	False

Fig. 8. Sample rows from the dataset

The usage of this data is also specified- different formats are used for regression and classification for feeding into the VW client. Subsequenty, a plaintext file compatible with VW is generated, which is fed to it as input (Fig. 9).

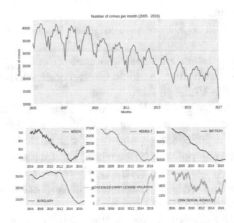

Fig. 9. Crime trends by year- overall and by crime type

3.3 Exploratory Data Analysis and Data Pre-processing

EDA is one of the crucial step in data science that allows us to achieve certain insights and statistical measure that is essential for the business continuity, stockholders and data scientists. It performs to define and refine our important features variable selection, that will be used in our model. Once EDA is complete and insights are drawn, its feature can be used for supervised and unsupervised machine learning modelling. The EDA is executed majorly by Uni-variate visualization, Bi-variate visualization, Multivariate Visualization and Dimensionality reduction. Other standard EDA techniques include Box Plots, Whisker Plots, Histograms, Pivot Tables, Parallel Coordinates etc.

In the case of the above dataset, a few generic questions such as number of crimes per month, their trend, to identify the different crime types and trends within their time series, their distribution across different days of the week and months of the year are explored to gain a preliminary insight into the data.

A lot of information can be inferred from the exploratory data analysis performed. Some crime types are actually increasing all along like homicide and deceptive practice. Other types started to increase slightly before 2016 like theft, robbery and stalking. Additionally, some crimes types are more likely to occur than other types depending on the place and time.We can also infer that crimes rates seem to peak at summer months.

Grouping the data by crime type clearly shows that certain types of crime, such as theft related, narcotics and criminal damage is more likely to occur than others such as obscenity and ritualism. The data we have is therefore skewed in terms of crime type. If we attempt to, therefore use crime type as a feature in any of our models, the performance of these will be affected. Therefore, types where the number of such occurrences are less than 10000 are pruned out. The resultant set is thus reduced to a set of 24 different types of crimes, from the original 31 (Fig. 10).

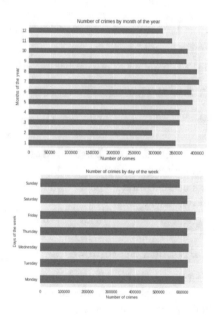

Fig. 10. Crime trends by month and day

The dataset is checked for and cleaned to remove null values and NaNs. he data is also filtered based crime types to ensure a minimal data skew. This final, polished data is fed to a Libsvm converter module which takes as input the data, prediciton/classification variable and the output file name to save into a format compatible with Vowpal Wabbit.

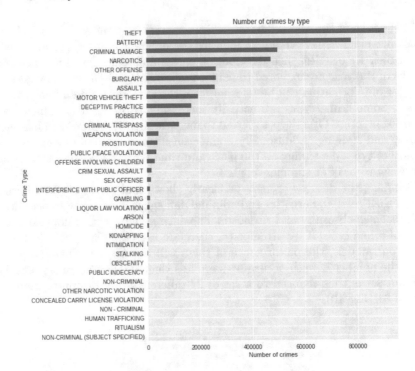

Fig. 11. Number of crimes per category

Table 1. Time taken for execution

Problem no	Algorithm	Offline mode	Online model stand alone	Online model cluster
1	Binary classifiaction logistic regression	63.45	8.72	4.32
2	Multi-class classification one-vs-all regression	85.4	11.23	5.31
3	Linear regression	34.3	6.87	4.23

3.4   Addressing the Problem Statements

The following are implemented in an ofline fashion using Python and in an online fashion using Vowpal Wabbit to compare and contrast the performace-time, accuracy and other performance metrics.

- The first question is a binary classification problem- to predict whether an arrest occours or not. We approach this binary classification problem using Logistic Regression with two different loss functions- HingeLoss and Logistic Loss functions.
- The second statement is a multi-class classification problem - to predict what types of crimes occour. Similar loss functions are used to classify and test the performace.

– The third statement is a use case for linear regression - the number of arrests is presented in terms of time-series data. Squared, Quantile and Poissons loss functions are used to analyse the performance (Fig. 11 and Table 1).

4   Performance and Results

Each model was run with both offline and online learning algorithms- the offline model implemented using Python and the online version using Vowpal Wabbit. To exploit the full potential of the VW package, the online version was also run in a distributed setup on a cluster. Their performance in terms of time and other metrics such as Log Loss, Jaccard Score, AUC-ROC curve, R-squared error etc are compared and contrasted (Fig. 12).

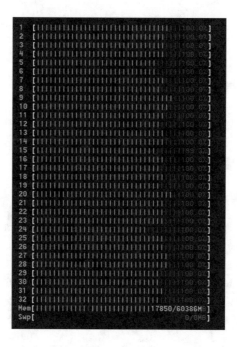

Fig. 12. VW running in a distributed mode: an 8-node 4 core setup

4.1   Performance Parameter- Time

The results clearly show the improvement in time when the online algorithm runs. This can be attributed to the fact that the algorithm does not require processing of the entire dataset, and requires only incremental data to be fed to the model. Thus, it does not need the entire dataset to be loaded into memory either- this reduces the inherent IO time.

Further, with Vowpal Wabbit supporting a distributed setup, the computation can be sped up. The above performance improvement was achieved by using a small cluster: an 8 node 4 core setup. As expected, there is an average of 40% decrease in the total time taken - or a 40% speedup.

4.2   Evaluation Metrics

4.2.1   Classification

To answer the question as to whether a certain type of crime, given information such as area and date leads to an arrest or not, binary classification using logistic regression is used. Two different loss functions

– HingeLoss
– LogisticLoss

were used. The resultant model was evalutated using the parameters-

– Accuracy
– Log Loss
– Mean Squared Error
– Jaccard Score
– AUC-ROC curve
– F1 score

The results are tabulated as follows:

Table 2. Binary classification

Evaluation metric	Hinge loss	Logistic loss	Offline model
Accuracy	0.884	0.832	0.916
Log loss	1.342	2.463	1.003
Mean suqared error	3.932	5.127	3.846
Jaccard score	0.846	0.887	0.902
AUC-ROC	0.862	0.834	0.878
F1 score	0.943	0.922	0.956

4.2.2   Prediction

Linear regression is used to address the final part of the problem statement- to predict the number of crimes that may occour during the next one month given the history of crimes in the past by area. Three loss functions were tested-

– Squared Loss
– Quantile Loss

Table 3. Multi-class classification

Evaluation metric	Hinge loss	Logistic loss	Offline model
Accuracy	0.814	0.786	0.906
Log loss	2.467	3.992	1.793
Mean suqared error	3.554	4.912	2.846
Jaccard score	0.826	0.763	0.881
AUC-ROC	0.843	0.797	0.850
F1 score	0.914	0.886	0.925

– Poisson Loss

and their performances were evaluated based on the following metrics:

– Accuracy
– MSE
– RMSE
– R-Squared Value

The results are tabulated as follows:

Table 4. Linear regression

Evaluation metric	Squared loss	Quantile loss	Poisson loss	Avg. offline model
Accuracy	0.926	0.945	0.886	0.947
MSE	5.588	6.031	14.564	5.126
RMSE	2.234	2.786	3.846	2.884
R-Squared	0.88	0.832	0.761	0.911

5   Conclusion

Most models above have performed either at par, or slightly lesser than the average offline models using the same loss functions. The increase in error can be attributed to the training methodology- the losses are calculated and minimised over every example against a batch of examples. Thus, the overall error is higher in the online learning setup than when compared to its respective offline model. Both binary (Table 2) as well as multi-class (Table 3) setups follow similar trends when compared to their native Python implementation. The same trend can be extended to prediction as well (Table 4) - Linear Regression has performed comparably to the existing offline models.

References

1. Strehl, A.L., Littman, M.L.: Online linear regression and its application to model-based reinforcement learning. In: NIPS (2007)
2. Kaelbling, L.P., Littman, M.L., Moore, A.W.: Reinforcement learning: a survey. J. Artif. Intell. Res. **4**, 237–285 (1996)
3. Kussul, E., et al.: Rosenblatt perceptrons for handwritten digit recognition. In: IJCNN 2001. Proceedings of the International Joint Conference on Neural Networks (Cat. No. 01CH37222), vol. 2. IEEE (2001)
4. Biehl, M., Schwarze, H.: Learning by online gradient descent. J. Phys. A **28**, 643–656 (1995)
5. Dekel, O., et al.: Optimal distributed online prediction using mini-batches. J. Mach. Learn. Res. **13**, 165–202 (2012)
6. Luo, H., et al.: Efficient second order online learning by sketching. In: Advances in Neural Information Processing Systems (2016)
7. Jain, A., Bhatnagar, V.: Crime data analysis using pig with Hadoop. In: International Conference on Information Security and Privacy (2015)
8. Almanie, T., Mirza, R., Lor, E.: Crime prediction based on crime types and using spatial and temporal criminal hotspots. Int. J. Data Min. Knowl. Manage. Process **5** (2015)
9. Caplan, J.M., Kennedy, L.W., Miller, J.: Risk terrain modeling: brokering criminological theory and GIS methods for crime forecasting. Just. Q. **28**, 360–381 (2011)
10. Ross, S., Mineiro, P., Langford, J.: Normalized online learning. In: Proceedings of the Twenty-Ninth Conference on Uncertainty in Artificial Intelligence, pp. 537–545 (2013)
11. Song, C.-B., Liu, J., Liu, H., Jiang, Y., Zhang, T.: Fully implicit online learning. CoRR (2018)
12. Humphrey, C., Jensen, S.T., Small, D.S., Thurston, R.: Urban vibrancy and safety in Philadelphia. Environ. Plan. B: Urban Anal. City Sci. **47**(9), 1573–1587 (2020)
13. Rakhlin, A., Sridharan, K.: Online learning with predictable sequences. COLT (2013)
14. Ruschendorf, L.: Solution of a statistical optimization problem by rearrangement methods. Metrika **30**(1), 55–61 (1983)
15. Skogan, W.G.: The changing distribution of big-city crime: a multi-city time-series analysis. Urban Affairs Q. **13**(1), 33–48 (1977)
16. Karampatziakis, N., Langford, J.: Online importance weight aware updates. In: Proceedings of the Twenty-Seventh Conference on Uncertainty in Artificial Intelligence, pp. 392-399 (2011)
17. Hoi, S.C.H., Sahoo, D., Lu, J., Zhao, P.: Online learning: a comprehensive survey. CoRR (2018)
18. Rakhlin, A., Sridharan, K.: A tutorial on online supervised learning with applications to node classification in social networks. CoRR (2016)
19. Langford, J.: Vowpal Wabbit Wiki (2015)

A Survey: Secure Indexing & Storage of Big Data

Poonam Kumari[✉], Amit Kr. Mishra, and Vivek Sharma

Department of Computer Science and Engineering,
Technocrats Institute of Technology, Bhopal, M.P., India
Kumari.pari836@gmail.com

Abstract. The sector of healthcare includes a lot of data related to the health of patients, where large number of people enrol into. Through any organization which is dealing with the large amount of data expected to increase its capacity in the next upcoming years where large amount of data is still unstructured, which is present in the silos as well as in the form of the images, medical notes of prescription, insurance claims data, EPR (Electronic Patient Records) etc. The main task is to integrate this data and then to generate better healthcare system which will capable of handling many aspects easily. Since the data is in the isolated form, unmanaged manner so there is a need for the personal health records system that will ultimately holds up the factors affecting the data and provides a better managing capability. Big data is making it simple and flexible because of its tools, its transformation factors. This paper gives an overall explanation of the big data services to the medical field where data plays a major role. Hence, an updated algorithm containing compress storage along with fast as well as secure access is also presented in this paper.

Keywords: Healthcare storage · Data security · Analytics · Big data · Cloud · Data management · Healthcare · Storage and accessing

1 Introduction

Big data is a field of computer science, it plays a major role in the sectors wherever there is a need to store large amount of data, manipulate that data, retrieve data etc. In this paper main focus of stream big data is on the medical science where large number of data in form of reports generated every day and that data is connected to the fast transition from the technologies which are digital give rise to the big data. The big data ia playing a major role in providing a breakthroughs in many fields with the large datasets. It collects the data which is quite complex and difficult to manage for the regular management tools or else data processing applications. It is also available in the structured, semi-structured, and unstructured format in petabytes and beyond.

Big Data Virtualization
It is a path for the collection of the data via different sources all in a single layer. The collected data layer is usually virtual. As compare to other methods, most of the data remains at a single place and is taken on demand directly from the source systems.

M. Tripathi and S. Upadhyaya (Eds.): ICDLAIR 2019, LNNS 175, pp. 37–45, 2021.
https://doi.org/10.1007/978-3-030-67187-7_5

1. Applications

It can be said that the demand of big data is increasing continuously because it increases the demand for the management of information for example- Software AG, Oracle Corporation, IBM, Microsoft, SAP, EMC, HP and Dell have spent more than $15 billion for the software's.

2. Government

The adoption of the big data in the government sectors tends to provide the effectiveness in terms of cost, innovation but does not contains any of the faults.

Data analysis may require several parts of the government organizations (central as well as local) for work in connection to the new innovations.

CRVS (Civil Registration and Vital Statistics) gathers all endorsements status from birth to death. CRVS is a wellspring of enormous information for governments.

3. Universal Development

Research on the successful use of data and correspondence advancements for improvement (otherwise called ICT4D) proposes that huge information innovation can make significant commitments yet in addition present one of a kind difficulties to International advancement. Headways in huge information examination offer financially savvy chances to improve basic leadership in basic advancement regions, for example, medicinal services, work, monetary efficiency, wrongdoing, security, and catastrophic event and asset the executives. Moreover, client created information offers new chances to give the unheard a voice. Nonetheless, longstanding challenges for creating districts, for example, insufficient innovative foundation and financial and human asset shortage worsen existing worries with enormous information, for example, protection, blemished philosophy, and interoperability issues.

4. Assembling

In view of TCS 2013 Global Trend Study, upgrades in supply arranging and item quality give the best advantage of enormous information for assembling. Enormous information gives a framework to straightforwardness in assembling industry, which is the capacity to unwind vulnerabilities, for example, conflicting segment execution and accessibility. Prescient assembling as a material approach toward close to zero vacation and straightforwardness requires tremendous measure of information and propelled forecast devices for an efficient procedure of information into valuable data. A theoretical structure of prescient assembling starts with information securing where distinctive kind of tangible information is accessible to procure, for example, acoustics, vibration, weight, current, and voltage and controller information. Huge measure of tactile information notwithstanding authentic information build the enormous information in assembling. The created enormous information goes about as the contribution to prescient devices and preventive procedures, for example, Prognostics and Health Management (PHM) [1–6].

2 Literature Review

The role of Big Data in several sectors such as banking, retail etc. has been witnessed but in the recent times even health care systems have shown the preparedness to bring the transformation and leverage on the benefits that Big data solutions provides for the health care systems.

[1] Hanlin Zhang introduced while recovering the EHR, the social insurance focus catches portions from incomplete cloud servers and recreates the EHRs. In the mean time, in all actuality, the remaking of a common EHR could be a lot of troublesome for a social insurance focus or a patient, we in this way propose a commonsense distributed storage plot which re-appropriates the recreation of a mutual EHR to a distributed computing specialist co-op. Such an answer can definitely help the proficiency of the proposed plan. Apparently, our plan is the first to characterize remaking redistributing idea in all distributed storage plans for EHRs dependent on mystery sharing, and the aftereffects of re-appropriating reproduction can be confirmed by social insurance focuses or patients in our plan. The hypothetical examination and test results additionally bolster that our proposed plan is secure and proficient.

[7] HarukaIshii, author presented a system of using wearable sensors capable for continuous monitoring of the elderly and forwarding the data to the big data systems. A big data system manages high volume, velocity and variety of information from different sources, so as to address this challenge a wearable sensor system with intelligent data forwarder was introduced adopting Hidden Markov model for human behaviour recognition for feeding meaningful data to the system. The sensor readings and states of a user are sent to Big Data analytics for improving and personalizing the quality of care.

[8] Shamsiah Abidin, author evaluates the ensembles design and combining different algorithms to develop novel intelligent ensemble health care and decision system to monitor the health using wearable sensors. New Novel Intelligent Ensemble method was constructed based on Meta classifier voting combining with three base classifiers J48, Random Forest and Random Tree algorithms.

[9] Walied Merghani, author in his work did analysis of the health care data using social network analysis, temporal analysis and text mining and higher order feature construction to understand how each of these areas contributes to understand the domain of healthcare. Temporal Analysis methods are used as they do not require trained classifier to identify anomalies and it can be used as a timely technique for detection of transient billing practices that are anomalous.

[10] Lorenzo Moreno Ruiz, author in his paper focuses on the performance limitation of the prediction engine CARE (Collaborative Assessment and Recommendation Engine). In order to solve the computation time issue of care algorithm two methods have been devised, in first method single patient version of CARE is taken to perform disease risk ranking on demand and with high degree of accuracy and second method is distributed computation of the care algorithm for nightly batch job on large patient datasets.

[11] Peter Triantafillou, author discusses how Big Data Analytics are beneficial to transform the rural healthcare by gaining insights from their clinical data and to effectively make the right decisions.

- Big data domain has two options, to use open source solution or the commercial solutions available. Some of the key products are: Hadoop based analytics, Data warehouse for operational insights, Stream computing software for real time analysis of streaming data. Apache Hadoop framework is an open source solution, followed by NoSQL databases such as Cassandra, MongoDB, DynamoDB, Neo Technologies and couch base [12–16].
- HDFS is Hadoop distributed file systems attempts to enable storage of large files and does this by distributing data among pools of data nodes.
- Map Reduce is a programming model that allows for massive job execution scalability against cluster of servers.
- Apache Hadoop YARN (Yet another Resource Negotiators) separates resource management and processing components.
- HBase is another example of a non-relational data management solution that distributes massive datasets over the underlying Hadoop framework. HBase derived from Google's BigTable, is a column-oriented data layout and provides fault tolerant storage solution.
- Hive is a SQL-like bridge that allows conventional BI applications to run queries against a Hadoop cluster.
- Zookeeper and Chukwa are used to manage and monitor distributed applications that run on Hadoop.
- PIG consists of a Perl-like language that allows for query execution over data stored on a Hadoop cluster, instead of a SQL-like language.
- Sqoop is used for transferring data between relational database and Hadoop
- Mahout is an Apache project used to generate free applications of distributed and scalable machine learning algorithms that support big data analytics on Hadoop framework.
- SkyTree is high-performance machine learning and data analytics platform focuses specifically on handling Big Data.

Table 1. Comparison among various works.

Authors	Year	Algorithms	Key features	Advancements performed	Limitations
Hanlin Zhang [1]	2018	EHRs based on secret sharing	Reconstruction outsourcing is a processing method of reconstruction in a cloud storage solution based on secret sharing	Secret sharing is performed in it	It is quite difficult to maintain the time and security as well at the same time

(*continued*)

Table 1. (*continued*)

Authors	Year	Algorithms	Key features	Advancements performed	Limitations
[7] Haruka Ishii	2018	Hidden Markov model	A system of using wearable sensors capable for continuous monitoring of the elderly and forwarding the data to the big data systems	The sensor readings and states of a user are sent to Big Data analytics for improving and personalizing the quality of care	Sometimes wearable sensors may not be able to read user data
[8] Shamsiah Abidin	2018	New Novel Intelligent Ensemble method	Develop novel intelligent ensemble health care and decision system to monitor the health using wearable sensors	Meta classifier voting combining with three base classifiers J48, Random Forest and Random Tree algorithms are also used with this algorithm	The random selection of the data may lead to the mismatch to the relevant outputs
[9] Walied Merghani	2018	temporal analysis and text mining analysis is done	Detect anomalies	Temporal Analysis methods are used as they do not require trained classifier to identify anomalies	Anomalies are quite difficult to find out
[10] Lorenzo Moreno Ruiz	2018	Collaborative Assessment and Recommendation Engine	Accuracy achieved by this method	First method single patient version of CARE is taken to perform disease risk ranking on demand and with high degree of accuracy	High degree of accuracy can be achieved only if the disease analysis is done in a proper manner
[11] Peter Triantafillou	2018	Big Data Analytics	Make accurate decisions	Big Data Analytics are beneficial to transform the rural healthcare by gaining insights from their clinical data and to effectively make the right decisions	This method is good in performing better results but the transformation sometimes bring the wrong decisions

3 Problem Definition

As per discussion of previous work, there ismultiple problem definition which arises and can be overcome in further implementation of proposed solution.

The following are problem formulation of existing solution. Handling large data complexity is lacking is swarm based technique.

1. Handling and configure kernel function while working with data optimization technique, optimizing it according to the requirement is a limitation to the existing work.
2. Fitness function optimization while working with document optimization using the genetic algorithm and other feature selection approach, computation to find fitness function value is observed challenge in genetic approach.
3. Existing solution has given the compression but having high SNR values which determine the low efficiency of approach.
4. The storage algorithm found less security over the document which is submitted over the server using big data process.
5. Accessing of the data and proper storage is needed refinement using which it can be handle over concurrent node using given system.
6. Short indexing keep high computation time while storage and accessing of the document which need to be minimized.

Thus the given problem formulation is needed to work out for providing effective algorithm for better prediction.

4 Proposed Work

In order to avoid the given problem definition, the following proposed work solution using the compressive sensing lexical indexing with enhance security algorithm is going to use with Map reduce environment of medical data.

A Secure Storage, Accessing indexing based compressing approach for Medical data is presented which consist of algorithm which make utilization of proper data sequencing for its optimized utilization.

Data accessing using the measures compressive data usage and sampling the available data in source folder for further processing. Data sampling and temporal storage is performed using the compressive sampling function computing the calculation, difference between the input file using further sampling equation.

A is the sensing matrix, which allows us to get from x to y (via random measurements, transformations, or a combination of the two).

The proposed work consist of enhance version of ECC technique after compressive optimization, where the technique is further applied with modified components and further it is applied at cloud server. The further work is compared with existing MAC and ECC based solution proposed by different author while performing cloud security and integrity verification [17].

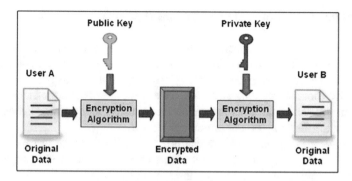

Fig. 1. Elliptic curve cryptography.

General working of ECC:

It is the most important step in which an algorithm is used to generate both public and private keys. Sender encrypts the message data with the help of receiver's public key and receiver decrypts the data using its private key.

Stage 1. The sender chooses an arbitrary number dA between the range [1, n − 1]. This is the private key of the sender.

Stage 2. At that point the sender creates the general population key PA utilizing the equation PA = dA*G.

Stage 3. Additionally recipient chooses a private key dB and produces its open key PB = dB*G.

Stage 4. The sender creates the security key K = dA*PB and the beneficiary additionally produces the security key K = dB*PA.

A proposed security approach with HECC hyper elliptic curve cryptography is going to perform for the secure storage of data.

Following is the given step for storage and accessing of the medical data works with the indexing storage using compressive sampling function.

- Proposed security approach will be used along with the lexical data storage and accessing mechanism.
- A pre-computed load and resource aware data will help in fast computation and immediate decision making.
- High throughput and low computation time is going to observe.
- Advancement of encryption technique make use of algorithm and provide efficient data sharing in between.
- A parallel dynamic search updating will save the current execution and run time execution.

Thus the given execution application steps help in completing the execution. This is the main advantage of processing data in multiple steps.

5   Expected Outcomes from Proposed Work

1. Bandwidth Costs The company with resource usage and multiple platform with bandwidth unlimited over the internet. Thus the user doesn't guarantee about the Bandwidth consumption, hence a heavy cost may be billed as per usage.
2. Reliability Cloud computing still does not always offer round the-clock reliability. The platform sometime faces issue with the industry, thus all time availability and reliability for that may be not possible for users.
3. Security As it is complete online platform, users not feel safe in several cases of data and passwords. Thus an security issue over the cloud need to be understand and challenge need to be solved by the providers.
4. Privacy The data accessing in between the several component of cloud and their communication may reveal some data in between the communication. Thus a privacy policy and discussion over the privacy issue may arise with cloud.

6   Conclusion

Health care segment generates huge data in the different forms such as medical reports, CT scan and other MRI reports. It contains large image and document data which need to be processed over the different entity. Big data is the technique which help in efficient storage and processing of the big data documents. Report storage and indexing required algorithm which can use for proper usage over the entity over the connected architecture. Secure storage is always a challenging issue which deals with security encryption algorithm with different key size. Data minimization is also a challenging issue which need optimization of input data. This paper deals with the requirement of big data technique in multiple sector. It also discuss about the algorithm which can help in optimize storage and then accessing it with the multiple node simultaneously. Security algorithm for the storage using the Map reduce is going to utilize which enhance the performance of medical record storage and accessing over multiple entity. Thus ECC is the cryptographic algorithm which provides security and authentication. Authentication to the data is provided with the help of smaller keys. The computational cost as well as the speed of this algorithm is comparatively better. It also makes use of the good exchange protocols giving another mark to the security.

Further providing a fully privacy preserving system with low computation time and efficiency is going to provide by the given algorithm.

References

1. Zhang, H.: Cloud storage for electronic health records based on secret sharing with verifiable reconstruction outsourcing. IEEE (2018)
2. Smorodin, G.: Internet of things: modem paradigm of health care. In: 2017 21st Conference of Open Innovations Association (FRUCT), pp. 311–320 (2017)

3. Lin, C.H., Huang, L.C., Chou, S.C.T., Liu, C.H., Cheng, H.F., Chiang, I.J., Lin, C.H.: Temporal event tracing on big healthcare data analytics. In: 2014 IEEE International Congress on Big Data, pp. 281–287 (2014)
4. Chen, M., Hao, Y., Hwang, K., Wang, L., Wang, L.: Disease prediction by machine learning over big healthcare data. IEEE Access 5(1), 8869–8879 (2017)
5. Chen, M., Zhou, P., Fortino, G.: Emotion communication system. IEEE Access 5, 326–337 (2017)
6. Chen, M., Ma, Y., Li, Y., Wu, D., Zhang, Y., Youn, C.: Wearable 2.0: enable human-cloud integration in next generation healthcare system. IEEE Commun. 55(1), 54–61 (2017)
7. Ishii, H., Kimino, K., Inoue, M., Arahira, M., Suzuki, Y.: Method of behaviormodeling for detection of anomaly behavior using hidden Markov model. In: International Conference on Electronics, Information and Communication, ICEIC 2018, 05 April 2018
8. Abidin, S., Xia, X., Togneri, R., Sohel, F.: Local binary pattern with random forest for acoustic scene classification. In: IEEE International Conference on Multimedia and Expo (ICME), 11 October 2018 (2018)
9. Merghani, W., Davison, A., Yap, M.: Facial micro-expressions grand challenge 2018: evaluating spatio-temporal features for classification of objective classes. In: 13th IEEE International Conference on Automatic Face & Gesture Recognition (FG 2018), 07 June 2018 (2018)
10. Ruiz, L.M., Nieves, D.C., Popescu-Braileanu, B.: Methodological proposal for automatic evaluation in collaborative learning. In: IEEE Global Engineering Education Conference (EDUCON), 24 May 2018 (2018)
11. Triantafillou, P.: Data-less big data analytics (towards intelligent data analytics systems). In: IEEE 38th International Conference on Distributed Computing Systems (ICDCS), 25 October 2018 (2018)
12. Zheng, Z., Zhu, J., Lyu, M.R.: Service-generated big data and big data-as-a-service: an overview. In: Proceedings of IEEE BigData, pp. 403–410, October 2013 (2013)
13. Bellogín, A., Cantador, I., Díez, F., et al.: An empirical comparison of social, collaborative filtering, and hybrid recommenders. ACM Trans. Intell. Syst. Technol. 4(1), 1–37 (2013)
14. Havens, T.C., Bezdek, J.C., Leckie, C., Hall, L.O., Palaniswami, M.: Fuzzy c-means algorithms for very large data. IEEE Trans. Fuzzy Syst. 20(6), 1130–1146 (2012)
15. Liu, X., Huang, G., Mei, H.: Discovering homogeneous web service community in the user-centric web environment. IEEE Trans. Serv. Comput. 2(2), 167–181 (2009)
16. Zielinnski, K., Szydlo, T., Szymacha, R., et al.: Adaptive SOA solution stack. IEEE Trans. Serv. Comput. 5(2), 149–163 (2012)
17. https://qvault.io/2020/09/17/very-basic-intro-to-elliptic-curve-cryptography/

Text Visualization Using t-Distributed Stochastic Neighborhood Embedding (t-SNE)

Chelimilla Natraj Naveen[✉] and Kamal Kumar

Department of Computer Science and Engineering, National Institute of Technology, Uttarkhand, Srinagar Garhwal 246174, Uttarakhand, India
{chelimilla.natraj.cse16,kamalkumar}@nituk.ac.in

Abstract. Data visualization is most important task to be done before classification and building a model. By data visualization we can easily know whether problem is directly classifiable or not. Below paper contains presentation of t-SNE visualization for donors choose data vectorized by different techniques like Term frequency-inverse document frequency, Bag of words, Tf-Idf weighted Word2Vec. Data visualization here means indirectly reducing dimensions of our data. Different data visualizations in below paper are done using different perplexity values of t-SNE. Pre-processing and vectorization are done accordingly for different features and numerical features are standardized before giving it to t-SNE.

Keywords: tsne · Term frequency · Bag of words · Inverse document frequency · Word2vector · Text visualization · Weighted word2vector

1 Introduction

Over the last few decades many of the data visualization and dimensionality reduction techniques have been proposed and out of them, t-SNE is one of the state of the art technique. It is so far the best method for dimensionality reduction. While constructing a model using modern machine learning algorithms is good but what if it takes days to train and about all the results we should take care of while building model, too much headache? Here where we can make the best use of data visualization.

In below paper visualization of donors choose data which is a mix of categorical, numerical, text features using t-SNE is presented. First, all categorical features are vectorized using one-hot encoding, text features are vectorized using different strategies like bag of words, tf-idf, tf-idf weighted word2vec. Numerical features are standardized using z-score mean normalization before visualization. First visualization of data containing all vectorized categorical features, all standardized numerical features merged with all bag of words vectorized text features is presented. Secondly, visualization of data containing all vectorized

M. Tripathi and S. Upadhyaya (Eds.): ICDLAIR 2019, LNNS 175, pp. 46–52, 2021.
https://doi.org/10.1007/978-3-030-67187-7_6

categorical features, all standardized numerical features merged with all term-frequency inverse document-frequency (aka tf-idf) vectorized text features is presented. Then visualization of data containing all vectorized categorical features, all standardized numerical features merged with all tf-idf weighted vectorized text features is presented. At last, visualization of data containing all vectorized categorical features, all standardized numerical features merged with all bag of words vectorized text features, all tf-idf vectorized text features, all tf-idf word2vec vectorized text features is presented.

2   Related Work

2.1   Parameters in t-SNE

t-SNE can be plotted using three parameters. The parameters are:

1. Step: Number of iterations
2. Perplexity: How many number of points needed to be considered in neighborhood while preserving local structure
3. Epsilon: How fast the structure need to be changed, can be said as learning rate.

   Out of these three parameters, Step and Perplexity are most important parameters [1,3,4].

2.2   Vectorization Techniques

Bag of Words. Bag of words is a feature vectorization technique. In this, we first find all unique words in whole text corpus (collection of all text documents which here are reviews) and initialize with a vector whose number of dimensions equal to the number of unique words where each dimension representing each word. The particular component of the vector is represented by some word which will be filled with a number of times that word occurred in that text document. So finally it converts each text document into some dimensional vector (which can be huge if there are large numbers of unique words). If we want to reduce the number of dimensions we introduce frequency condition which means that it will consider a word only if it occurs given a number of times in whole text corpus and so some of the unique words will not be considered and so the number of dimensions will be reduced. Generally, we should not try to take a large number of dimensions because higher dimensions will be affected by the curse of dimensionality [8,14].

Term Frequency-Inverse Document Frequency (Tf-Idf). Same as a bag of words it first finds all unique words in whole text corpus and initializes with a vector whose number of dimensions equal to the number of unique words where each dimension representing each word for each text document. The only difference is that here each component of vector is filled with tf(word w_i)*idf(word w_i). Term frequency of a particular word w_i in a review r_i is number of times the word w_i occurred in the review r_i/total number of words in the review r_i [9].

Tf-Idf Weighted Word2Vector. Word2Vector is an algorithm that converts each word in the text corpus to some specified dimensional vector. Similar words will get vector values in such a way they are nearer to each other in distance. In Tf-Idf weighted word2vector we do the sum of Tf-Idf(word)*word2Vector(word) for each review [2].

3   Proposed Work

In our dataset, we have three types of features like categorical features, numerical features, text features. In this generally while building a model or visualizing data we generally ignore categorical features. So instead of ignoring them which will result in data loss, we can use one-hot encoding on them and use it for our data visualization because we might get some better results or we might retrieve good conclusions with visualization on such data. Now about the other one generally while visualizing or building a model for data if there are text features we just vectorize it using one-strategy and use that data for training. But, why can't we make use of all vectorizations of data and use it for our visualization because the main problem in the real world about vectorization is that we don't know which vectorization technique is best for our data until we get results? So what I am saying is we will take data containing all text features vectorized with all strategies. Proudly, I got better results when I applied this technique. Data visualization using my technique is like a small group of clusters which can be useful in making some conclusions [Fig. 24]. Now the other technique I have used in the below paper is instead of using just word2vector I have used tf-idf weighted word2vector but unfortunately, I didn't get better results [Fig. 16].

3.1   Vectorizations of Different Features

Text data:

1. project_resource_summary - text data
2. project_title - text data
3. project_resource_summary - text data

For vectorizing text data we use different vectorization techniques like bag of words, tf-idf, average word2vec, tf-idf weighted word2vec. For vectorizing categorical data we do one hot encoding using count vectorizer. For numerical data we perform mean-std standardization and min-max standardization as per situation [5, 11, 13].

3.2   Text Pre-processing

In our text reviews, there may be unnecessary words like 'and', 'the' called stop words which are useless words and we cannot draw any information from them and so we need to remove such words from reviews. The words like won't, can't, should all be replaced with will not, cannot and so it will be helpful retrieving information and similarity between reviews. The other unnecessary things that should be removed in reviews are HTTP links, HTML tags, words with numbers, special characters [6,7,12].

4   Results

All categorical features and numerical features are vectorized and standardized accordingly and this data is merged with a respective vectorized text features. This total data is given to t-sne which will find similarity between features and returns lower dimensional data. The plot below is a scattered plot of the lower dimensional data returned by t-SNE. The number of data points in the below plot is 6000.

Below plots are t-SNE visualizations at different perplexity values. In the below plots, the green points are approved data points and red points are rejected data points.

4.1   t-SNE of Bag of Words Representation

(See Fig. 1)

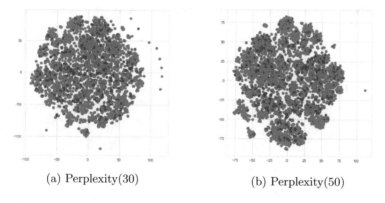

(a) Perplexity(30)   (b) Perplexity(50)

Fig. 1. t-SNE plot for merged data with bow vectorized text at different perplexities

Observations: The t-SNE visualization of data containing a bag of words vectorized text is like a group of clusters. Even at different perplexity values, there is a huge overlap of data points. There is no much difference in the structure of data visualization at different perplexities. The structure is better at higher perplexities than lower perplexities [8].

4.2   t-SNE of Tf-Idf Representation

(See Fig. 2)

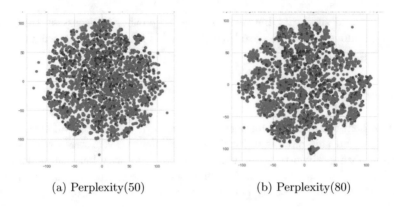

(a) Perplexity(50)   (b) Perplexity(80)

Fig. 2. t-SNE plot for merged data with tf-idf vectorized text at different perplexities

Observations: The t-SNE visualization of data containing tf-idf vectorized text is better than a bag of words text features as it is trying to make a small group of clusters. The structure is better at higher perplexities than lower perplexities as it was trying to form a small group of clusters at higher values from which we can retrieve some useful information [9].

4.3   t-SNE of Tf-Idf Weighted Word2Vector Representation

(See Fig. 3)

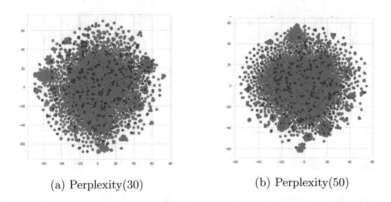

(a) Perplexity(30)   (b) Perplexity(50)

Fig. 3. t-SNE plot for merged data with tf-idf weighted word2vec vectorized text at different perplexities

Observations: The t-SNE visualization of data containing tf-idf weighted word2vec vectorized text is like some random cluster of data points. The visualizationa of bag of words and tf-idf are far better than this because we cannot retrieve any information from the structure which is like a random cluster of data points [10].

4.4   t-SNE of Combined Vectorizations

(See Fig. 4)

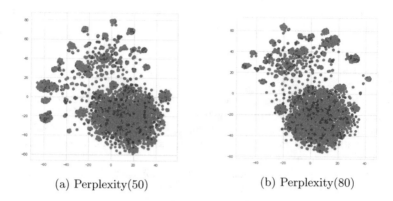

(a) Perplexity(50)   (b) Perplexity(80)

Fig. 4. t-SNE plot for merged data with tf-idf weighted word2vec vectorized text at different perplexities

Observations: The t-SNE visualization of data containing a combination of all vectorized text is like a group of one big and small clusters. It's trying to make clusters but not able to do it perfectly. Even at different perplexities, the structure of data is almost the same and from this, we can say that this data visualization is trying to draw some information from data.

5   Conclusions About Data Visualization Using t-SNE

The green and red points in the above scatter plots represent projects approved and projects rejected. Bag of Words, Tf-Idf are better than tf-idf weighted word2vec vectorizations because of forming some small group of clusters with less overlap of overall data when compared to others. Higher perplexity values seem better in data visualization because of less overlap of data than others. None of the techniques are useful for classification because of the huge overlap of data. It is not a separable problem in 2-dimensions but it may be separable in higher dimensions. Then all data visualizations the one with a combination of all text features of all vectorizations is far better than others because of the much better clustering of data where we can draw some good conclusions about the data. This data visualization can be useful in understanding about how some of the data points are related by seeing clusters because if there is cluster forming containing some data points then it means that there is some relationship between these data points and if there is some relation we can use that information in constructing a model for the problem. Our problem when vectorized contains more than 10000 features (dimensions) and by using t-sne we have reduced it into 2-dimensions and represented graphically. Almost in every plot, there is a huge overlap of data points and this might not be the case when we have taken some higher dimensions but we cannot represent that visually. The only way to represent such higher-dimensional plots is by using scatter plots of every combination of two features which will be huge if there are more number of features.

References

1. van der Maaten, L., Hinton, G.: Visualizing Data using t-SNE, November 2008
2. Sarkar, S.: Benefits of sentiment analysis for businesses, May 2018
3. Kurita, K.: Paper dissected - "visualizing data using t-SNE" explained
4. Olah, C.: Visualizing MNIST - an exploration of dimensionality reduction, September 2018
5. DonorsChoose.org : DonorsChoose.org application screenings, April 2018
6. McAuley, J., Leskovec, J.: From amateurs to connoisseurs: modeling the evolution of user expertise through online reviews, March 2013
7. Shayaa, S., Jaafar, N.I., Bahri, S., Sulaiman, A., Wai, P.S., Chung, Y.W., Piprani, A., Al-Garadi, M.A.: Sentiment analysis of big data: methods, applications, and open challenges, June 2018
8. Zhang, Y., Jin, R., Zhou, Z.-H.: Understanding bag-of-words model: a statistical framework, December 2010
9. Qaiser, S., Ali, R.: Text mining - use of TF-IDF to examine the relevance of words to documents, July 2018
10. Mikolov, T., Sutskever, I., Chen, K., Corrado, G., Dean, J.: Distributed representations of words and phrases and their compositionality (2013)
11. Guo, G., Wang, H.: KNN model-based approach in classification, August 2004
12. Wong, A.Y.L., Warren, S., Kawchuk, G.: A new statistical trend in clinical research - Bayesian statistics, October 2010
13. Agarwal, A., Xie, B., Vovsha, I., Rambow, O., Passonneau, R.: Sentiment analysis of Twitter data, June 2011
14. Mntyl, M.V., Graziotin, D., Kuutila, M.: The evolution of sentiment analysis - a review of research topics, venues, and top cited papers, November 2016

Detecting Fake News with Machine Learning

Nagender Aneja[1](✉) and Sandhya Aneja[2]

[1] Institute of Applied Data Analytics, Universiti Brunei Darussalam,
Jalan Tungku Link, Brunei Darussalam
nagender.aneja@ubd.edu.bn
[2] Faculty of Integrated Technologies, Universiti Brunei Darussalam,
Jalan Tungku Link, Brunei Darussalam
sandhya.aneja@ubd.edu.bn

Abstract. Fake news is intentionally written to influence individuals and their belief system. Detection of fake news has become extremely important since it is impacting society and politics negatively. Most existing works have used supervised learning but given importance to the words used in the dataset. The approach may work well when the dataset is huge and covers a wide domain. However, getting the labeled dataset of fake news is a challenging problem. Additionally, the algorithms are trained after the news has already been disseminated. In contrast, this research gives importance to content-based prediction based on language statistical features. Our assumption of using language statistical features is relevant since the fake news is written to impact human psychology. A pattern in the language features can predict whether the news is fake or not. We extracted 43 features that include Parts of Speech and Sentiment Analysis and shown that AdaBoost gave accuracy and F-score close to 1 when using 43 features. Results also show that the top ten features instead of all 43 features give the accuracy of 0.85 and F-Score of 0.87.

Keywords: Fake news · Machine learning · AdaBoost · Decision tree · Naive Bayes · K-nearest neighbors · Stochastic gradient descent · Support vector machine

1 Introduction

One of the challenging problems for traditional news media and social media service providers in Natural Language Processing (NLP) is to detect fake news due to its social and political impact on individuals [1]. It is also important since the capability of humans to detect deceptive content is minimal, especially when the volume of false information is high. Although fake news has been around from a long time as propaganda, however, it has grown exponentially in recent years. Privately owned websites and social media users/groups have

© The Author(s), under exclusive license to Springer Nature Switzerland AG 2021
M. Tripathi and S. Upadhyaya (Eds.): ICDLAIR 2019, LNNS 175, pp. 53–64, 2021.
https://doi.org/10.1007/978-3-030-67187-7_7

amplified the distribution of fake news since anyone can create a website or social media page and claim as news media. Social media has advantages to sharing information informally, however, this feature has been misused by a few people or organizations to distribute unverified content. The content which is well documented but fake is being distributed for political or other malicious purposes. The objective of fake content writers is to influence beliefs and thus to impact users' decisions. Social media has become a place for campaiging misinformation that affects the credibility of the entire news ecosystem.

Another issue that prevents social media users from seeing both sides of the coin is the informational separation caused by filtration of information through news aggregators [2]. Newsfeed of a user is most likely to contain posts of his friends who have the same attitude, and thus belief of the user is influenced by such posts. While on the other hand, information about different point of view doesn't reach to a user. This is more a rational and social issue, wherein, the users may be algorithmically advised if the news feed of a user represents one view only.

Fake news is defined as a piece of news, which is stylistically written as real news but is entirely or partially false [3]. Undeutsch hypothesis also [4] states that a fake statement differs in writing style and quality from a true one. Recent techniques are based on content, however, the fundamental theories in social and forensic psychology have not played a significant role in these techniques. Research efforts have been to automate the detection of fake news so that a user is informed about the content even if his or her friends share it. Fully automatic detection is still a research topic, however, supervised machine approaches that identify patterns in the fake news are being explored.

2   Problem Statement

Identification of fake news is a binary classification problem since there are two classes, fake and real. Mathematically, the problem may be stated as follows.

Let $N = \{n_1, n_2, n_3, ..., n_M\}$ be a collection of M news items and $L = \{l_1, l_2, l_3, ..., l_M\}$ be their corresponding labels of news items such that label l_i of news item n_i is either 1 or 0 depending on if the news item n_i is fake or real.

We need a machine-learning algorithm that can predict the accurate label of news item $n_z \notin N$.

Most of the current approaches are based on a dictionary that is developed from news items $n_i \in N$. In other words, a dictionary $D = \{w_1, w_2, w_3, ..., w_K\}$ of K words is such that the all words of new items $words_{n_i} is \subset D \ \forall \ n_i \in N$ Thus, performance of the model for a news item $n_z \notin N$ will be based on similarity of n_z with N.

Instead of a dictionary-based approach, we want to evaluate if the language features can be used to detect fake news. The assumption may work since there is clear intention to write fake news, and the content in fake news is writtenbased on human psychology to influence social belief system. In this research, we extracted numerical features from all news items and used the machine learning binary classification algorithm to detect fake news. Since we don't give importance to the similarity of words within the dataset, this approach may generalize well on other datasets also. The following section describes the proposed algorithm.

3   Related Work

This section describes a survey of recent prior work published in the area of fake news detection.

Guacho et al. [5] proposed semi-supervised content-based method that uses tensor-based article embeddings to construct a k-nearest neighbor graph of news articles that captures similarity in a latent, embedding space. The authors showed 75:43% accuracy using 30% of labels of one dataset and 67:38% accuracy using 10% labels of another dataset. Additionally, the method attains 70:92% accuracy using only 2% labels of the dataset.

Oshikawa et al. [6] presented a review on natural language processing solutions for automatic fake news detection and developed a textual content-based method on multi-class fake news detection based on natural language processing.

Zhou et al. [7] investigated news content at various levels: lexicon-level, syntax-level, semantic-level and discourse-level. The authors observed that the current techniques capture non-latent characteristics e.g. word-level statistics based on Term Frequency-Inverse Document Frequency (TF-IDF) Pérez-Rosas et al. [8], n-gram distribution Pérez-Rosas et al. [8] and/or utilize Linguistic Inquiry and Word Count (LIWC) features Pennebaker et al. [9]. Recently neural networks have also been used using the latent characteristics within news content Volkova et al. [10], Wang et al. [11].

Gravanis et al. [12] exploited the use of linguistic-based features in combination with Machine Learning methods to detect news with deceptive content. The proposed features combined with ML algorithms obtained the accuracy of up to 95% with the AdaBoost as first in rank and SVM & Bagging algorithms to be next in ranking but without statistically significant difference. The authors used Linguistic features as proposed by Burgoon et al. [13], Newman et al. [14], Zhou et al. [15].

Reis et al. [16] proposed that recent work in fake news detection identify pattern after these have been disseminated. Thus, it is difficult to gauge the potential that supervised models trained from features proposed in recent studies can be used for detecting future fake news. The authors used Language Features, Lexical Features, Psycholinguistic Features, Semantic Features, and Subjectivity. The results have shown that Random Forest and XGBoost perform better.

4   Proposed Method

We propose two phase-process that includes extracting numerical features in the first phase and then using the numerical features to predict the label of the news item using machine learning classifiers in the second phase.

4.1   Datasets

To determine the feature set that can help to predict news items as fake or real, we consider two datasets, namely dataset of new items labeled fake and dataset of new items labeled real.

Fake news dataset was taken from Kaggle website [17] and Real news dataset was downloaded from Guardian website [18]. We only considered the text of news and ignored other metadata. The reason for using text is to find language style of authors in writing fake content vs. real content. Our final dataset included 12249 fake news items and 9725 real news items after considering only news items with the English Language.

4.2   Preprocessing and Features Extraction

In the preprocessing step, first, we cleaned the news items so that there is no special character and in case there is a special character we split the word at the special character, e.g., refugees/immigrants was split into words refugees and immigrants. After cleaning the news items, we tokenized each news items using the tokenizer function of the nltk library and filtered the stopwords using nltk corpus.

To extract features of all news items, we applied Parts of Speech (POS) pos tag and Vader sentiment of nltk on filtered words representing news items. Table 1 provides 43 features that were extracted from pos tagging and sentiment analysis in addition to counting unique words. The description of POS tags have been explained in Table 2.

Table 1. Features set

POS Tags (39)	$, "", ., :, CC, CD, DT, EX, FW, IN, JJ, JJR, JJS, MD, NN, NNP, NNPS, NNS, PDT, POS, PRP, PRP$, RB, RBR, RBS, RP, SYM, TO, UH, VB, VBD, VBG, VBN, VBP, VBZ, WDT, WP, WP$, WRB
Sentiment (3)	'positive', 'neutral', 'negative'
Miscellaneous (1)	'unique'

We applied sentiment analysis to extract three features namely number of positive words, negative words, and neutral words from filtered words after converting the words to its base form using lemmatization. The process of lemmatization converts a word to meaningful base form while still maintaining the

Table 2. Description of NLTK POS Tags used in the present dataset

#	POS Tag	Description	POS Tag	Description
1	$	dollar	""	quotes
2	.	Dot	:	colon
3	CC	conjunction, coordinating	CD	numeral, cardinal
4	DT	determiner	EX	existential there
5	FW	foreign word	IN	preposition or conjunction, subordinating
6	JJ	adjective or numeral, ordinal	JJR	adjective, comparative
7	JJS	adjective, superlative	MD	modal auxiliary
8	NN	noun, common, singular or mass	NNP	noun, proper, singular
9	NNPS	noun, proper, plural	NNS	noun, common, plural
10	PDT	pre-determiner	POS	genitive marker
11	PRP	pronoun, personal	PRP$	pronoun, possessive
12	RB	adverb	RBR	adverb, comparative
13	RBS	adverb, superlative	RP	particle
14	SYM	symbol	TO	"to" as preposition or infinitive marker
15	UH	interjection	VB	verb, base form
16	VBD	verb, past tense	VBG	verb, present participle or gerund
17	VBN	verb, past participle	VBP	verb, present tense, not 3rd person singular
18	VBZ	verb, present tense, 3rd person singular	WDT	WH-determiner
19	WP	WH-pronoun	WP$	WH-pronoun, possessive
20	WRB	Wh-adverb		

context instead of stemming that removes a few characters from the end. For example, lemmatization of caring is care, while stemming of caring is car. The word set that we get after lemmatization is the words that represent a particular news item for Sentiment Analysis that gives the number of positive, neutral, negative words and miscellaneous feature that provides the unique number of words.

Finally, we divided all features of POS tags, Sentiment Features, and Number of Unique words with the total number of words so that the value of the feature lies between 0 to 1. This is in contrast to other published work in which features were divided by the number of sentences. Next section describes the application of Machine Learning Algorithms on the feature set.

Table 3. Dataset

Fake news items	12249
Real news items	9725
Total news items	21974
Features extracted for each news item	43

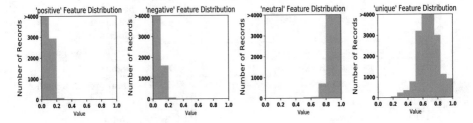

Fig. 1. Sentiment feature distribution before normalization

4.3   Supervised Learning

In this research, we employed several supervised machine learning algorithms to model the fake and real news items accurately. We determined the best candidate algorithm from preliminary results and further optimized the algorithm to best model the data.

Data Exploration. Table 3 provides the information that the dataset includes 12249 fake news and 9725 real news items. Total 43 features were extracted as explained in Table 1.

Figure 1 describes Feature Distribution (before normalization) for sample features {Positive, Negative, Neutral, Unique words} wherein the values lies between 0 to 1. The features were normalized so that the mean value of each feature is 0, and the standard deviation is 1. Figure 2 shows Features Distribution for sample features {Positive, Negative, Neutral, Unique words} after normalization.

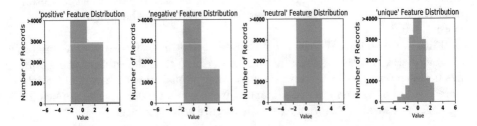

Fig. 2. Sentiment feature distribution after normalization

The dataset of 21974 news items was divided into a training set and testing set using sklearn train_test_split function so that 80% of the data is used for training and 20% of the data is used for testing. Thus, the training set included 17579 samples, and the testing set included 4395 samples.

Evaluating Model Performance. We implemented Logistic Regression (LR), Stochastic Gradient Descent Classifier (SGDC), Support Vector Machines, K-Nearest Neighbors (KNeighbors), Gaussian Naive Bayes (GaussianNB), and Decision Trees in addition to Naive Predictor that always predict news item as not fake. To evaluate model performance, accuracy may be appropriate, however, predicting a piece of real news as fake may be a concern. Thus, we used a metric based on precision and recall.

Accuracy measures the correct output, whether correctly predicted fake or correctly predicted real from total news items from the test dataset. Precision, as shown in Eq. 1, measures the proportion of news items that the system classified as fake were fake. Recall, as shown in Eq. 2, tells as what proportion of fake news items were identified as fake. Thus,

$$Precsion = \frac{TruePositive}{TruePositive + FalsePositive} \tag{1}$$

$$Recall = \frac{TruePositive}{TruePositive + FalseNegative} \tag{2}$$

A model's ability to precisely predict fake is more critical than the model's ability to recall. We may use F-beta score ($\beta = 0.5$), as shown in Eq. 3, as a metric that considers both precision and recall.

$$F_\beta = (1 + \beta^2) \cdot \frac{precision \cdot recall}{(\beta^2 \cdot precision) + recall} \tag{3}$$

Naive Predictor. The purpose of Naive Predictor is to show what a base model without intelligence would perform. A model that predicts all news items as Fake gives Accuracy score of 0.5574 and F-score of 0.6116.

Supervised Learning Models. We implemented the following six machine learning models.

1. Ada Boost Classifier
2. Decision Trees Classifier
3. Gaussian Naive Bayes (GaussianNB)
4. K-Nearest Neighbors (KNeighbors)
5. Stochastic Gradient Descent Classifier (SGDC)
6. Support Vector Machine

AdaBoost trains multiple weak learners and combines the weak learners. The algorithm is very fast and able to boost performance. The weakness of AdaBoost is that it is sensitive to noise or outliers since a week learner may increase contributions of noise or outliers.

Decision Trees are non-parametric methods used for classification and regression. It is a tree of decisions that predicts the target variable based on decision rules. The deeper the curve, more complex decision rules and better fitting. A decision tree is a good candidate algorithm since there should be few features that will contribute more to predict and thus a decision tree is an easy tool for this type of problem. It also has no significant impact on outliers.

Naive Bayes is a good algorithm for working with text classification. The relative simplicity of the algorithm and the independent features assumption of Naive Bayes make it a good candidate for classifying texts. Further, Naive Bayes works best when training data set and features (dimensions) is small. In case of a huge feature list, the model may not give better accuracy, because the likelihood would be distributed and may not follow the Gaussian or other distribution. Naive Bayes works best if the features are independent of each other, which looks like the case when we plotted the data using a scatter diagram.

K-Nearest Neighbors (KNeighbors) provides functionality for the supervised and unsupervised model. Supervised nearest neighbors can be used for classification and regression problems. KNN finds a predetermined number of samples closest in the distance to a point to predict the label. The samples can be k in numbers or based on the local density of points (radius-based neighbor learning). Although, it may take time the KNN algorithm is simple to visualize and can easily find similarity/patterns in the data.

Stochastic Gradient Descent Classifier (SGDC) is generally useful when number of features and samples are large. SGDClassifier and SGDRegressor fit linear models for classification and regression using different loss functions. with log loss SGDClassifier fits Logistic Regression (LR) and with hingle loss it fits a Linear Support Vector Machine. SGD classifier implements regularized linear models using SGD (GD - full dataset; SGD - 1 sample to update weight; MBGD - mini batch to update weights).

Support Vector Machines SVM is used for classification, regression, and outliers detection and is effective in high dimension spaces and even when the number of dimensions is greater than samples. Since it uses a subset of training points (supporting vectors) in decision function so is memory efficient. It is versatile since different Kernel Functions can be used in the decision function.

The initial performance of the above models was tested using default configurations based on three training sets considering (i) 1% of training data, (ii) 10% of training data, and (i) 100% of training data.

Figure 3 shows results for the time taken, accuracy score, F-Score for both phases, namely Training and Testing. Training time at 100% of the dataset was highest for Support Vector Machines, followed by Ada Boost Classifier. However, the testing time of K-Nearest Neighbors was observed highest. Most

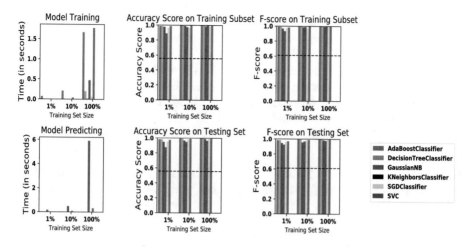

Fig. 3. Initial results with 1%, 10%, and 100% of training data

Table 4. Initial performance with 100% of training data

Algorithm	Test accuracy	Test F-Score
AdaBoostClassifier	**0.9989**	**0.9993**
DecisionTreeClassifier	0.9980	0.9983
GaussianNB	0.9902	0.9867
KNeighborsClassifier	0.9747	0.9822
SGDClassifier	0.9859	0.9851
SVC	0.9966	0.9968

of the algorithms performed well and gave close to 1 accuracy and F-score when 100% of the training data used.

Table 4 shows that results close to 1 but AdaBoost performed well in case of Test Accuracy and Test F-Score. The default base estimator of AdaBoost Classifier is DecisionTreeClassifier with max_depth as 1.

Adaboost (Adaptive Boosting) combines classifiers with poor performance, also known as, weak learners, into a bigger classifier with much higher performance. Adaboost starts with a week learner that classifies some points correctly but misclassify few. This week learner is added into the list and train another week learner that gives more weight to classify the previous mis-classify points and also may mis-classify few other points. This week learner is saved and the algorithm continues to train another week learner that gives more weights to classify the misclassified points of the previous learner. This approach is continued up to a threshold, and the final model is a combination of the different week learners. Thus, each classifier focus on the mistakes of the previous classifier. AdaBoost is particularly useful when it is easy to create a simple classifier, e.g., in this case, we know that few features can predict the style of news content to predict the correct label. Adaboost can create simple classifiers and optimally combine them. Once a model has been trained, it can easily predict whether the news is fake or not.

Results also indicate the performance of AdaBoost is better than others even when reduced training data is used. The F-score increases with an increase in training sample size and is maximum when 100% training data is used. The training time of AdaBoost is higher when 100% training data is used but should be excellent due to negligible prediction time and higher F-Score.

Improving Results by Model Tuning. Since AdaBoost performed well in the initial results, so we fine-tuned AdaBoost using GridSearchCV module of sklearn. We considered the following parameters for GridSearch.

1. *n_estimators*: 50, 100, 150, 175, 200
2. *base_estimator*: Decision Tree Classifier with max depth as 1, 2, and 3
3. scorer: F-Score with beta 0.5
4. cv: 5

The default value of n_estimators is 50 that defines a maximum number of estimators at which boosting is terminated. In case of a perfect fit, the learning procedure is stopped early. The default value of base estimator from which the boosted ensemble is built Decision Tree Classifier with max depth 1. The cv stands for cross-validation that defines cross-validation splitting strategy. The default value of cv is 3 specifies the number of folds in Stratified KFold.

Grid Search provided the best parameters with max depth 3 and the number of estimators as 175. AdaBoost with optimized parameters gave accuracy and F-Score equal to 1.0000 that was an improvement from the accuracy of 0.9989 and F-score of 0.9993. Figure 4 shows confusion matrix for AdaBoost with best tuned parameters.

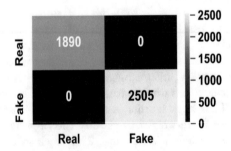

Fig. 4. Confusion matrix based on AdaBoost test results

Feature Importance. The section describes the importance of particular features for the predictive power to help understand the crucial features for determination of fake news. Figure 5 shows the most crucial top ten features are NN (noun, common, singular or mass); CD (numeral, cardinal); VBP (verb, present tense, not 3rd person singular); VBG (verb, present participle or gerund); positive (positive sentiment); NNP (noun, proper, singular); JJ (adjective or numeral, ordinal); IN (preposition or conjunction, subordinating); VBN (verb, past participle); and unique (unique words). These top ten features cover around 60% of features domain.

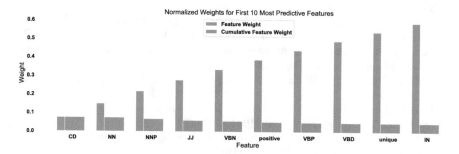

Fig. 5. Importance of features

Additionally, to test the impact of top features, we again trained the model only on these essential ten features. AdaBoost with the best classifier provided accuracy of 0.8578 and F-Score of 0.8753 in comparison to the accuracy of 1.0000 and F-score on the full feature set. Thus, we can say that features NN (noun, common, singular or mass); CD (numeral, cardinal); VBP (verb, present tense, not 3rd person singular); VBG(verb, present participle or gerund); positive (positive sentiment); NNP (noun, proper, singular); JJ (adjective or numeral, ordinal); IN (preposition or conjunction, subordinating); VBN (verb, past participle); and unique (unique words) are highly essential to predict fake news.

5   Conclusion and Future Work

This research implemented AdaBoost classifier; DecisionTreeClassifier; GaussianNB; KNeighborsClassifier; SGDClassifier; and SVC to predict whether a piece of particular news is fake or real. Results show that AdaBoost Classifier with base estimator as Decision Tree of maximum depth 3 and 175 estimators performs best and provides accuracy close to 1 when 43 features were considered. Features NN, CD, VBP, VBG, positive, NNP, JJ, IN, VBN, and unique were found top predictive features that provided accuracy of 0.85 and F-score of 0.87. In future work, we will implement this algorithm on other datasets.

References

1. Lazer, D.M.J., Baum, M.A., Benkler, Y., Berinsky, A.J., Greenhill, K.M., Menczer, F., Metzger, M.J., Nyhan, B., Pennycook, G., Rothschild, D., et al.: The science of fake news. Science **359**(6380), 1094–1096 (2018)
2. Zhuk, D., Tretiakov, A., Gordeichuk, A.: Methods to identify fake news in social media using machine learning. In: Proceedings of the 22st Conference of Open Innovations Association FRUCT, p. 59. FRUCT Oy (2018)
3. Sukhodolov, A.P.: The phenomenon of "fake news" in the modern media space. Eurasian Coop.: Humanit. Aspects, 36 (2017)
4. Undeutsch, U.: The development of statement reality analysis. In: Credibility Assessment, pp. 101–119. Springer (1989)
5. Guacho, G.B., Abdali, S., Shah, N., Papalexakis, E.E.: Semi-supervised content-based detection of misinformation via tensor embeddings. In: 2018 IEEE/ACM International Conference on Advances in Social Networks Analysis and Mining (ASONAM), pp. 322–325. IEEE (2018)
6. Oshikawa, R., Qian, J., Wang, W.Y.: A survey on natural language processing for fake news detection. arXiv preprint arXiv:1811.00770 (2018)
7. Zhou, X., Jain, A., Phoha, V.V., Zafarani, R.: Fake news early detection: a theory-driven model. arXiv preprint arXiv:1904.11679 (2019)
8. Pérez-Rosas, V., Kleinberg, B., Lefevre, A., Mihalcea, R.: Automatic detection of fake news. arXiv preprint arXiv:1708.07104 (2017)
9. Pennebaker, J.W., Boyd, R.L., Jordan, K., Blackburn, K.: The development and psychometric properties of liwc2015. Technical report (2015)
10. Volkova, S., Shaffer, K., Jang, J.Y., Hodas, N., Separating facts from fiction: linguistic models to classify suspicious and trusted news posts on twitter. In: Proceedings of the 55th Annual Meeting of the Association for Computational Linguistics (Volume 2: Short Papers), pp. 647–653 (2017)
11. Wang, Y., Ma, F., Jin, Z., Yuan, Y., Xun, G., Jha, K., Su, L., Gao, J.: Eann: event adversarial neural networks for multi-modal fake news detection. In: Proceedings of the 24th ACM SIGKDD International Conference on Knowledge Discovery & Data Mining, pp. 849–857. ACM (2018)
12. Gravanis, G., Vakali, A., Diamantaras, K., Karadais, P.: Behind the cues: a benchmarking study for fake news detection. Expert Syst. Appl. **128**, 201–213 (2019)
13. Burgoon, J.K., Blair, J.P., Qin, T., Nunamaker, J.F.: Detecting deception through linguistic analysis. In: International Conference on Intelligence and Security Informatics, pp. 91–101. Springer (2003)
14. Newman, M.L., Pennebaker, J.W., Berry, D.S., Richards, J.M.: Lying words: predicting deception from linguistic styles. Pers. Soc. Psychol. Bull. **29**(5), 665–675 (2003)
15. Zhou, L., Burgoon, J.K., Nunamaker, J.F., Twitchell, D.: Automating linguistics-based cues for detecting deception in text-based asynchronous computer-mediated communications. Group Decis. Negot. **13**(1), 81–106 (2004)
16. Reis, J.C.S., Correia, A., Murai, F., Veloso, A., Benevenuto, F., Cambria, E.: Supervised learning for fake news detection. IEEE Intell. Syst. **34**(2), 76–81 (2019)
17. Risdal, M.: Getting real about fake news (2017). https://www.kaggle.com/mrisdal/fake-news
18. Guardian News and Media Limited. The Guardian Open Platform (2017). http://open-platform.theguardian.com/

Handwritten Gujarati Character Detection Using Image Processing and Recognition Using HOG, DHV Features and KNN

Rupali N. Deshmukh$^{(\boxtimes)}$, Tiwari Saurabh, Sant Shambhavi, Bhopale Jayant, and Kachroo Mokshi

Department of Information Technology, Fr C. Rodrigues Institute of Technology, Navi-Mumbai, India
rupali7580@gmail.com

Abstract. This paper presents a compendious system which can be used to detect and recognize Gujarati handwritten scripts using image processing techniques and machine learning concepts. It focuses on the essential steps that are involved in the entire process. The complexity of the entire process is on account of variety of handwritings and the curves of letters involved in Gujarati scripts. This paper presents the essential steps of image processing including thresholding, noise removal, character detection and comprehensive steps of machine learning including training and testing of the system for recognition purpose in order to deal with the difficulties and challenges possessed by the script. This paper can be used by researchers and technology enthusiasts to develop systems for Gujarati script recognition. The paper aims to present and deal with special properties associated with Gujarati script.

Keywords: Clustering · Feature extraction · Grayscale · Histogram · Histogram of oriented gradients · Noise removal · OCR · Pixel density · DHV features · KNN · Thresholding · Unicodes

1 Introduction

Optical character recognition (OCR) is an intricate process involving the detection and recognition of handwritten as well as machine-printed characters. Although the first concrete work in this field was done in 1900 by Tyurin, its roots were visible in 1870 itself in the form of aid for the visually handicapped [1]. In 1914, Emanuel Goldberg developed a machine that read characters and converted them into standard telegraph code (early OCR) [2]. Since then it has continuously flourished and today it's getting used in various sectors such as education, navigation, postal processing, security, authentication and language identification. In recent years, OCR has been able to recognize Chinese as well as Japanese characters which were once considered as an extremely difficult task [4, 5]. Proceeding in the same direction, In this paper, we primarily focus on the detection and recognition of Gujarati characters.

© The Author(s), under exclusive license to Springer Nature Switzerland AG 2021
M. Tripathi and S. Upadhyaya (Eds.): ICDLAIR 2019, LNNS 175, pp. 65–72, 2021.
https://doi.org/10.1007/978-3-030-67187-7_8

2 Related Work

Hardly any work can be seen in literature towards detecting Gujarati characters but some amount of work has been done towards recognizing Gujarati characters. Antani et al. [6] have tried to recognize Gujarati characters using minimum Hamming distance, KNN classifier and managed to achieve an accuracy of 67%. Prasad et al. [7] have divided Gujarati characters into 6 classes, each class contains characters having a similar shape and then neural network, template matching was used to classify the characters belonging to a particular set. Naik et al. [8] have tried to recognize Gujarati numerals using a support vector machine(SVM) with linear, RBF and polynomial kernel. The study has made use of chain code as a feature to train the SVM model. SVM with Polynomial kernel gave the best accuracy of 95% with 10 numerals, also one common trend was observed with all kernels that as the number of numeral increases classification accuracy decreases.

3 System Details

The system takes the input in the form of image. The extension of the image can be .jpg, .jpeg or .png etc. The image is first converted to grayscale format therefore the pixel intensity will now be ranged from 0 to 255, where pixel with intensity 0 is the black pixel and pixel with intensity 255 is the white pixel. Further, thresholding is performed which helps to remove some noise. Next the major step of segmentation is done so as to detect characters. For detecting the individual characters clustering is used, such that each character is an individual cluster. Image processing algorithms used are such that implementation is irrespective of the font style, thickness, size and color of text. Once the characters are detected, an intensive training will be performed on the system. The learning involved is a supervised one. For every detected character, that is for every cluster a unique label is given and system is trained with that data. Later on for testing, unknown samples are fed to the system and system would be able to recognize character. The recognition is in terms of labels which will then be displayed on screen using unicodes.

3.1 System Architecture

Figure 1 shows a sequence of steps followed to generate a set of Unicode corresponding to an image of a Gujarati script. The system takes the Gujarati script as an input, then preprocessing is done to remove noise from the image. Thereafter, segmentation is done to detect individual characters. Once the characters are detected then features are extracted and passed to the KNN classifier that classifies and gives Unicode as the final output for every detected character.

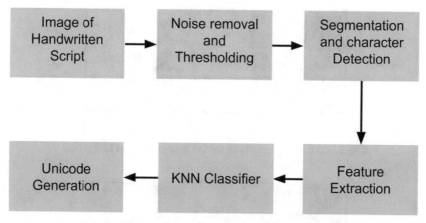

Fig. 1. Flow of handwritten script recognition system [9].

4 Methodology

4.1 Image Thresholding

Image thresholding is implemented on the gray scale image to separate the image from clearly as background and foreground portion. A certain threshold value for intensity between 0 to 255 allows us to keep the pixel of our interest by. Different algorithms such as binary thresholding, adaptive thresholding [9] and Otsu thresholding gives different results for different values. Binary thresholding classifies the grayscale intensities into two values- 0 or 255. That is either the pixel will be treated as black or white. In adaptive thresholding the central tendency for intensities of neighbourhood pixels are also considered for a pixel and treated as threshold for that pixel and the class is determined accordingly. The results of experiments show that effectiveness of Otsu's algorithm is significantly higher for **character** detection.

4.1.1 OTSU Thresholding Algorithm

Otsu's thresholding method involves iterating through all the possible threshold values and calculating a measure of variance for the pixel levels each side of the threshold, the pixels that either fall in foreground or background. The aim is to find the threshold value where intra variance is minimum and inter variance is maximum. Example- Consider grayscale image below [10] (Figs. 2 and 3).

4.2 Segmentation

After document binarization a top-down segmentation approach is applied. First lines of the documents are detected, then words are extracted and finally words are segmented in characters [11].

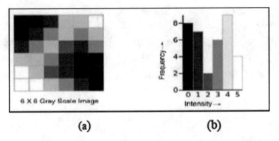

Fig. 2. (a) 6 × 6 grayscale image (b) histogram of an Image [12]

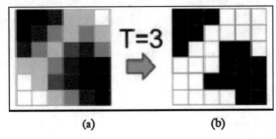

Fig. 3. (a) 6 × 6 grayscale image (b) Result of image after implementation of OTSU thresholding [12].

4.3 Character Detection

Character detection is performed by calculating the relative spacing between words and each character. This helps to obtain vowels and consonants into different clusters. Even if the combination of both is obtained as single cluster, it can be recognized by proper use of training set.The average of these is compared with each of the character and the categorization is done by plotting rectangles around the desired data. Example- Detection of line, words and characters (Figs. 4 and 5).

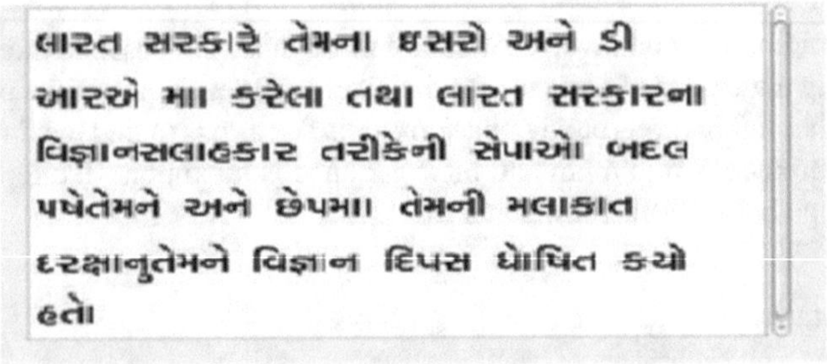

લારત સરકારે તેમના ઇસરો અને ડી
આરએ માા કરેલા તથા લારત સરકારના
વિજ્ઞાનસલાહકાર તરીકેની સેપાઓ બદલ
પધેતેમને અને છેપમાા તેમની મલાકાત
દરક્ષાનુતેમને વિજ્ઞાન દિપસ ધોષિત ક્યે
હતો

Fig. 4. Character detection (digital sample)

Fig. 5. Character detection (handwritten sample)

4.4 Feature Extraction

Feature extraction [11] is the crucial phase of any character recognition system. Features means the useful information that describes an object uniquely. For example the size of edges of a 2-D geometrical object can be used to determine its shape. Similarly, for characters the distribution of intensity values is used to distinguish a character from other characters. For feature extraction Histogram of oriented gradient (HOG) has been used.

4.4.1 Histogram of Oriented Gradients

Histogram of oriented gradients [12] is a very popular way of feature extraction that has been used from 1986 onwards. The orientation and gradient magnitude of every pixel is calculated. This value ranges between 0 to 180 degrees. Generally HOG makes use of 9 bins. Every bin represents discrete orientation value i.e. 0, 20, 40, 60, 80, 100, 120, 140, 160. Orientation value of 180 is equivalent to 0. The orientation of every pixel is resolved along two neighbouring discrete valued orientations. Suppose a pixel has a orientation of 50 degrees then the gradient magnitude of this pixel is resolved along 40 and 60 degree. Thus a histogram of oriented gradients is calculated.

4.4.2 HOG and Character

Orientation of gradients i.e. orientation of change in intensity for an image depends upon two factors: intensity values and shape of objects in an image. Every character has different shape and different pattern of intensity distribution [13]. Thus orientation of gradients for character is unique and hence the histogram of oriented gradient(HOG) for every character is also unique. This property of a character can be used to distinguish between characters and also to recognize them.

4.4.3 HOG Calculation for a Character

Step 1 - The character detected in the character detection phase is resized such that its width and height are integral multiple of 8.

Step 2 - The resized character image is divided into blocks and these blocks are further divided into cells.

Step 3 - For all the pixels of a cell within the block G and θ [13] are calculated as-

$$G = \sqrt{(g\_x)^2 + (g\_y)^2} \tag{1}$$

$$\theta = \tan(g\_y/g\_x) \tag{2}$$

Where G is the gradient magnitude, θ is the orientation, g_x and g_y are the gradient along horizontal and vertical direction respectively.

Step 4 - Using θ, the G is resolved along the discrete orientation valued bins.

Step 5 - Repeat step 3 and step 4 until HOG is calculated for all the pixels of cells within the blocks of a characters.

The output of the phase will be the unicodes[14] of the characters. These unicodes will be used to display all the characters of an image in a digitized form.

5   Results

5.1   DHV Features

DHV features refer to diagonal-horizontal-vertical features of a cluster along the respective lines.On every detected cluster two diagonal lines, two vertical line at a distance 25% and 75% and two horizontal lines at a distance 25% and 75% refer to these lines. Step 1 - Draw the lines over the segment chosen Step 2 - Consider continuous chunks of black pixels as a single sub-cluster. Step 3 - A threshold of 4 is fixed wherein counting of sub-clusters is terminated when threshold is achieved. Step 4 - The coordinates of the clusters are recorded for every line and that makes the core of DHV features.

5.2   Recognition Using KNN

Character recognition algorithm classify a character into a particular class based on their features. KNN algorithm is used for character recognition.Knn (K Nearest Neighbour) is one of the simplest supervised machine learning algorithm. KNN stands for KNearest Neighbours where k is a hyper parameter. It is the simplest character recognition algorithm. It makes use of a voting mechanism for classification. For an unclassified character it conducts a voting wherein nearest K classes participate. The class assigned to unclassified character is the class with maximum of votes (Tables 1 and 2).

Table 1. Accuracy summary results without DHV

K-Value	5	7	9	11	13
No of correct classification	425	455	443	429	420
No of incorrect classification	132	102	114	128	137
No of untrained incorrect classification	54	54	54	54	54
Accuracy (Ignoring untrained characters)	84.49%	90.46%	88.07%	85.28%	83.49%

Table 2. Accuracy summary results with DHV

K-Value	5	7	9	11	13
No of correct classification	455	449	438	419	413
No of incorrect classification	102	108	119	138	144
No of untrained incorrect classification	54	54	54	54	54
Accuracy (Ignoring untrained characters)	90.46%	89.26%	87.07%	83.30%	82.11%

The above table shows a summary for digital image inputs. Maximum accuracy obtained is 90.46%. When the same algorithm is implemented for handwritten samples 75% accuracy is obtained (Figs. 6 and 7).

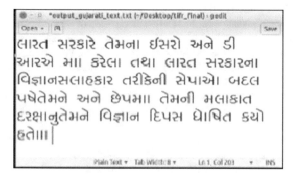

Fig. 6. Recognized characters (digital sample)

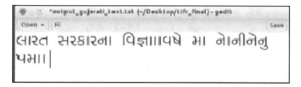

Fig. 7. Recognized character (handwritten sample)

6 Conclusion

The existing Gujarati character script recognition systems have lower accuracy percentages due to complexity involved in the handwritten Gujarati characters. These characters involve several curves and discontinuities which makes detection a difficult process. The system aims to improve the segmentation step, wherein entire character or vowel gets clustered as a separate entity. Later after extensive training vowel or consonant can be easily recognized. The advantage of this system is that every consonant or vowel is segmented as a separate cluster, hence recognition can give better accuracy. With the use of

DHV features, accuracy for digital samples has increased from 84.49% to 90.46%. This success of this system will be a useful contribution in this field of research. One future task of this research could be to incorporating prediction systems which can improvise the accuracy and hence the results.

References

1. Govindan, V.K., Shivaprasad, A.P.: Character recognition - a review. Pattern Recogn. **23**(7), 671–683 (1990)
2. https://history-computer.com/Internet/Dreamers/Goldberg.html
3. Charles, P.K., Harish, V., Swathi, M., Deepthi, C.H.: A review on the various techniques used for optical character recognition. Int. J. Eng. Res. Appl. **2**(1), 659–662 (2012)
4. Zhu, B., Nakagawa, M.: On-line handwritten Japanese characters recognition using a MRF model with parameter optimization by CRF. In: 2011 International Conference on Document Analysis and Recognition (2011)
5. Liu, C.-L., Jaeger, S., Nakagawa, M.: Online recognition of Chinese characters: the state-of-the-art. Online Recognit. Chin. Characters: the state-of-the-art. **26**(2) (2004)
6. Antani, S., Agnihotri, L.: Gujarati character recognition. In: Proceedings of the Fifth International Conference on Document Analysis and Recognition. ICDAR 1999 (Cat. No. PR00318) (1999)
7. Prasad, J.R., Kulkarni, U.V., Prasad, R.S.: Template matching algorithm for gujrati character recognition. In: 2009 Second International Conference on Emerging Trends in Engineering & Technology (2009)
8. Naik, V.A., Desai, A.A.: Online handwritten Gujarati numeral recognition using support vector machine. Int. J. Comput. Sci. Eng. **6**(9), 416–421 (2018)
9. https://docs.opencv.org/3.2.0/d7/d4d/tutorial_py_thresholding.html
10. https://computervisionwithvaibhav.blogspot.in/2015/10/otsu-thresholding.html
11. Patel, C., Desai, A.: Gujarati handwritten character recognition using hybrid method based on binary tree classifier and k-nearest neighbour (2013)
12. Bhopale, J., Deshmukh, R., Dugad, S., Kachroo, M., Sant, S., Tiwari, S.: Character detection using image processing and recognition using HOG. In: International Conference on Recent Trends in Electronics, Communication and Computing (NCRTEEC 2018) (2018)
13. https://www.learnopencv.com/histogram-of-oriented-gradients/
14. https://jrgraphix.net/r/Unicode/0A80-0AFFSS

Fake and Live Fingerprint Detection Using Local Diagonal Extrema Pattern and Local Phase Quantization

Rohit Agarwal[1(✉)], A. S. Jalal[1], Subhash Chand Agrawal[1], and K. V. Arya[2]

[1] GLA University, Mathura 281406, UP, India
{rohit.agrwal,asjalal,subhash.agrawal}@gla.ac.in
[2] ABV-IIITM, Gwalior, MP, India
kyarya@iiitm.ac.in

Abstract. Fingerprint is widely used physical human trait for uniquely identification and verification of human but use of fingerprint is becoming very challenging due to spoofing attacks. In this paper, we present new combination of local diagonal extrema pattern and local phase quantization descriptors to extract the features to generate feature vector for training and testing fingerprint images. Local diagonal extrema pattern finds the relationship between center pixels with diagonal neighboring pixels and local phase quantization extracts the local phase information by using short-term Fourier transform. LDEP reduces the dimensionality problem and forms a good combination of feature descriptor with LPQ. Combined extracted features of training and testing images using both descriptors are passed to Support Vector Machine for discriminating live and fake fingerprints. Experiments have been performed on LivDet2009 dataset and results show the effective and efficient performance of the proposed system. The proposed system achieved good accuracy and very less error rate in comparison to the different descriptors.

Keywords: Biometric · Fingerprint · Liveness detection · Descriptor

1 Introduction

In recent decade, use of biometrics is being increased day by day for human identification and verification. Biometrics has advantages over conventional authentication systems such as PIN, and password based systems due to its uniqueness, permanence, convenience of use, implausibility of stolen. There are various biometric traits (fingerprint, iris, palm print, finger vein, finger knuckle, face, ear, hand geometry etc.) available for identification and verification but fingerprint is widely used due to its easy user acceptance, highly distinctive for the twins, high accuracy, low cost sensors. In spite of emerging as a prominent choice for the security concerning to person authentication, fingerprint is vulnerable to spoof attacks. Espinoza et al. [1] concluded that current state-of-art sensors can be deceived using spoof attacks, created with or without user cooperation. This motivates us for the development of a novel method to anti-spoofing system. The

© The Author(s), under exclusive license to Springer Nature Switzerland AG 2021
M. Tripathi and S. Upadhyaya (Eds.): ICDLAIR 2019, LNNS 175, pp. 73–81, 2021.
https://doi.org/10.1007/978-3-030-67187-7_9

remaining sections of this paper are structured as follows: Sect. 2 introduces the state-of-art, Sect. 3 discusses the proposed approach, Sect. 4 presents the experimental results and Sect. 5 has conclusion of this paper.

2 Related Work

In past decades, important researches have been done to study the vulnerabilities of biometric system to indirect and direct attacks. Direct attack made to sensor by using artificial biometric traits such as gummy fingers or high quality iris printed images while indirect attack is done against inner module of the system. Many authors have presented their work to identify fake fingerprints.

Ojala et al. [2] introduced a local binary pattern as a feature descriptor for texture classification, which has adopted to make distinction between fake and live fingerprint [3]. Ghiani et al. [4] adopted blurred insensitive texture classification Local Phase Quantization (LPQ) [5] to detect the fingerprint liveliness. Nikam et al. [6] applied Grey-Level Co-occurrence Matrix (GLCM) and computed ten texture features-entropy, maximum probability, contrast, angular second moment etc., to detect live and fake fingerprint. In 2011, Galbally et al. [7] presented a ridge-feature based and minutiae based system to detect real and fake fingerprints. In 2012, Galbally et al. [8] proposed a novel finger parameterization using quality related features and experiments performed on ATVS DB and LIVDET 2009 datasets.

Curse of dimensionality is the main problem with local binary pattern which is caused by the increasing number of local neighboring pixels. Local Binary Pattern computes the relationship of central pixels with its neighboring pixels.

To handle the dimensionality problem, we have adopted Local Diagonal Extrema Pattern [9] to identify fake and liveness of a fingerprint. LDEP (Local Diagonal Extrema Pattern) is very effective and efficient to discriminate the fake and live fingerprints by creating relationship between central pixels and local diagonal extremas. We have also adopted Local Phase Quantization which is insensitive to the blurring effects; and combined with LDEP to enhance the detecting capability of Live and Fake Fingerprints. The proposed LDEP and LPQ combined approach outperforms over LBP [3, 10], LPQ [4], LDEP [9], LBP + LPQ [4], CLBP + WT [12] and SMD [13] in fake and live fingerprint detection.

3 Proposed System

In this section, we have proposed Local Diagonal Extrema Pattern and Local Phase Quantization combined features sets for the fake and live images. LDEP has been adopted as a feature descriptor which computes relationship between central pixels and its local diagonal neighboring pixels. LPQ uses local phase information extracted using short-term Fourier Transform; makes capable for finding spectrum differences between fake and live fingerprints. Figure 1 depicts the proposed system. In this, we have read the fingerprint images and dimension is reduced to 256×256. After this, LPQ + LDEP feature set is generated for the test and train fingerprint images. Then, this feature sets are passed to the Support Vector Machine to classify the test fingerprint images into fake and live fingerprint images. Proposed system consists of the following steps:

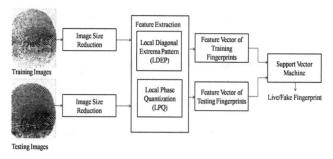

Fig. 1. Proposed system for live and fake fingerprint detection

3.1 Image Size Reduction

In preprocessing step, we have reduced the dimension of each image to 256 x 256.

3.2 Feature Extraction

A) Local Diagonal Extrema Pattern: First step is to compute the local diagonal minima and maxima using first order local diagonal derivative of any center pixel. Here, diagonal neighbors have been used to reduce the dimension of feature vector. We consider a gray scaled image M of dimension $m \times n$ where m is no. of rows and n is no. of columns. Suppose $P^{i,j}$ is a center pixel value at i^{th} row and j^{th} column and $P_t^{i,j}$ is t^{th} diagonal neighbor pixel of any center pixel where $t \in [1, 4]$. Let us consider $I_t^{i,j}$ is intensity value of $P_t^{i,j}$ and $I^{i,j}$ is the intensity of $P^{i,j}$. There are three diagonal derivatives computed to find relationship between each diagonal with other remaining diagonals which helps to find local diagonal extremas. Then, following formula is used to compute relation between local diagonal extrema and center pixels:

$$\overrightarrow{I}_{t,\gamma}^{i,j} = I_{(1+\text{mod}(\ t+\gamma,4))}^{i,j} - I_t^{i,j} \tag{1}$$

Here, $t \in [1, 4]$, $\gamma \in [0, 2]$ *and* $\text{mod}(\lambda 1, \lambda 2)$ remainder of division of $\lambda 1$ *by* $\lambda 2$. We consider $\tau_{\max}$ *and* $\tau_{\min}$ are the indexes for the central pixel $P^{i,j}$ whose intensity is $I^{i,j}$ *and* $I_{\tau \max}^{i,j}$ *and* $I_{\tau \min}^{i,j}$ are the local diagonal maxima and minima values respectively.

We assume $LDEP^{i,j}$ is the binary pattern for local diagonal extrema pattern of $P^{i,j}$ when R is the distance of local diagonal neighbors. $LDEP^{i,j}$ is generated using following formula:

$$LDEP^{i,j} = (LDEP_1^{i,j}, LDEP_2^{i,j}, \ldots LDEP_{\dim}^{i,j}) \tag{2}$$

Where dim is dimension length and $LDEP_k^{i,j}$ is computed using following formula:

$$LDEP_k^{i,j} = \begin{cases} 1, & \text{if } k = (\tau \max + 8\delta) \text{ or } k = (\tau \min + 4 + 8\delta) \\ 0 & \text{otherwise} \end{cases} \tag{3}$$

Where δ is an extrema center relation factor which is computed according to the given formula:

$$\delta = \begin{cases} 1, \text{ if } (sign\,(\Delta^{i,j}_{max}) = 0 \text{ and } sign(\Delta^{i,j}_{min}) = 0) \\ 2, \text{ if } (sign\,(\Delta^{i,j}_{max}) = 1 \text{ and } sign(\Delta^{i,j}_{min}) = 1) \\ 3, \text{ otherwise} \end{cases} \tag{4}$$

Where $\Delta^{i,j}_{max}$ and $\Delta^{i,j}_{min}$ are computed as following formula:

$$\Delta^{i,j}_{max} = I^{i,j}_{max} - I^{i,j} \tag{5}$$

$$\Delta^{i,j}_{min} = I^{i,j}_{min} - I^{i,j} \tag{6}$$

Finally, we consider LDEP over the image M is as follows:

$$LDEP = (LDEP_1, LDEP_2, ...LDEP_k) \tag{7}$$

Where $LDEP_k$ is the kth element of LDEP which can be computed using given formula:

$$LDEP_k = \frac{1}{(m-2R)(n-2R)} \sum_{i=R+1}^{m-R} \sum_{j=R+1}^{n-R} LDEP^{i,j}_k \tag{8}$$

B) Local Phase Quantization: Local Phase Quantization is blur invariance which uses local phase information. This information is extracted Over $m \times m$ neighborhood Nx at each pixel position x of the image f(x) is defined as:

$$F(u, x) = \sum_{y \in Nx} f(x - y)\, e^{-j2\pi\, u^T y} = w_u^T fx \tag{9}$$

Where w_u is considered as a basis vector of 2-Dimensional DFT at u frequency, and $f(x)$ is another vector containing all $m \times m$ image samples from N_x. In this LPQ, we considered four coefficients corresponding to 2-Dimensional frequencies $u1 = [a, 0]^T$, $u2 = [0, a]^T$, $u3 = [a, a]^T$ and $u4 = [a, -a]^T$ where a scalar frequency is.

$$Let \ F^c_x = [F(u1, x),\ F(u2, x),\ F(u3, x),\ F(u4, x)] \tag{10}$$

$$and \ F_x = [Re\{F^c_x\},\ Im\{F^c_x\}]^T \tag{11}$$

where $Im\{.\}$ and $Re\{.\}$ gives imaginary and real part of a complex number. The corresponding $8\ m \times m$ transformation matrix is as follows:

$$W = [Re\{w_{u1}, w_{u2}, w_{u3}, w_{u4}\},\ Im\{w_{u1}, w_{u2}, w_{u3}, w_{u4}\}]^T \tag{12}$$

$$so \ that \ \ Fx = Wfx \tag{13}$$

After this process, Gaussian distribution Gx is calculated for all image positions i.e. $X \in \{X1, X2, X3, X4, \ldots\ldots XN\}$ and we quantize to resulting vectors using a single scalar quantizer:

$$qj = \begin{cases} 1, & if\ gj \geq 0 \\ 0, & otherwise \end{cases} \tag{14}$$

Where g_j is the jth component of Gx. We represent quantized coefficients between 0–255 as integer constants using binary coding:

$$b = \sum_{j=1}^{8} q_j 2^{j-1} \tag{15}$$

Finally, we get *256* dimensional feature vectors after composing histograms of all integer values from all positions of image.

C) Classification: The advantage of SVM over the other learning algorithms is that it can be analyzed theoretically using the concepts from computational learning theory, and at the same time it provides good performance for real world problem. Proposed approach is two class classification problems. We have adopted SVM (Support Vector Machine) to classify fingerprints into live and fake categories. We pass the extracted features set (training feature set and testing feature set) to the SVM classifier where SVM classify to the test fingerprint images into fake and live category efficiently.

4   Experiment Results and Discussion

In this section, we present the experiment results of the proposed system. The experiments have been performed on LivDet2009 [11] dataset to detect fake and live fingerprint images. We have compared our proposed approach I.E. LPQ + LDEP with LBP [10], LPQ [4], LDEP [9], LBP + LPQ [4], CLBP + WT [12] and SMD [13] descriptors to show its discriminative ability.

LivDet2009 Dataset: This dataset consists of 17993 fingerprint images which have been captured through three sensors-Biometrika, Crossmatch and Identix. Each sensor has been used to produce fingerprint using different materials such as gelatin, playdoh and silicon. The data set for the final evaluation is constituted of three sub-sets, which contain live and fake fingerprint images from three different optical sensors. Table 1 shows the details of sensors of LivDet2009 dataset used in proposed method and Table 2 describe training and testing samples used in proposed approach.

We have extracted feature vectors of training and testing images of all type sensors. We trained SVM with feature vector of training images and performed 3-fold and 10-fold cross validation for testing images. We have evaluated performance of LBP [10], LPQ [4], LDEP [9], LBP + LPQ [4], CLBP + WT [12] and SMD [13] descriptors for the live and fake fingerprint detection to compare with proposed approach LPQ + LDEP. Table 3 shows the performance of all the above mentioned descriptors for 3-fold cross validation. We can see accuracy of all descriptors with proposed combined descriptor in Table 3 where proposed system outperforms to LBP [10], LPQ [4], LDEP [9], LBP + LPQ [4], CLBP + WT [12] and SMD [13 descriptors. We have also evaluated the

Table 1. LivDet2009 dataset

DATASET	Scanners	Model No.	Resolution (dpi)	Image size
Dataset #1	Crossmatch	Verifier 300LC	500	480 x 640
Dataset #2	Identix	DFR2100	686	720 x 720
Dataset #3	Biometrika	FX2000	569	312 x 372

Table 2. LivDet2009 dataset sampling for training and testing

Sensor	Training(Live/Fake)	Test(Live/Fake) Subhead
Biometrika	520 A + 520 S	1473 A + 1480 S
CrossMatch	1000A + 344G + 346P + 310S	3000A + 1036G + 1034P + 930S
Identix	750A + 250G + 250 P + 250S	2250A + 750G + 750P + 750S

Note: A-Alive, G-Gelatin, P-Playdoh and S-Silicon

combination of LBP + LDEP which outperformed to LBP [10], LPQ [4], LDEP [9], CLBP + WT [12] and SMD [13 descriptors individually but could not outperformed to the combination of LBP + LPQ [4] descriptors.

Table 3. Performance on LIVDET2009 by the 3-fold Cross Validation

Descriptors/Sensors	Biometrika	CrossMatch	Identix
LBP [3, 10]	80.52	91.60	74.44
LPQ [4]	93.69	91.53	80.04
LDEP [9]	81.71	86.15	90.57
LBP + LPQ [4]	92.14	92.95	97.95
LBP + LDEP	82.26	92.06	96.20
CLBP + WT [12]	87.54	93.2	86.7
SMD [13]	88.3	90.8	87.9
Proposed (LPQ + LDEP)	**94.65**	**93.12**	**98.00**

To evaluate the performance of all above mentioned descriptors on the basis of less no. of testing images; we have applied 10-fold cross validation and presented accuracy of all descriptors in Table 4 which again depicts that the proposed system performs well in comparison to all listed descriptors in the table.

The performance of the proposed system and other descriptors are evaluated in terms of ACE (Average Classification Error) [12]. ACE can be defined as per given formula 16:

$$ACE = (FLR + FFR)/2 \qquad (16)$$

Where, False Living Rate (FLR) can be defined as the percentage of fake fingerprints misclassified as real fingerprint class and False Fake Rate (FFR) can be defined as percentage of real fingerprints misclassified as fake fingerprint class.

Table 4. Performance on LIVDET2009 by the 10-fold cross validation

Descriptors/Sensors	Biometrika	CrossMatch	Identix
LBP [3, 10]	80.55	91.81	74.49
LPQ [4]	94.11	91.46	79.33
LDEP [9]	81.46	87.00	90.95
LBP + LPQ [4]	91.24	92.73	97.93
CLBP + WT [12]	87.54	93.2	86.7
SMD [13]	89.8	91.7	88.24
LBP + LDEP	83.07	92.35	96.37
Proposed (LPQ + LDEP)	**94.64**	**93.20**	**98.13**

Table 5. Average Classification Rate (ACE) of proposed system and other listed descriptors on Livdet2009 dataset

Descriptors/Sensors	Biometrika	CrossMatch	Identix
LBP [3, 10]	17.00	08.18	22.59
LPQ [4]	05.98	08.47	17.33
LDEP [9]	18.19	13.66	09.21
LBP + LPQ [4]	07.83	07.25	02.06
CLBP + WT [12]	4.21	5.29	3.27
SMD [13]	3.82	6.81	2.87
LBP + LDEP	16.05	07.93	03.79
Proposed (LPQ + LDEP)	**5.11**	**6.79**	**1.86**

Table 5 clearly proves the performance of the proposed system and outperforms to other descriptors. Average classification error rate of proposed system is 5.11, 6.79 and 1.86 respectively. These ACE rates of Biometrika, CrossMatch and Identix sensors of proposed system are lesser than the other descriptors LBP [10], LPQ [4], LDEP [9], LBP + LPQ [4], LBP + LDEP, CLBP + WT [12] and SMD [13].

5 Conclusion

In this paper, we have presented an effective and efficient anti-spoofing system which works well to detect live and fake fingerprints. We have combined two efficient descriptors i.e. Local Diagonal Extrema Patterns and Local Phase Quantization to extract the features of fingerprint images and passed to Support Vector Machine to detect live and fake fingers. Proposed system yields good accuracy and less Average Classification Error Rate. Experimental results proved the superiority of our proposed system which outperforms to different descriptors of state-of-art. In order to develop new optimum fingerprint recognition which will work with any variety of information from the variability of fingerprint databases, the neural network approaches are often employed in the future work.

References

1. Espinoza, M., Champod, C., Margot, P.: Vulnerabilities of fingerprint reader to fake fingerprints attacks. Forensic Sci. Int. **204**(1–3), 41–49 (2011)
2. Ojala, T., Pietikäinen, M., Mäenpää, T.: Multiresolution gray-scale and rotation invariant texture classification with local binary patterns. IEEE Trans. Pattern Anal. Mach. Intell. **7**, 971–987 (2002)
3. Kim, W.: Fingerprint liveness detection using local coherence patterns. IEEE Sig. Process. Lett. **24**(1), 51–55 (2016)
4. Ghiani, L., Marcialis, G.L., Roli, F.: Fingerprint liveness detection by local phase quantization. In: Proceedings of the 21st International Conference on Pattern Recognition (ICPR 2012), pp. 537–540. IEEE, November 2012
5. Ojansivu, V., Heikkilä, J.: Blur insensitive texture classification using local phase quantization. In: International Conference on Image and Signal Processing, pp. 236–243. Springer, Heidelberg, July 2008
6. Nikam, S.B., Agarwal, S.: Wavelet energy signature and GLCM features-based fingerprint anti-spoofing. In: 2008 International Conference on Wavelet Analysis and Pattern Recognition, vol. 2, pp. 717–723. IEEE, August 2008
7. Galbally, J., Fierrez, J., Alonso-Fernandez, F., Martinez-Diaz, M.: Evaluation of direct attacks to fingerprint verification systems. Telecommun. Syst. **47**(3–4), 243–254 (2011)
8. Galbally, J., Alonso-Fernandez, F., Fierrez, J., Ortega-Garcia, J.: A high performance fingerprint liveness detection method based on quality related features. Future Gen. Comput. Syst. **28**(1), 311–321 (2012)
9. Dubey, S.R., Singh, S.K., Singh, R.K.: Local diagonal extrema pattern: a new and efficient feature descriptor for CT image retrieval. IEEE Sig. Process. Lett. **22**(9), 1215–1219 (2015)
10. Kulkarni, S.S., Patil, H.Y.: A fingerprint spoofing detection system using LBP. In: 2016 International Conference on Electrical, Electronics, and Optimization Techniques (ICEEOT), pp. 3413–3419. IEEE, March 2016
11. Marcialis, G.L., Lewicke, A., Tan, B., Coli, P., Grimberg, D., Congiu, A., ... & Schuckers, S.: First international fingerprint liveness detection competition—LivDet 2009. In: International Conference on Image Analysis and Processing, pp. 12–23. Springer, Heidelberg (2009)

12. Kundargi, J., Karandikar, R.G.: Fingerprint liveness detection using wavelet-based completed LBP descriptor. In: Proceedings of 2nd International Conference on Computer Vision & Image Processing, pp. 187–202. Springer, Singapore (2018)
13. Ri, G.I., Kim, M.C., Ji, S.R.: A Stable Minutia Descriptor based on Gabor Wavelet and Linear Discriminant Analysis. arXiv preprint arXiv:1809.03326(2018)

A Survey on Human Action Recognition in a Video Sequence

A. Bharathi$^{(\boxtimes)}$ and M. Sridevi

National Institue of Technology, Trichy, Tiruchirappalli, India
bharathisriramd@gmail.com, msridevi@nitt.edu

Abstract. In the recent decade, many new technologies were developed to store and process huge amounts of data. Because of this development, there was a significant increase in video cameras usage and analysis. And also, rapid advancement of machine learning and computer vision led video analysis from recognizing the present state to future state prediction. Such tasks are human action recognition (present state) based on complete actions and human action prediction (future state) based on incomplete actions. These tasks can be used in many areas like surveillance, self driving vehicles, monitoring traffic and entertainment. In last few decades, many robust frame works has been developed for human action recognition and prediction. In this paper, the state-of-art techniques, action datasets, and challenges has been surveyed and presented.

Keywords: Activity recognition · Activity prediction · Video sequence · Deep architectures

1 Introduction

Every action performed by human has some purpose, no matter how difficult it is. For instance, a person is doing exercise means, he/she is interacting to environment with legs, hands, torsos or bodies. The human eye can easily identify the actions performed by human. It is difficult to appoint human to monitor these actions in a video surveillance. The major question is "can a machine predict these actions as human?". One of the major goals in the field of computer vision is to build a machine that can accurately predict human actions as human eye does.

1.1 Human Action Recognition - Overview

Human activity recognition plays a significant role in human to human interaction and inter personal relations. Because it provides information about the identity of the person, their personality, and psychological states. It is difficult to extract. The human ability to recognize another person's activities is one of the main subjects of study of the computer vision and machine learning. As a result of this research, many applications, including video surveillance systems,

M. Tripathi and S. Upadhyaya (Eds.): ICDLAIR 2019, LNNS 175, pp. 82–93, 2021.
https://doi.org/10.1007/978-3-030-67187-7_10

human computer interaction and robotics for human behavior characterization require a multi-activity recognition system.

When attempting to recognize human activities, one must determine the kinetic states of a person, so that the computer can efficiently recognize this activity. Human activities such as walking and running arise naturally in daily life and are relatively easy to recognize. On the other hand, more complex activity such as doing more than one activity at the same time (For example: watching TV and talking to friend), interpret similar activities in different ways (For example: opening refrigerator door may considered as a cooking or cleaning operation), activities overlapped to each other (For example: while person is cooking and the phone rings he are she will stop cooking for a while) are more difficult to identify. Complex activities may be decomposed into other simpler activities which are generally easier to recognize.

1.2   Decomposition of Human Actions

In a field of computer vision research, the word "human action" ranges from simple movement of limb to complex movement of many limbs. This is a dynamic process and usually represented in video. The formal definition of human action in computer vision area is difficult to frame, so examples has been provided in Fig. 1. Example for human actions decomposition are listed below

  i. **Individual action:** the simple actions performed in day to day life such as walking, running, clapping, etc.
 ii. **Human interaction:** the actions can be performed between two humans such as pushing, handshake, etc.
iii. **Human object interaction:** the actions can be performed by human with the use of objects such as actions in sports namely weight lifting, archery, etc.
 iv. **Group action:** the actions can be performed by group of individual and each human can perform same action or different one.
  v. **RGB-d sensor action:** the actions can be captured by using RGB – D sensor.
 vi. **Multi-view action:** the actions can be captured from multiple view points of camera

| i | ii | iii | iv | v | vi |

Fig. 1. Frames from video action dataset

1.3   Tasks in Human Action

There is more advancement in technology which leads to automatic understanding of human actions carried by machine. There are two fundamental tasks in the computer vision community and is represented in Fig. 2.

1) **Action recognition:** recognize a human action from a video containing complete action execution.
2) **Action prediction:** reason a human action from temporally incomplete video data.

a. Action recognition   b. Action prediction

Fig. 2. Fundamental tasks

The most important difference in action recognition and prediction is "when a decision can be made". Action recognition is useful in non-urgent situations. Action prediction infers the actions without observing the full video and is more important in some situations like a system can infer a fall prediction before it happens. In this survey, the recent advances in recognition and prediction are discussed in Sect. 2. The dataset available for action recognition is provided in Sect. 3. In Sect. 4, Experimental results are discussed from various papers. The challenges in HAR are mentioned in Sect. 5. Finally the paper is concluded in Sect. 6.

2   Action Classifiers Used in Human Action Recognition

The action classifier classifies the video into different action classes based on the training samples. The action classifiers are classified into following categories and shown in Fig. 3.

2.1   Direct Classifiers

This type of approaches summarize an action video into a feature vector, and then directly recognize actions using off-the-shelf classifiers such as Support Vector Machine (SVM) [1,38,42], K-Nearest Neighbor (K-NN) [2,40,41], etc. In these methods, action dynamics is characterized in a holistic way using action

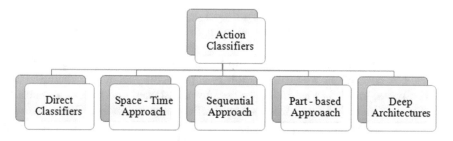

Fig. 3. Action classifiers

shape [2], or using the so-called bag-of-words model, which encodes the distribution of local motion patterns using a histogram of visual words [1,2,38–40].
As shown in Fig. 4, these approaches first detect local salient regions using the
spatiotemporal interest point detectors [1,43]. Finally, an action can be represented by a histogram of visual words, and can be classified by a classifier such
as support vector machine. The bag-of-words approaches have been shown to
be insensitive to appearance and pose variations. However, they do not consider
the temporal characteristics in human actions, as well as the structural information of human actions, which is addressed by sequential approaches [27,37] and
space-time approaches [42], respectively.

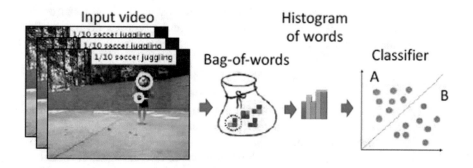

Fig. 4. Bag – of – words models

2.2   Space-Time Approaches

Although direct approaches have shown promising results on some action
datasets [1,39], they do not consider the spatiotemporal correlations between
local features, and do not take the potentially valuable information about the
global spatio-temporal distribution of interest points into account. This problem
was addressed in [34], which learns a global Gaussian Mixture Model (GMM)
using the relative coordinates features, and uses multiple GMMs to describe

the distribution of interest points over local regions at multiple scales. The spatiotemporal distribution of interest points is described by a Directional Pyramid Co-occurrence Matrix in (DPCM) [35]. DPCM characterizes the co-occurrence statistics of local features as well as the spatio-temporal positional relationships among the concurrent features. Graph is a powerful tool for modeling structured objects, and it is used in [36] to capture the spatial and temporal relationships among local features. A novel family of context-dependent Graph Kernels (CGKs) is proposed in [36] to measure the similarity between the two-graph models. Although the above methods have achieved promising results, they are limited to small datasets as they need to model the correlations between interest points which are explosive on large datasets.

2.3   Sequential Approaches

This line of work captures temporal evolution of appearance or pose using sequential state models such as hidden Markov Models (HMMs) [28–30], Conditional Random Fields (CRFs) [25,26,31,32] and structured support vector machine (SSVM) [27,33]. These approaches treat a video as a composition of temporal segments or frames. The work in [28] considers human routine trajectory in a room, and use a two-layer HMMs to model the trajectory. Recent work in [37] shows that representative key poses can be learned to better represent human actions. This method discards a number of non-informative poses in a temporal sequence, and builds a more compact pose sequence for classification. Nevertheless, these sequential approaches mainly use holistic features from frames, which are sensitive to background noise and generally do not perform well on challenging datasets.

2.4   Part-Based Approaches

Part-based approaches consider motion information from both the entire human body as well as body parts. The benefit of this line of approaches is that it inherently captures the geometric relationships between body parts, which is an important cue for distinguishing human actions. A constellation model was proposed in [23], which models the position, appearance and velocity of body parts. Inspired by [23], a part-based hierarchical model was presented in [24], in which a part is generated by the model hypothesis and local visual words are generated from a body part as shown in Fig. 5.

2.5   Deep Architectures

Action features learned by deep learning techniques has been popularly investigated [9–21] in recent years. The two major variables in developing deep networks for action recognition are convolution operation and temporal modeling, leading to a few lines of networks. Convolution operation is one of the fundamental components in deep networks for action recognition, which aggregates pixel values

Fig. 5. Constellation model for part prediction

in a small spatial (or spatiotemporal) neighborhood using a kernel matrix. 2D convolution over images is one of the basic operations in deep networks, and thus it is straightforward to use 2D convolution on video frames as shown in Fig. 6(a). The work in [30] presented a single-frame architecture based on 2D CNN model, and extracted a feature vector for each frame. Such a 2D convolution network (2D ConvNet) also enjoys the benefit of using the networks pretrained on large-scale image datasets such as ImageNet. However, 2D ConvNets do not inherently model temporal information, and requires an additional aggregation or modeling of such information. As multiple frames are presenting in videos, 3D convolution is more intuitive to capture temporal dynamics in a short period of time and it is given in Fig. 6(b). Using 3D convolution, 3D convolutional networks (3D ConvNets) directly create hierarchical representations of spatio-temporal data [12, 17, 21, 22]. However, the issue is that they have many more parameters than 2D ConvNets, making them hard to train. In addition, they are prevented from enjoying the benefits of ImageNet pre-training.

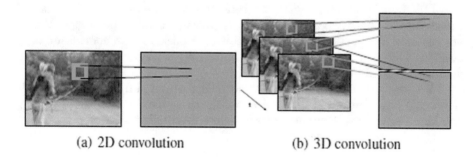

(a) 2D convolution   (b) 3D convolution

Fig. 6. Convolution on 2D and 3D

3 Datasets

The various action recognition dataset are available in video format. These videos are captured in both uncontrolled and controlled environment. These datasets are widely used for recognition with diverse algorithms.

3.1 Controlled Environment

This has datasets captured in controlled environment action performed by individual or by two or more people.

a. Individual Action Recognition Dataset

Weizmann dataset [2] has 90 video sequences with a resolution of 180 * 144. These videos are classified into 10 action classes such as walk, bend or jump, etc. This dataset has been collected by a static camera from 9 different individuals and each person performs all the 10 actions in a simple background.

KTH dataset [1] has 2391 video sequences with a resolution of 160*120. The video has been captured by camera in a homogenous background with 25 individuals. Each individual performs action several times and is classifies into 6 different categories such as "walk", "clap", etc. These videos are captured in 4 different scenarios such as outdoor, indoor, outdoor with variation in scale, outdoor with different cloth.

NRIA Xmas Motion Acquisition Sequences (IXMAS) [5] is a view-invariant dataset for human action recognition. The videos are collected from 11 individuals and each repeats the same action 3 times. The actions are classified into 13 categories such as "scratch, "wave", "bend", etc.

b. Group Action Recognition Dataset

UT-Interaction dataset [6] denotes interaction between humans and consists of 60 videos. These videos have 2 different sets captured in parking lot and a lawn. It has 6 different action such as "handshake", "punching", etc and each set has 10 different videos for each action category.

BIT-Interaction dataset [4] has a total of 400 videos and classified into 8 classes such as "bow", "boxing", "high-five, etc with 50 videos per action. Videos are captured in practical scenes with cluttered backgrounds, moving objects, occluded body parts, and variations in human appearance, variations in scale, illumination condition and viewpoint.

TV Interaction dataset [3] has 300 video clips captured from 20 different TV shows and classified into 4 different actions such as hand shake, high five, hug and kiss. Each frame is annotated by discrete head orientation, upper body of people and person's interaction label.

3.2   Uncontrolled Environment

The controlled environment datasets are not predicting actions in real world scenario. To overcome this problem, the videos are gathered from various internet source and compiled by researcher.

UCF101 dataset [8] has 13320 videos and classified into 101 action classes. These videos are collected from youtube and provide huge variation in actions. The videos also have huge variation of camera motion, viewpoint, pose, appearance and cluttered background.

HMDB51 dataset [7] has 7000 videos and classified into 51 different action classes. Each class has minimum 101 videos. Adding to that, each video is annotated with action and meta-label which denotes various properties such as motion, body parts, viewpoint, quality, and number of individual.

Sports -1M dataset [9] is the largest action dataset has total of 1,133,158 URLs and annotated with action and has 487 action categories. This dataset is very challenging to researcher because of its huge variation of videos.

The detailed information about the video action dataset is given in Table 1.

Table 1. Video action datasets

Dataset	#Subjects	#videos	Views	Year	Environment	# Actions
KTH [1]	25	599	1	2004	Controlled	6
Weizmann [2]	9	90	1	2005	Controlled	10
INRIA AMAS [5]	10	390	5	2006	Controlled	13
UT-Interaction [6]	10	60	6	2010	Controlled	6
BIT-Interaction [4]	50	400	–	2012	Controlled	8
TV Interaction [3]	–	300	–	2010	Controlled	4
UCF101 dataset [8]	–	13320	–	2012	Uncontrolled	101
HMDB51 dataset [7]	–	7000	–	2011	Uncontrolled	51
Sports -1M [9]	–	1,133,158	–	2014	Uncontrolled	487

4   Experimental Result Analysis

Action recognition approaches are used in various datasets. Each dataset has own set of actions and these approaches are developed based on the dataset to recognize the action more precisely and accurately. The experimental results in our survey paper are discussed in terms of accuracy as a measure. The following Table 2 shows the accuracy of action recognition approach in %.

Table 2. Experimental results

Approach used	UCF101	Sports 1M	UCF + Sports	HMDB51
Transfer learning [9]	–	60.9	–	–
Two stream network [12]	89.3	–	–	–
LRCN [15]	82.9	–	–	–
Two stream fusion [22]	92.5	–	–	65.4
Dense trajectory [27]	–	–	88.2	–
LSTM + Conv Pooling [14]	–	73.1	–	–
C3D on SVM [13]	–	–	–	62.2
Optical flow Network [11]	–	–	–	62

Fig. 7. Variations in different viewpoints of camera

5   Challenges in Human Action Recognition

The major challenges of Human action recognition are listed below.

i. Application Domain
Based upon application, the activities and the fine details will vary. For example, in a traffic monitoring system, the main interest is monitoring crowd and in surveillance system, the main interest is finding strange behavior.

ii. Inter-class and Intra-class variations
For one action category, each people will behave differently. For action "running", people can run slow, fast or jump and run. Many different styles can be followed for one category of action. In addition to that same action can be captured from different viewpoints of camera as in Fig. 7 i.e. from top, front and side shows variations in appearance. All these factors results in intra-class variations and pose variations.

iii. Occlusion
Occluded portion has been occurred due to self or other objects and has a major impact in action recognition. Due to loss of occluded portion, the action can recognized wrongly by the system.

iv. Cluttered Background

Many action recognition systems works well for controlled (indoor) environment than uncontrolled (outdoor) environment. Because of the background noise and dynamic background, the performance of system is degraded.

6 Conclusion

The availability of big data and powerful models diverts the research focus about human actions from understanding the present to reasoning the future. This paper presented a complete survey of state-of-the-art techniques for action recognition and prediction from videos. These techniques became particularly interesting in recent decades due to their promising and practical applications in several emerging fields focusing on human movements. Several aspects of the existing attempts including handcrafted feature design, models and algorithms, deep architectures and datasets are investigated and discussed in this survey.

References

1. Schuldt, C., Laptev, I., Caputo, B.: Recognizing human actions: a local SVM approach. In: IEEE ICPR (2004)
2. Blank, M., Gorelick, L., Shechtman, E., Irani, M., Basri, R.: Actions as space-time shapes. In: Proceedings of ICCV (2005)
3. Patron-Perez, A., Marszalek, M., Zisserman, A., Reid, I.: High five: recognising human interactions in tv shows. In: Proceedings of British Conference on Machine Vision (2010)
4. Kong, Y., Jia, Y., Fu, Y.: Learning human interaction by interactive phrases. In: Proceedings of European Conference on Computer Vision (2012)
5. Weinland, D., Ronfard, R., Boyer, E.: Free viewpoint action recognition using motion history volumes. In: Computer Vision and Image Understanding, vol. 104, no. 2–3 (2006)
6. Ryoo, M.S., Aggarwal, J.K.: UT-Interaction Dataset. ICPR contest on semantic description of human activities (SDHA) (2010). http://cvrc.ece.utexas.edu/SDHA2010/HumanInteraction.html
7. Kuehne, H., Jhuang, H., Garrote, E., Poggio, T., Serre, T.: Hmdb: a large video database for human motion recognition. In: ICCV (2011)
8. Khurram Soomro, A.R.Z., Shah, M.: UCF101: a dataset of 101 human action classes from videos in the wild (2012). cRCV-TR-12-01
9. Karpathy, A., Toderici, G., Shetty, S., Leung, T., Sukthankar, R., Fei-Fei, L.: Large-scale video classification with convolutional neural networks. In: CVPR (2014)
10. Yang, Y., Shah, M.: Complex events detection using data-driven concepts. In: ECCV (2012)
11. Wang, K., Wang, X., Lin, L., Wang, M., Zuo, W.: 3D human activity recognition with reconfigurable convolutional neural networks. In: ACM Multimedia (2014)
12. Taylor, G.W., Fergus, R., LeCun, Y., Bregler, C.: Convolutional learning of spatio-temporal features. In: ECCV (2010)
13. Sun, L., Jia, K., Chan, T.-H., Fang, Y., Wang, G., Yan, S.: DL-SFA: deeply-learned slow feature analysis for action recognition. In: CVPR (2014)

14. Pötz, T., Hammerla, N.Y., Olivier, P.L.: Feature learning for activity recognition in ubiquitous computing. In: IJCAI (2011)
15. Le, Q.V., Zou, W.Y., Yeung, S.Y., Ng, A.Y.: Learning hierarchical invariant spatio-temporal features for action recognition with independent subspace analysis. In: CVPR (2011)
16. Heilbron, F.C., Escorcia, V., Ghanem, B., Niebles, J.C.: ActivityNet: a large-scale video benchmark for human activity understanding. In: CVPR (2015)
17. Ji, S., Xu, W., Yang, M., Yu, K.: 3D convolutional neural networks for human action recognition. In: ICML (2010)
18. Hasan, M., Roy-Chowdhury, A.K.: Continuous learning of human activity models using deep nets. In: ECCV (2014)
19. Bengio, Y., Courville, A., Vincent, P.: Representation learning: a review and new perspectives. IEEE Trans. Pattern Anal. Mach. Intell. **35**, 1798–1828 (2013)
20. Simonyan, K., Zisserman, A.: Two-stream convolutional networks for action recognition in videos. In: NIPS (2014)
21. Ji, S., Xu, W., Yang, M., Yu, K.: 3D convolutional neural networks for human action recognition. IEEE Trans. Pattern Anal. Mach. Intell. **35**, 221–231 (2013)
22. Tran, D., Bourdev, L., Fergus, R., Torresani, L., Paluri, M.: Learning spatiotemporal features with 3D convolutional networks. In: ICCV (2015)
23. Fanti, C., Zelnik-Manor, L., Perona, P.: Hybrid models for human motion recognition. In: CVPR (2005)
24. Niebles, J.C., Fei-Fei, L.: A hierarchical model of shape and appearance for human action classification. In: CVPR (2007)
25. Morency, L.-P., Quattoni, A., Darrell, T.: Latent-dynamic discriminative models for continuous gesture recognition. In: CVPR (2007)
26. Sminchisescu, C., Kanaujia, A., Li, Z., Metaxas, D.: Conditional models for contextual human motion recognition. In: International Conference on Computer Vision (2005)
27. Shi, Q., Cheng, L., Wang, L., Smola, A.: Human action segmentation and recognition using discriminative semi-Markov models. IJCV **93**, 22–32 (2011)
28. Duong, T.V., Bui, H.H., Phung, D.Q., Venkatesh, S.: Activity recognition and abnormality detection with the switching hidden semi-Markov model. In: CVPR (2005)
29. Rajko, S., Qian, G., Ingalls, T., James, J.: Real-time gesture recognition with minimal training requirements and on-line learning. In: CVPR (2007)
30. Ikizler, N., Forsyth, D.: Searching video for complex activities with finite state models. In: CVPR (2007)
31. Wang, S.B., Quattoni, A., Morency, L.-P., Demirdjian, D., Darrell, T.: Hidden conditional random fields for gesture recognition. In: CVPR (2006)
32. Wang, L., Suter, D.: Recognizing human activities from silhouettes: motion subspace and factorial discriminative graphical model. In: CVPR (2007)
33. Wang, Z., Wang, J., Xiao, J., Lin, K.-H., Huang, T.S.: Substructural and boundary modeling for continuous action recognition. In: CVPR (2012)
34. Wu, X., Xu, D., Duan, L., Luo, J.: Action recognition using context and appearance distribution features. In: CVPR (2011)
35. Yuan, C., Li, X., Weiming, H., Ling, H., Maybank, S.J.: Modeling geometric-temporal context with directional pyramid co-occurrence for action recognition. IEEE Trans. Image Process. **23**(2), 658–672 (2014)
36. Wu, B., Yuan, C., Hu, W.: Human action recognition based on context-dependent graph kernels. In: Proceedings of the IEEE Conference on Computer Vision and Pattern Recognition, pp. 2609–2616 (2014)

37. Raptis, M., Sigal, L.: Poselet key-framing: a model for human activity recognition. In: CVPR (2013)
38. Marszałek, M., Laptev, I., Schmid, C.: Actions in context. In: IEEE Conference on Computer Vision & Pattern Recognition (2009)
39. Laptev, I., Marszalek, M., Schmid, C., Rozenfeld, B.: Learning realistic human actions from movies. In: CVPR (2008)
40. Laptev, I., Perez, P.: Retrieving actions in movies. In: ICCV (2007)
41. Tran, D., Sorokin, A.: Human activity recognition with metric learning. In: ECCV (2008)
42. Ryoo, M., Aggarwal, J.: Spatio-temporal relationship match: video structure comparison for recognition of complex human activities. In: ICCV, pp. 1593–1600 (2009)
43. Klaser, A., Marszalek, M., Schmid, C.: A spatio-temporal descriptor based on 3d-gradients. In: BMVC (2008)

Analysis of Optimal Number of Cluster Heads on Network Lifetime in Clustered Wireless Sensor Networks

Vipin Pal[1][(✉)], Anju Yadav[2], and Yogita[1]

[1] National Institute of Technology Meghalaya, Shillong, India
vipinrwr@gmail.com, thakranyogita@gmail.com
[2] Manipal University Jaipur, Jaipur, India
anju.anju.yadav@gmail.com

Abstract. For the resource constraint wireless sensor network, clustered architecture has been admitted as energy efficient approach. Clustered architecture reduces the energy consumption of sensor nodes with extra advantage of scalability and fault tolerance to the network. Low Energy Adaptive Clustering Hierarchy (LEACH) and based clustering techniques for wireless sensor networks optimize the number of cluster heads in a round but because of probabilistic nature of the techniques does not give an assurance of optimal number of clusters in each round. Work of this paper addresses that issue and compares the results of optimal number of cluster heads in each round with probabilistic cluster heads in each round. Analysis of simulation results advocates that selection of optimal number of cluster heads in each round is better than random number of cluster head selection.

Keywords: Wireless sensor networks · Cluster · Network lifetime · LEACH

1 Introduction

Wireless sensor networks, collection of various resource constrained sensor nodes deployed over a region of interest, have been well entertained by the research community over past few years because of well wide spread application areas and consequently has been well received by market [6]. Functioning of wireless sensor network depends on the in-collaborative nature of sensor nodes. Sensor nodes sense the region of interest collectively and send the information to the base station. Authentic user can access the information by means of any communication strategy collected at base station [11].

Literature of wireless sensor network have the viewpoint that the region of interest for wireless sensor network to monitor is remote in nature and offer very less interference of human after deployment of sensor nodes, for this reason it is quite difficult to recharge or replace the limited on-board battery unit of sensor nodes [19]. Consequently, economical energy consumption of sensor nodes

M. Tripathi and S. Upadhyaya (Eds.): ICDLAIR 2019, LNNS 175, pp. 94–102, 2021.
https://doi.org/10.1007/978-3-030-67187-7_11

has been the prime design for wireless sensor network from circuitry of sensor boards to communication protocols [3]. Various strategies have been employed to address the above issue of efficient energy consumption and out of these clustering approaches have been implemented rigorously [1]. In clustered wireless sensor network, sensor nodes are grouped into independent clusters with one representative of each cluster, named cluster head. Member nodes of a cluster send information to the respective cluster head and cluster head sends the collected data to base station after performing data aggregation to the collected data that reduces the volume of data to be communicated [1]. To support the load balance of network, the role of cluster head is rotated among the nodes. Clustering approach reduces the overall communication distance of the network that results in reduced energy consumption of sensor nodes [1].

Low Energy Adaptive Clustering Hierarchy (LEACH) [9] protocol has been appraised as break through point for clustering protocols in wireless sensor networks. LEACH and based protocols [4,17] work in rounds and have optimized the number of cluster heads for the rounds that depends on the number of sensor nodes, dimensions of region of interest, round number and base station positioning. The probabilistic nature of these protocols for selecting the cluster head does not assure that in each round optimal numbers of cluster heads are selected. Work of this paper analyses the effect of optimal number cluster heads and compares the network lifetime for optimal cluster head clustering approach and random number cluster head approach. The simulation results suggest that selection of optimal number of cluster heads in each round approach is better than random number of cluster heads in each round.

Rest of the paper is organized in 6 sections. Literature review of clustered wireless sensor network has been presented in Sect. 2. Section 3 and 4 discuss the network Model and problem statement. Analysis and Results have been showcased in Sect. 5. Conclusion of the work has been presented in Sect. 6.

2   Literature Review

Clustering approach has been implemented for energy conservation in wireless sensor network [1]. Clustering approaches have been categorised as Distributed and Centralized in the literature. In distributed clustering approaches selection of cluster heads and formation of clusters is performed locally by the sensor nodes and in oppose of that, in centralized clustering approaches selection of cluster heads and formation of clusters is performed by a centralized unit mostly by base station.

Low Energy Adaptive Clustering Hierarchy (LEACH) is fully distributed clustering approach which is considered as basis of most of clustering approaches proposed in the literature for wireless sensor network. LEACH performs the assigned task in rounds and selects clusters heads in round probabilistically. Nodes select a random number between 0 and 1 and a threshold is calculated that depends upon the predefined optimal number of cluster heads and current round number as in Eq. (1).

$$
T(n) = \begin{cases} \dfrac{p_{opt}}{N - p_{opt} \times (r \, mod \, \frac{N}{p_{opt}})} & \text{if } n \in G; \\ 0 & \text{otherwise.} \end{cases}
\tag{1}
$$

If the selected number by the node is less than threshold for the current round, the node is selected as cluster head. Remaining nodes select the nearest cluster head and constitute the clusters. The probabilistic cluster head selection method ensures that all nodes are selected as cluster head once in the epoch. The nodes send the sense data to cluster head according to the time slots assigned in TDMA schedule.

LEACH protocol has been appraised as quantum leap for energy efficient clustered wireless sensor network [4, 17] and various approaches in the literature extend the work of LEACH. V. Pal et al. in [13] constitutes the balanced size clusters, clusters of same number of member nodes, to have balanced frame length for each cluster. M. Elshrkawey et al. in [7] addressed the problem of uneven frame length in TDMA schedule for uneven size clusters and put the member nodes of smaller nodes in sleep node more time to load balance the network.

LEACH-C [10] a centralized variant of LEACH selects cluster head with the help of simulated annealing. Nodes send the information of location and remaining energy to the base station. Base station selects the p_{opt} number of cluster heads and broadcast the information to the network. J. C. Cuevas-Martinez et al. in [5] applied type-2 fuzzy algorithm for efficient cluster head selection for longer network lifetime. Fuzzy system selects cluster head with consideration of residual energy of the nodes, relative distance to the base station, historical contribution as a Cluster Head and efficiency. [2, 12, 16] applied fuzzy system for better load balanced wireless sensor network considering as LEACH as the basis.

SEP [18] and DEEC [14] protocols successfully embedded the concept of node heterogeneity to increase the network lifetime. SEP protocol introduced two types on sensor nodes – *Normal nodes* and *Advance Nodes*. An advance node has more battery power as compared to normal nodes and plays the role of cluster head more times than the normal nodes in an epoch. DEEC protocol also includes the remaining energy concept for cluster head selection. Both the protocols are well referred approaches for heterogeneous clustering approaches.

3   Network Model

A wireless sensor network with 100 sensor nodes deployed randomly over an area of 100×100 m^2 has been considered for simulation. All sensor nodes are homogeneous in nature in the capacity of battery power, computational capability. Location of base station is outside the application area. The communication model as shown in Fig. 1 [10] has been considered for network simulation.

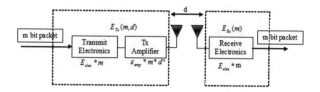

Fig. 1. Communication model

Energy has been consumed by both sender and receiver node during the communication of m-bits information. The energy consumption of sender node depends upon the distance between sender node and receiver node. The energy consumption is according to *Two-Ray Ground model* for distance less than $d_{crossover}$ and *Free-Space model* has been considered otherwise. The energy consumption of sender and receiver node is accordance of Eq. (3) and (4) respectively [15].

$$E_{Tx}(m, d) = E_{Tx-elec}(m) + E_{Tx-amp}(m, d) \tag{2}$$

$$E_{Tx(m,d)} = \begin{cases} m \times E_{elec} + \left(m \times E_{fs} \times d^2 \right) & d < d_{crossover} \\ m \times E_{elec} + \left(m \times E_{two-ray} \times d^4 \right) & d \geq d_{crossover} \end{cases} \tag{3}$$

$$E_{Rx}(m) = m \times E_{elec} \tag{4}$$

where, ϵ_{fs} and $\epsilon_{two-ray}$ signifies the amplification coefficients for the free space communication and two-ray communication models, respectively. For running the electronic circuit of the sender and receiver the amount of energy consumed is E_{elec}. Value of $d_{crossover}$ has been set to 87 m [9]. Table 1 provides the various parameters and corresponding value taken for simulation.

Matlab2010 is used for simulation. 50 different runs of simulation have been performed for each approach and then average of these simulation runs is considered for analysis.

4   Problem Statement

LEACH and based protocols consider that the optimal number of cluster heads has been selected in each round of clustering but the probabilistic nature of cluster head selection does not ensure the selection of optimal number of cluster heads in each round and only ensures that all nodes are selected as cluster head

Table 1. Network parameters and values

Parameter	Value
Number of nodes (N)	100
Network area	$100 \times 100\,\mathrm{m}^2$
Base station location	75, 150
Initial energy	0.5 J
Packet header size	25 bytes
Data packet size	500 bytes
E_{elec}	$50\,\mathrm{nJ/bit}$
E_{fs}	$100\,\mathrm{pJ/bit/m}^2$
$E_{two-ray}$	$0.0013\,\mathrm{pJ/bit/m}^4$

once in the epoch. LEACH protocol has been simulated for $p_{opt} = 5$ and $p_{opt} = 10$. Figure 2 and Fig. 3 shows number of selected cluster heads over the rounds for $p_{opt} = 5$ and $p_{opt} = 10$ respectively. It can be seen from both the figures that the LEACH protocol does not guarantee the selection of optimal number of cluster heads in each round as claimed earlier.

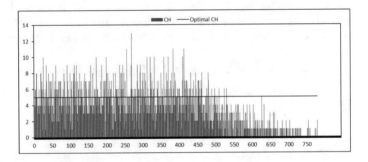

Fig. 2. Number of selected cluster heads over the rounds for $p_{opt} = 5$

It can also be analysed from Table 2 and Table 3. Number of rounds with CHs more than p_{opt}, number of rounds with CHs less than p_{opt} and number of rounds with CHs equal to p_{opt} has been shown in Table 2 and Table 3 for $p_{opt} = 5$ and $p_{opt} = 10$ respectively. It can be analysed that selection of optimal cluster head happened in very few rounds.

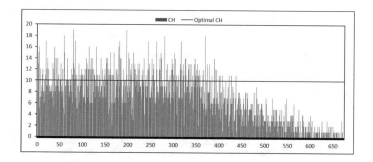

Fig. 3. Number of selected cluster heads over the rounds for $p_{opt} = 10$

Table 2. Cluster heads selected for $p_{opt} = 5$

	No. of rounds	Percentage of rounds
No. of rounds with CHs more than p_{opt}	181	23
No. of rounds with CHs less than p_{opt}	503	65
No. of rounds with CHs equal to p_{opt}	95	12

Table 3. Cluster heads selected for $p_{opt} = 10$

	No. of rounds	Percentage of rounds
No. of rounds with CHs more than p_{opt}	165	24
No. of rounds with CHs less than p_{opt}	465	69
No. of rounds with CHs equal to p_{opt}	45	7

5   Results and Analysis

For simulation and analysis, same network topology with $p_{opt} = 5$ and $p_{opt} = 10$ has been considered for both the approaches. Two network metrics namely *Nodes Alive* and *Network Lifetime* have been examined for performance analysis. Nodes Alive metric defines the stable region [8] which should be lengthened for energy constraint wireless sensor network. Network Lifetime metric defines how much the network can sustain and work in the region of interest.

Figure 4 and Fig. 5 show the number of nodes alive over the clustering rounds for $p_{opt} = 5$ and $p_{opt} = 10$ respectively. It can be seen from the figures that Mod approach has extended stable region as compared to LEACH in both the cases. There is increase of 7% and 11% for stable region in case of Mod approach over LEACH approach for $p_{opt} = 5$ and $p_{opt} = 10$ respectively.

Network lifetime has been measured for First Node Death (FND), time of first node death, Half Node Death (HND), time of half node death in the network, and Last Node Death (LND), time of last node death. In our performance analysis, we consider LND as the time till there are number of alive nodes more than p_{opt}. Figure 6 and Fig. 7 demonstrate the FND, HND and LND over the clustering

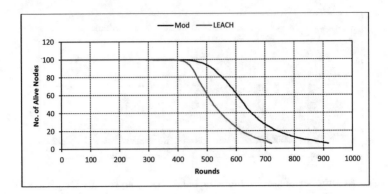

Fig. 4. Number of Nodes Alive over the rounds for $p_{opt} = 5$

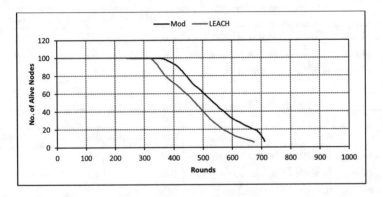

Fig. 5. Number of Nodes Alive over the rounds for $p_{opt} = 10$

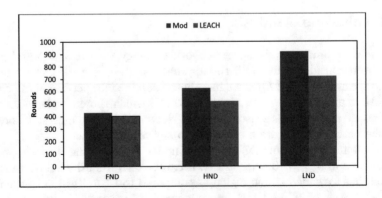

Fig. 6. Network Lifetime for $p_{opt} = 5$

rounds for $p_{opt} = 5$ and $p_{opt} = 10$ respectively. There is increase of 7 %, 20 % and 27 % for the number of clustering rounds for FND, HND and LND respectively in case of Mod approach over LEACH for $p_{opt} = 5$. In the same scenario, there is increase of 11 %, 13 % and 6 % for the number of clustering rounds for FND, HND and LND respectively in case of Mod approach over LEACH for $p_{opt} = 10$.

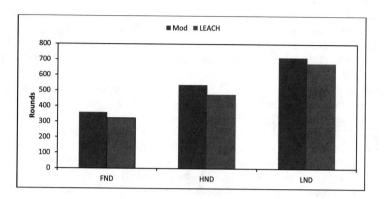

Fig. 7. Network Lifetime for $p_{opt} = 10$

6   Conclusion

Clustering algorithm has been proved as energy efficient mechanism for wireless sensor network. LEACH and based clustering approaches select cluster heads with probabilistic approaches which did not assure selection of predefined optimal number of cluster heads in each round. Work of the paper investigated the effect of probabilistic cluster head selection on network performance. Analysis of simulation results suggested that network performs better for selection of optimal number of cluster heads in each round as compared to probabilistic cluster head selection in each round.

Acknowledgement. The work of the paper has been supported by NATIONAL MISSION ON HIMALAYAN STUDIES (NMHS) sanctioned project titled "Cloud-assisted Data Analytics based Real-Time Monitoring and Detection of Water Leakage in Transmission Pipelines using Wireless Sensor Network for Hilly Regions" (Ref. No.: GBPNI/NMHS-2017-18/SG21).

References

1. Afsar, M.M., Tayarani-N, M.H.: Clustering in sensor networks: a literature survey. J. Netw. Comput. Appl. **46**, 198–226 (2014)
2. Agrawal, D., Pandey, S.: FUCA: fuzzy-based unequal clustering algorithm to prolong the lifetime of wireless sensor networks. Int. J. Commun. Syst. **31**(2), e3448 (2018)

3. Anastasi, G., Conti, M., Di Francesco, M., Passarella, A.: Energy conservation in wireless sensor networks: a survey. Ad Hoc Netw. **7**(3), 537–568 (2009)
4. Arora, V.K., Sharma, V., Sachdeva, M.: A survey on leach and other's routing protocols in wireless sensor network. Optik - Int. J. Light Electron Opt. **127**(16), 6590–6600 (2016)
5. Cuevas-Martinez, J.C., Yuste-Delgado, A.J., Triviño-Cabrera, A.: Cluster head enhanced election type-2 fuzzy algorithm for wireless sensor networks. IEEE Commun. Lett. **21**(9), 2069–2072 (2017)
6. Dargie, W., Poellabauer, C.: Fundamentals of Wireless Sensor Networks: Theory and Practice (2011)
7. Elshrkawey, M., Elsherif, S.M., Wahed, M.E.: An enhancement approach for reducing the energy consumption in wireless sensor networks. J. King Saud Univ. - Comput. Inf. Sci. **30**(2), 259–267 (2018)
8. Handy, M.J., Haase, M., Timmermann, D.: Low energy adaptive clustering hierarchy with deterministic cluster-head selection. In: 4th International Workshop on Mobile and Wireless Communications Network, pp. 368–372 (2002)
9. Heinzelman, W., Chandrakasan, A., Balakrishnan, H.: An application-specific protocol architecture for wireless microsensor networks. IEEE Trans. Wirel. Commun. **1**(4), 660–670 (2002)
10. Heinzelman, W.B.: Application-specific protocol architectures for wireless networks. Ph.D. thesis, Cambridge, MA, USA (2000). aAI0801929
11. Karl, H., Willig, A.: Protocols and Architectures for Wireless Sensor Networks. Wiley, Hoboken (2007)
12. Logambigai, R., Kannan, A.: Fuzzy logic based unequal clustering for wireless sensor networks. Wirel. Netw. **22**(3), 945–957 (2016)
13. Pal, V., Singh, G., Yadav, R.P.: Balanced cluster size solution to extend lifetime of wireless sensor networks. IEEE Internet Things J. **2**(5), 399–401 (2015)
14. Qing, L., Zhu, Q., Wang, M.: Design of a distributed energy-efficient clustering algorithm for heterogeneous wireless sensor networks. Comput. Commun. **29**(12), 2230–2237 (2006)
15. Rappaport, T.: Wireless Communications: Principles and Practice, 2nd edn. Prentice Hall PTR, Upper Saddle Rive (2001)
16. Shokouhifar, M., Jalali, A.: Optimized sugeno fuzzy clustering algorithm for wireless sensor networks. Eng. Appl. Artif. Intell. **60**, 16–25 (2017)
17. Singh, S.K., Kumar, P., Singh, J.P.: A survey on successors of leach protocol. IEEE Access **5**, 4298–4328 (2017)
18. Smaragdakis, G., Matta, I., Bestavros, A.: SEP: a stable election protocol for clustered heterogeneous wireless sensor networks. In: Second International Workshop on Sensor and Actor Network Protocols and Applications (SANPA 2004), Boston, MA (2004)
19. Yan, J., Zhou, M., Ding, Z.: Recent advances in energy-efficient routing protocols for wireless sensor networks: a review. IEEE Access **4**, 5673–5686 (2016)

Frequent Pattern Mining Approach for Fake News Detection

S. Pranave, Santosh Kumar Uppada$^{(\boxtimes)}$, A. Vishnu Priya, and B. SivaSelvan

Computer Science and Engineering, IIITDM Kancheepuram, Kancheepuram, India
{coe15b003,coe18d005,coe15b015,sivaselvanb}@iiitdm.ac.in

Abstract. Spreading of Fake news is not a new problem, many people have been using News or online social media for propaganda or to influence for centuries. The rise of web-generated news on social media makes fake news a more powerful force that challenges traditional journalistic norms.The extensive spread of fake news has the potential for extremely negative impacts on individuals and society. Therefore, fake news detection has recently become an emerging research that is attracting tremendous attention. To help mitigating the negative effects caused by fake news - both to benefit the public and the news ecosystem, it is critical to develop methods to automatically detect fake news on social media. This paper has analyzed the existing approaches to Fake news detection such as Naive Bayes Classifier, Decision tree and has proposed a novel approach for Fake news detection by implementing Association rule based classification ARBC). Experimental results has indicated notable improvement in detection accuracy.

Keywords: Data mining · Twitter · Fake news · Natural language processing · Association rule based classification

1 Introduction

1.1 Data Mining

Data Mining is an interdisciplinary sub-field of computer science with an overall goal to extract information (with intelligent methods) from a data set and transform the information into a comprehensible structure for further use [1]. The above figure illustrates different domains in which Data Mining has been treated as vital. Domains include statistics, machine learning techniques and algorithm analysis, pattern recognition given with extracted and useful features, visualization of the obtained patterns, Databases and warehouses for being repository of data, Information retrieval etc. [2]. In Data Mining, five perspectives were observed by Naren Ramakrishnan et al. which are Compression, Search, Induction, Approximation, and Querying [3] (Fig. 1).

Data mining involves in six major task class types like Anomaly detection, Associative rule mining, Clustering, Classification, Regression and Summarization. World Wide Web has given a huge impact of amount of data that has been generated and laid path for the dissemination of data of different types. with the increase in the use of internet,

M. Tripathi and S. Upadhyaya (Eds.): ICDLAIR 2019, LNNS 175, pp. 103–118, 2021.
https://doi.org/10.1007/978-3-030-67187-7_12

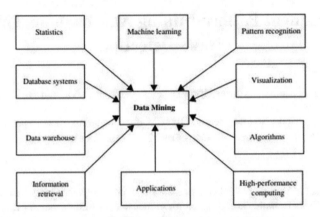

Fig. 1. Data mining techniques for various domains

there has been a major drift in the type of data that to be shared. Dimension of data that is being shared across network has increased.

The major dimensions of data mining are data, knowledge, technologies, and applications. As a general technology, data mining can be applied to any kind of data as long as the data are meaningful for a target application. The explosive growth of data in various fields like industry and academia and the rapidly increasing computing power are the driving forces to make data mining, an important area of study.

A few of the mining techniques such as Classification, Clustering, Association Rule Mining(ARM) may be explored from the fake news detection perspective. With the advent increase in multimedia data, there is an immediate need of discovering patterns and knowledge from this data. Further on par of knowledge discovered, Data Mining tasks are classified into either Descriptive, that deals with characterization or Predictive, wherein hidden knowledge and interestingness patterns are derived.

1.2   Social Media

Social media is the collection of online communications channels dedicated to community-based input, interaction, content-sharing and collaboration. Social media provides a platform for quick and seamless access to Information. Social media helps in improving individual with different communities and act as an effective way of communication. Bots and Cyborgs are typically used for collecting information from different social media. Social networks defines the generation of huge amount of data that is treated as social media. Social networks are important sources of online interactions and contents sharing, subjectivity, assessments, approaches, evaluation, influences, observations, feelings, opinions and sentiments expressions borne out in text, reviews, blogs, discussions, news, remarks, reactions, or some other documents. Social networks are being used for many purposes like social interactions, business collaborations, educational summits etc. Social media has become part of everyone's life.

The increase in types of Social Networks has showm a clear evidence for the"Small world phenomenon", typically termed as"six degrees of seperation", which states that the whole world can be connected in nearly six hops, as given by Stanley Milgram [4, 5]. Examples of social network platforms include Facebook, Twitter, Linkedin, while media-sharing networks include sites like Instagram, Snapchat and YouTube. With the increase in amount of data that has been generated, quality of data has become most challenging task in recent days.

Social bots and chatbots are typically used for mimicing human behaviour patterns like commenting on posts, following and unfollowing, liking etc. Cyborgs are typically claimed to spread more fake news to create market buzz [6]. Deception detection in online social media has took a greater attention as it has a clear influence on business, law enforcement, political and cultural issues and National Security. Fake news has a strong impact on the customer behaviour and marketing patterns also [7, 8]. Online social networks or social media does not only gain you relations, but also provides anxiety and depression with psychological distress. Cyber - bullying is another problem that has increased drastically with anonymous friends and publicly available data in many social media sites [9]. This cyberbullying not only causes deep pyschological scar but may also lead to depression and suicide tendency in many cases. FOMO, "Fear of Missing Out" is another psychological anxiety problem that people has been observed in Facebook users. Phishing, Vishing, Pretexting, Baiting, Tailgaiting, "Quid Pro Quo", Compromising accounts are some of the major problems in online social networks. The spread of fake news on social media, on celebrities or on some social conditions, is yet another aspect that has become problematic which even leads to killing of people or social unbalance [10].

Fake news detection models have been classified as either News Content Models or Social Context Models, basing on the input sources. [11]. Origin of news, Proliferation and tone of the news are some of the vital attributes that could predict the news published to be fake or real. Proliferation is observed more in case of unverified users compared to verified users on twitter. With over 2 billion active users Facebook holds the majority market share. The number of social media users worldwide in 2018 is 3.196 billion, up 13% year-on-year [12]. People tend to share thier views through social media and hence the number of users are exponentially increasing over period of time. Facebook, Twitter, Instragram are being popularly used for social interactions, whereas sites like LinkedIN are being used for professional network findings. The number of users depending on such platforms is increasing and thus became a branch of study (Figs. 2 and 3).

Thus, social media plays a profound role in influencing the social, economic and political domains of everyday decision making.

Social network sites worldwide ranked by number of active users (in millions, as of January 2017,)

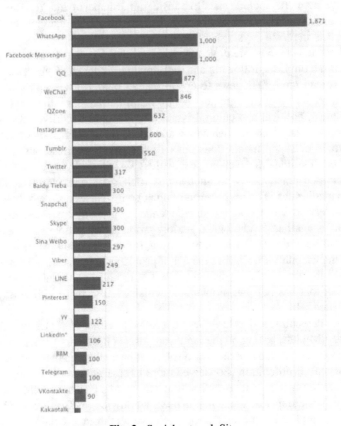

Fig. 2. Social network Sites

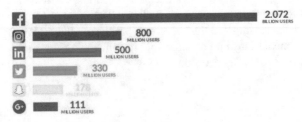

Fig. 3. User data of social networks

2   Related Work

Dave Chaffey has proposed a method of detecting opinion spams and fake news using text classification [13]. Spammers write false opinions to influence others, this is also known as Opinion spam mining. A detection model that combines text analysis using n-gram features and terms frequency metrics and machine learning classification has been proposed.

Hadeer Ahmed et al., has proposed an Automatic Online Fake News Detection Combining Content and Social Signals. A novel machine learning fake news detection method is proposed which combines news content and social context features and outperforms existing methods in the literature [14]. Combining content-based and social-based approaches for prediction and classification tasks is a solution that has been successfully applied in many fields.

Marco L. Della vedova et al., has propoased a method of detecting Fake News with Machine Learning Methods [15]. Different Machine learning methods like Naive Bayes, Neural Network and Support Vector Machine (SVM) are used to classify the believable and unbelievable messages on twitter. Supanya et al., has given a method for detecting Fake Account in Twitter basing on Minimum Weighted Feature set. Feature based detection approach is being implemented for monitoring the behavior of the user such as number of tweets, retweets, friends, etc. this is based on the confidence that humans usually have differently than fakers or simply fake account holders, therefore, detecting this behavior will lead to the revealing of the fake accounts for some extent [16].

Cody Buntain and Jennifer Golbeck proposed a method for automatic identification of Fake news from Twitter threads. Credibility of twitter stories is taken as a metric to classify the upcoming news into either of accurate or inaccurate classes. Using this mechanism, they have assessed the accuracy of social media stories. Here accuracy measures are deployed basing on the features extracted from crowd-sourced workers which have outperformed the accuracy obtained from models trained on journalists. Here the accuracy analysis performed using crowd-source is more influenced by network issues whereas the latter is more influenced by tweet content and language [17].

Saranya Krishnan and Min Chen have proposed a prediction method, which identifies the current news or pictures that are posted in twitter and cross-check with existing resources. This proposed framework uses a reverse image/text search, where the present photo or text is checked with different information sources from the past. URLs that are posted are checked with resources that stores fake news URLs, to check if it from a fake website. Repository will be updated periodically if any of the sources are proved as fake sources [18].

Marco L. Della Vedova et al., have proposed a method that detects fake news with the combination of news and social content features of any post. Social content features consists of other features like user-id, when user has created account, number of followers, tweets and re-tweets that he makes in case of Twitter etc. Machine learning algorithms are used for detection of fake news with respect to the user data and content posted. Further, chatbots have been implemented which automatically identifies fake accounts and fake news [15].

Stefan Helmstetter, and Heiko Paulhein claimed that the identification of fake news can be treated as binary problem. They claimed that it is always tedious to obtain enough content for training the system and instead of taking a small dataset of suitable size; data is collected at large-scale noisy training dataset. Initially the tweets are labelled automatically by its source, i.e., Trustworthy or Untrustworthy. Ensemble learning mechanisms like Random Forest and XG-Boost are used for more accuracy [19].

Terry Traylor et al., suggested the usage of NLP in detecting the fake news as generation of fake news has been claimed to have excessive use of unsubstantiated hyperbole and non-attributed quoted content. Textblob, NLP, SciPy toolkits are used on the top of Bayesian classifier for finding the likelihood of the tweet or content posted to be fake news. This method is termed as influence mining which not only identifies news to be fake news but also identifies the direction of propaganda of such fake news [20].

3   Problem Statement

Fake news, also termed as Pseudo-news is typically treated as a type of yellow media or propaganda which creates false news and spread them over traditional media to create hoax. Checkbook journalism is considered as the main reason for this spread, whose intention is to deviate or mislead people to become popular. Fake news spread is generally targeted to a person, society to damage their reputation, financial and social status. Clickbait stories, especially on YouTube is carried to increase the view count and to get easily popular. People generally use anonymous social networks to spread these types of news as the origin of the news cannot be easily tracked. Fake news is treated as neologism often used to create fabricated news. Claire Wardle has categorised this fake news into seven different categories like satire, false connection, misleading content, false content, impostor content, manipulated content and fabricated content. Fake news information is often treated as polluted information which is a collection of mis-information, dis-information and mal-information.

There have been many cases where spread of fake news has created hoax in society even leading to killing of innocent persons. With the spread of internet throughout the world, spread of information has become very simple and at the same time, origin of information has been drastically increasing. There has been a tremendous increase in the number of sources spreading news, especially over internet. Credibility of the information or sources spreading such information is more important. Online social networks made information passing a very easy task and spreads in no time. It has been proved many times that fake news generally spreads more than the real news, for instance during Boston Marathon blast more than 7.8 million tweets occurred. Fake content got popular in less than 30 min and around 30 thousand new accounts are created within few hours of blast. Only after 95 h of blasts, real news spread became equal to fake news. The same has been the case with Hurricane sandy, that occurred in 2012, where fake photos like shark on waterlogged highway, fake pictures of clouds, Flood in McDonalds at Virginia Beach etc. have been spread over twitter. In case of identifying fake pictures from the real ones, tweet based features are generally used for effective classification. For text based news, Number of characters in a tweet can also be considered as most influential for determining the credibility of the tweet made [6, 7].

In both of the above cases, Twitter has been the key-door for spreading information. It is termed that Twitter has been used for the first time to spread news about Boeing 737 being crashed in Florida River. There have been many cases where this type of news has disturbed social tolerance in society. Spread of manipulated or fabricated data will always create problems. Therefore, people should be cautious in believing any information or spreading this information.

While dealing about identifying fake news, certain information about the authors background, type of vocabulary that has been used, number of tweets and re-tweets made for unit time could be considered. For twitter data, features can be extracted from the tweets made and machine learning techniques can be employed further for analysis. In general twitter data is classified into either of fake or real news. However if associations could be derived between different phrases or words used in the tweet, it would be easy to categorize such data. The present work deals about extracting tweets from credible sources, extracting patterns that are required, establishing associations among different patterns selected and then use classification for determining whether the tweet is fake or real. Supervised learning techniques are used to classify tweets or posts into its specific categories. As feature extraction and selection are vital in the process of feature extraction, the initial step would be in the representation of data into feature vector. Present methodology uses both count vectorizer and TF-IDF for obtaining specific features. Nave Bayes and ARBC; Association Rule based Classification; are used to classify the tweets into required categories like Fake or Real news.

4 Data Set

4.1 Fake or Real News

This data set is taken from Amazon AWS and consists of 6,334 posts related to US Politics. The attributes in this dataset include unique id, author, text and title of news article and label of news article, that is, fake or real.

4.2 Twitter Extracted Data

.

4.3 Social Media API

Social Media like Twitter provide Application Programming Interfaces shortly called as API's to give access to their vast and dynamic data. Twitter has Streaming API and Facebook has Graph API. The response will be in JSON format. setup twitter oauth() command from twitteR package in R sets up the authentication credentials for a Twitter session. Consumer key, Consumer secret, Access token, Access secret are the parameters to setup twitter oauth() command and are unique to the application created by the user on dev.twitter.com. searchTwitter() from twitteR Package is used to retrieve tweets relevant to search word. Twitter has API limit of 180 calls every 15 min.

4.3.1 Few Examples of SearchTwitter ("IndiaElections")

- "KatAdeney: Alliance politics is all the rage in #India'Internal assessments that predicted a tough contest in the Hindi heart… https://t.co/St5rEELRab"

- "OmmcomNews: The much-awaited #2019generalelection and assembly election dates are likely to be declared next week… https://t.co/DEKIjtS1Ed"
- "tanvimor28: It seems the #dynamics of friendships are changing in the heat of upcoming #IndiaElections and #Pulwamaattacks. https://t.co/G588YsGRzH".

4.3.2 Attributes in Twitter Extracted Data

The fifteen attributes returned by searchtwitter() command [?] include the actual text of the status update, favorited indicates whether this tweet has been liked by the user or not, the number of times this tweet has been liked by Twitter users, reply to screen name, tweet created time (UTC), truncated indicates whether the text was truncated or not(for tweets exceeding max. tweet length), unique identifier of the tweet, utility used to post the tweet, screen name of the user in twitter, number of times this tweet has been retweeted, isRetweet indicates whether the tweet is a retweet by the authenticating user, Retweeted indicates whether this Tweet has been Retweeted by the authenticating user, location of the user, language used by the user, URL of the user's profile image.

5 Data Preprocessing Techniques

5.1 Language

Data set has text content from different languages. It has been cleaned so that only English texts are there in the training data and extra space is also removed.

5.2 URL Extraction

Extracted the URLs present in the text and title of each news article and stored in a separate column. Example:
  Input - this is fake proof at https://www.foobar.com
  url- https://www.foobar.com and Output - this is fake proof at

5.3 Tokenization

Tokenization describes the general process of breaking down a text corpus into individual elements that serve as input for various natural language processing algorithms. Usually, tokenization is accompanied by other optional processing steps, such as the removal of stop words and punctuation characters, stemming or lemmatizing, and the construction of n-grams. Example: Input - this is fake news and

  - ['this', 'is', 'fake', 'news']

5.4 Removal of Stop Words

Stop words are the words like "a", "an", "the", "is". These type of words are present in documents in greater frequencies, many programs that work on natural language such as search engines will ignore such type of words. Example: Input - this is fake news and Output - fake news.

6 Feature Extraction Techniques

One of the most important sub-tasks in pattern classification are feature extraction and selection. Prior to fitting the model and using machine learning algorithms for training, we need to best represent a text document as a feature vector. First comes the creation of the vocabulary which is the collection of all different words that occur in the training set and each word is associated with a count of how it occurs.

6.1 Count Vectorizer

We define a fixed length vector where each entry corresponds to a word in our pre-defined dictionary of words. The size of the vector equals the size of the dictionary. Then, for representing a text using this vector, we count how many times each word of our dictionary appears in the text and we put this number in the corresponding vector entry [?].

Example: If our dictionary contains the words MonkeyLearn, is reached, which contains the prediction or the outcome of is, the, not, great, and we want to vectorize the text "MonkeyLearn is great", we would have the following vector: (1, 1, 0, 0, 1).

6.2 TF-IDF Vectorizer

Term Frequency-Inverse Document Frequency and is a very common algorithm to transform text into a meaningful representation of numbers [21].

6.2.1 Term Frequency

A better representation would be to normalize the occurrence of the word with the size of the corpus and is called term-frequency. Numerically, term frequency of a word is defined as follows:

tf(w) = doc.count(w)/total words in corpus

6.2.2 Inverse Document Frequency

But, there would be certain words which are so common across documents that they may contribute very little in deciding the meaning of it. Term frequency of such words for example 'the', 'a', 'in', 'of' etc. might suppress the weights of more meaningful words. Therefore, to reduce this effect, the term frequency is discounted by a factor called inverse document frequency.

idf(w) = log(total number of documents/number of documents containing word w).

6.2.3 Term Frequency - Inverse Document Frequency

As a result, we have a vector representation which gives high value for a given term if that term occurs often in that particular document and very rarely anywhere else. TF-IDF is the product of term-frequency and inverse document frequency.

Tf-idf(w) = tf(w) * idf(w)

7 Application of Algorithms

7.1 Naive Bayes Classification

The probabilistic model of naive Bayes classifiers is based on Bayes theorem, and there is an assumption that the features in a dataset are mutually independent. Empirical comparisons provide evidence that the multinomial model works well for data which can easily be turned into counts, such as word counts in text [22, 23]. However, the performance of machine learning algorithms is highly dependent on the appropriate choice of features [24, 25].

7.2 Decision Tree

A decision tree is one of most frequently and widely used supervised machine learning algorithms that can perform both regression and classification tasks. For each attribute in the dataset, the decision tree algorithm forms a node, where the most important attribute is placed at the root node. For evaluation we start at the root node and work our way down the tree by following the corresponding node that meets our condition or "decision". This process continues until a leaf node is reached, which contains the prediction or the outcome of the decision tree [26].

7.3 Association Rule Based Classification

Association Rule Based Classification integrates the task of mining association rules with the classification task to increase the efficiency of the classification process [?]. Many studies have shown that associative classification (AC) achieves greater accuracy than other traditional approaches. Several AC-based studies have recently presented classification based on association (CBA), classification based on multiple association rules (CMAR), and classification based on predictive association rules (CPAR). An AC-based approach typically consists of three phases: rule generation, rule pruning, and classification [?]. In CBA, the system initially executes the Apriori algorithm to progressively generate association rules that are satisfied with a user-defined minimum support and confidence threshold. One subset of the generated classification rules becomes the final classifier [?].

Example for Association Rule Based Classifier. Let us suppose that rules are to be derived from the text given for classifying the text into either of sports or non-sports class type. This process initiates with the process of extracting the words from the text given, termed as tokenization. The below table depicts the text that should be classified into specific class type. Once tokens are generated from the text, stemming and bagging is performed to remove any stop words or phrases that has no significance. The below table depicts the process of extracting the required patterns from the text (Tables 1 and 2).

- Tokens including class labels are: great, game, election, over, clean, match, forgettable, close, sports, not sports. In order to create interestingness patterns, Association rule

Table 1. Categorizing tweets into different classes

Text	Category
A great game	Sports
The election was over	Not sports
Very clean match	Sports
A clean but forgettable game	Sports
It was a close election	Not sports

Table 2. Extracting Tokens from Tweets

A great game	Great, game
The election was over	Election, over
Very clean match	Clean, match
A clean but forgettable game	Clean, forgettable, game
It was a close election	Close, election

based classification technique is being employed which initially generates candidate item-sets of different sizes. Minimum support is taken to be 2, which means that any phrase or words that occurred less than 2 times will be discarded and any further combinations with that phrase or word would be directly considered infrequent and discarded directly. It has been observed from the table that no rules can be generated with more than 2 item-set combinations. Once the combinations are obtained, rules are generated from the patterns chosen.

– Association rules ending with class labels (count $\xi = 2$)

8 Results

8.1 ARBC on Fake News Data Set

Collection of Text Documents to a Matrix of the Counts/Frequency: Fake news data has been processed and ARBC is employed on it. The data is classified as two-class problem where the data falls in either Fake news category or Real news. As showm in Table 6, out of all the samples that are considered, around 4537 samples are classified correctly, whereas 1798 samples are misclassified, therefore accuracy in this aspect will be given with the formula (Tables 3 and 4)

$$Accuracy = (TP + TN)/(TP + FP + TN + FN) \qquad (1)$$

where True positive (TP) implies the number of tweets that are correctly classified as Fake, True Negative (TN) is the number of Non-Fake tweets classified as Non-Fake, False Positive (FP) implies Non-fake (Real) tweets to be misclassified as Fake and False Negative (FN) is the number Fake tweets misclassified as Non-Fake (Real) as shown in Table 6. Therefore for the dataset that has been considered, ARBC has shown an accuracy of 71.61%.

Collection of Text Documents to a Matrix of the TF-IDF Scores: Table 7, corresponds to the confusion matrix that has been generated by applying ARBC on Twitter Fake news data. Term frequency - Inverse document frequency has been considered further for analysis and confusion matrix is obtained as depicted in the below table. Table shows that even though True Positive (TP) cases are less when compared to the previous approach, there is a huge change in True Negative (TN) samples. Here out of 6335 samples, around 4876 samples are correctly classified and 1459 are misclassified, therefore here accuracy is obtained, as per Formula (1), to be around 76.9%.

Table 3. Candidate 1-itemset generation

great	1
game	2
election	2
over	1
clean	2
match	1
forgettable	1
close	1
sports	3
not sports	2

⇒

game	2
election	2
clean	2
sports	3
not sports	2

Table 4. Candidate 2-itemset generation

game, election	0
game, clean	1
game, sports	2
game, not sports	0
election, clean	0
election, sports	0
election, not sports	2
clean, sports	2
clean, not sports	0

⇒

game, sports	2
election, not sports	2
clean, sports	2

Table 5. Association rules generated from patterns

game → sports
election → not sports
clean → sports

Table 6. Confusion matrix - ARBC on fake news for count/frequency measure

Prediction	Truth	
	FAKE	REAL
FAKE	2892	1526
REAL	272	1645

8.2 ARBC on Twitter Extracted Data Set

ARBC usually works well for data with more attributes. For this experiment, around 1923 sample tweets have been extracted from Twitter (Table 5).

Feature Vector Includes Only Twitter Attributes: Once worked with fake news data, data has been extracted from Twitter. ARBC is applied on this collected data. It is to be noted that ARBC works well with data having more attributes. For the dataset in which the feature vector selected has only twitter attributes, the total samples collected are 1923, when ARBC is applied on this database, around 1525 samples are correctly classified, belonging to True Positive and False Negative cases, whereas 398 samples are wrongly classified, belonging to False Positive and False Negative cases as shown in Table 8. Therefore the accuracy is calculated to be 79.3%.

Feature Vector Includes Twitter Attributes and Some Other Attributes: Once ARBC has been worked on the Twitter data, for experiment purpose, some more attributes have been added to Twitter attributes. Experiment has shown that, out of 1923 samples, around 1627 samples are correctly classified and 296 samples are wrongly classified as shown in Table 9, thus leading to an accuracy of 84.6% of accuracy.

Table 7. Confusion matrix - ARBC on fake news for TF-IDF measure

Prediction	Truth	
	FAKE	REAL
FAKE	2050	345
REAL	1114	2826

8.3 Comparison of Algorithms

Table 10 depicts the comparison of different algorithms with respect to different measures like Accuracy, Precision, Recall, Specificity, F1-score etc. it has been proved that specific attributes like Accuracy, Recall, F1 Score, Negative predictive value, False positive rate are higher compared to other algorithms like Nave Bayes and Decision tree.

Table 8. Confusion matrix - ARBC on Twitter data with only Twitter attributes

Prediction	Truth	
	FAKE	REAL
FAKE	995	384
REAL	14	530

Table 9. Confusion matrix - ARBC on Twitter data with Twitter data combined with additional attributes

Prediction	Truth	
	FAKE	REAL
FAKE	971	258
REAL	38	656

Table 10. Comparison of measures in Naive Bayes, decision tree and ARBC

Measure	Naive bayes	Decision tree	ARBC
Accuracy	0.8173	0.7654	0.8461
Precision	0.8624	0.7731	0.7901
Recall	0.7741	0.7801	0.9623
F1 score	0.8159	0.7766	0.8677
Specificity	0.8647	0.7492	0.7177
Negative predictive value	0.7774	0.7567	0.9452
False positive rate	0.1353	0.2508	0.2823
False discovery rate	0.1376	0.2269	0.2099
False negative rate	0.2259	0.2199	0.0377

9 Inferences

The proposed Association Rule Based Classification (ARBC) algorithm has shown a significant improvement in performance, by detecting Fake News with an accuracy of 84.61%, which is better than other algorithms like Nave Bayes and Decision Tree. Recall, which specifies the completeness or sensitivity of the classifier, has shown a significant improvement in the proposed algorithm with 96.23%. This specifies that there is significantly less chance for the Fake news to be predicted as Real. That is, False positives (Real tweets that are flagged as fake) are more acceptable than false negatives (Fake tweets that are flagged as real). It is observed that precision of the proposed algorithm is better than Decision Tree. Improving recall can often decrease precision because it gets increasingly harder to be precise as the sample space increases. It is also observed that even if the proposed algorithm is less significant in predicting false positive samples, the overall F1 score, which is the weighted average of Precision and Recall, is relatively higher compared to the other algorithms with a value of 86.77%. Sensitivity and specificity have an inverse relationship; hence increasing one would always decrease other. Our classifier is highly sensitive, not specific. There is a notable reduction in False Negative Rate when compared to other algorithms. That is, classifier is sensitive in predicting fake tweets. It is also observed that even though the False Positive and False Discovery rate is lower than Decision Tree, it is still higher when compared to Naive Bayes.

10 Conclusion and Future Work

With the increasing popularity of social media, more and more people consume news from social media instead of traditional news media. However, social media has also been used to spread fake news, which has strong negative impacts on individual users and broader society. In this paper, we explored the fake news problem by reviewing existing literature work and proposed a new paradigm, Association Rule based classification(ARBC) to identify Fake News in Twitter media. We have used two data sets, one collected from Twitter which is profiled with sixteen attributes and other from Amazon AWS containing posts related to US Politics. Our proposed approach outperformed the existing approaches like Naive Bayes and Decision Tree with better accuracy. This work can be extended by implementing ARBC on prioritised attributes, Identification of fake accounts in social networks. We also hope to broaden the scope of our data collection and try to apply our method in a more general way in the future.

References

1. Berland, M., Baker, R.S., Blikstein, P.: Educational data mining and learning analytics: applications to constructionist research. Technol. Knowledge and Learning **19**(1–2), 205–220 (2014)
2. Han, J., Pei, J., Kamber, M.: Data Mining: Concepts and Techniques. Elsevier, Amsterdam (2011)
3. Ramakrishnan, N., Grama, A.Y.: Data mining: from serendipity to science. Computer **32**(8), 34–37 (1999)
4. Watts, D.J.: Networks, dynamics, and the small-world phenomenon. Am. J. Sociol. **105**(2), 493–527 (1999)
5. Kleinberg, J.: The small-world phenomenon: an algorithmic perspective. Technical report, Cornell University (1999)
6. Zhang, C., Sun, J., Zhu, X., Fang, Y.: Privacy and security for online social networks: challenges and opportunities. IEEE Netw. **24**(4), 13–18 (2010)
7. Girgis, S., Amer, E., Gadallah, M.: Deep learning algorithms for detecting fake news in online text. In 2018 13th International Conference on Computer Engineering and Systems (ICCES), pp. 93–97. IEEE (2018)
8. Shin, D.-H.: The effects of trust, security and privacy in social networking: a security-based approach to understand the pattern of adoption. Interact. Comput. **22**(5), 428–438 (2010)

9. Madden, M.: Privacy management on social media sites. Pew Internet Report, pp. 1–20 (2012)
10. Valkenburg, P.M., Peter, J.: Social consequences of the internet for adolescents: a decade of research. Curr. Direct. Psychol. Sci. **18**(1), 1–5 (2009)
11. Shu, K., Sliva, A., Wang, S., Tang, J., Liu, H.: Fake news detection on social media: a data mining perspective. ACM SIGKDD Explor. Newsl. **19**(1), 22–36 (2017)
12. Garton, L., Haythornthwaite, C., Wellman, B.: Studying online social networks. J. Comput.-Mediat. Commun. **3**(1), JCMC313 (1997)
13. Chaffey, D.: Global social media research summary 2016. Smart Insights: Social Media Marketing (2016)
14. Ahmed, H., Traore, I., Saad, S.: Detecting opinion spams and fake news using text classification. Secur. Priv. **1**(1), e9 (2018)
15. Vedova, M.L.D., Tacchini, E., Moret, S., Ballarin, G., DiPierro, M., de Alfaro, L.: Automatic online fake news detection combining content and social signals. In: 2018 22nd Conference of Open Innovations Association (FRUCT), pp. 272–279. IEEE (2018)
16. Aphiwongsophon, S., Chongstitvatana, P.: Detecting fake news with machine learning method. In: 2018 15th International Conference on Electrical Engineering/Electronics, Computer, Telecommunications and Information Technology (ECTI-CON), pp. 528–531. IEEE (2018)
17. Buntain, C., Golbeck, J.: Automatically identifying fake news in popular twitter threads. In: 2017 IEEE International Conference on Smart Cloud (SmartCloud), pp. 208–215. IEEE (2017)
18. Krishnan, S., Chen, M.: Identifying tweets with fake news. In: 2018 IEEE International Conference on Information Reuse and Integration (IRI), pp. 460–464. IEEE (2018)
19. Helmstetter, S., Paulheim, H.: Weakly supervised learning for fake news detection on twitter. In 2018 IEEE/ACM International Conference on Advances in Social Networks Analysis and Mining (ASONAM), pp. 274–277. IEEE (2018)
20. Traylor, T., Straub, J., Snell, N., et al.: Classifying fake news articles using natural language processing to identify in-article attribution as a supervised learning estimator. In: 2019 IEEE 13th International Conference on Semantic Computing (ICSC), pp. 445–449. IEEE (2019)
21. Zhang, W., Yoshida, T., Tang, X.: A comparative study of tf* idf, lsi and multi-words for text classification. Expert Syst. Appl. **38**(3), 2758–2765 (2011)
22. Rennie, J.D., Shih, L., Teevan, J., Karger, D.R.: Tackling the poor assumptions of naive bayes text classifiers. In: Proceedings of the 20th International Conference on Machine Learning (ICML-03), pp. 616–623 (2003)
23. Raschka, S.: Naive bayes and text classification i-introduction and theory. arXiv preprint arXiv:1410.5329) (2014)
24. McCallum, A., Nigam, K., et al.: A comparison of event models for naive bayes text classification. In: AAAI - 98 workshop on learning for text categorization, vol. 752, pp. 41–48. Citeseer (1998)
25. Kibriya, A.M., Frank, E., Pfahringer, B., Holmes, G.: Multinomial naive bayes for text categorization revisited. In: Australasian Joint Conference on Artificial Intelligence, pp. 488–499. Springer (2004)
26. Safavian, S.R., Landgrebe, D.: A survey of decision tree classifier methodology. IEEE Trans. Syst. Man Cybern. **21**(3), 660–674 (1991)

A Hybrid Method for Intrusion Detection Using SVM and k-NN

Abhishek Singh$^{(\boxtimes)}$, Maheep Singh, and Krishan Kumar

Department of Computer Science, National Institute of Technology Uttarakhand,
Srinagar, India
abhishek.cse16@nituk.ac.in

Abstract. The growing amount of data has necessitated the use of Intrusion Detection Systems in the modern days. The performance of an IDS is determined by the feature selection and classifiers. The traditional IDS fail to give satisfactory performance in today's world of growing data. We, in this paper, have propose a hybrid, two step approach for intrusion detection. In the first step, the data is classified into different classes with the help of Support Vector Machines. After the first steps, the records whose classification is not certain are passed on to the second step in which we use k-Nearest Neighbor method to further classify the incoming request into its respective class. Then we show how our proposed model compares with some recent approaches. We observe the proposed model to evaluate better on parameters like accuracy and precision.

Keywords: Support vector machines · k-Nearest Neighbors · Intrusion detection

1 Introduction

In this world of ever growing information, Network Security holds utmost importance. Network Intrusion detection is the field of network security which attempts to detect any unwanted intrusion in the network [3], which may be in the form of an attack or any attempt to steal or misuse the information [7].

There are two types of Intrusion Detection Systems

- Signature-Based Intrusion Detection System
- Anomaly Based Intrusion Detection System

The former take reference from a database of intrusions identified attacks and checks if the new request matches any of them [9]. While the latter [4] tries to detect anomalies in the behaviour of the incoming request and flags them as an intrusion if they seem inappropriate.

As the amount of data grows and takes varied form, the first type fails at providing satisfactory results in classification [12]. This necessitates the development of new Anomaly Based Intrusion Detection approaches [5]. Over the

M. Tripathi and S. Upadhyaya (Eds.): ICDLAIR 2019, LNNS 175, pp. 119–126, 2021.
https://doi.org/10.1007/978-3-030-67187-7_13

past few years it has been found out that SVM can come in handy at such kind of classification and various past works have proposed models based on the same [14]. With a culmination of SVM and k-NN we propose a reliable two step approach for intrusion detection in this work.

1.1   Support Vector Machines

A Support Vector Machine is a definitive classifier that, on being given a labeled data set, produces an optimum hyperplane that maximizes the margin between the given classes [8]. The vectors which define this hyperplane are termed as Support Vectors.

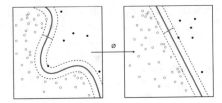

Fig. 1. Classification of example dataset using linear SVM [Support-vector machine, Wikipedia]

In the given Fig. 1 the black dots and the circles represent the two classes and the red hyperplane acts as a divisor among the classes.

1.2   k-Nearest Neighbor

The k-Nearest Neighbor is yet another method for classification and regression [6]. In this method a current data point is classified into a class depending upon the class which is most common among it's k nearest neighbors (Fig. 2).

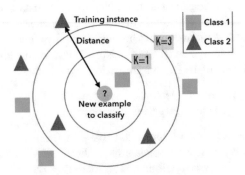

Fig. 2. Classification of a new entity based on k-NN for k = 1 and k = 3 [A Quick Introduction to K-Nearest Neighbors Algorithm]. Retrieved May 3, 2019 from https://blog. usejournal.com/a-quick-introduction-to-k-nearest-neighbors-algorithm-62214cea29c7

The rest of this paper is organized as follows. Section 2 gives a description of the datasets utilised for experimentation purposes. Section 3 provides for an evaluation criteria on which we measure the performance of the model. Section 4 provides the method itself and the inner functioning of the model in detail. Section 5 shows the various experiments and analysis performed on the dataset compare the proposed model with already existing models. Section 6 concludes the work and provides the future scope and improvements in the proposed work.

2 Dataset Used

In order to evaluate the proposed model, we have chosen the NSL-KDD dataset. It is a benchmark dataset for checking the performance of Intrusion Detection Systems. The NSL-KDD Dataset is an improvement over the KDD'99 data set. In each of the record in the dataset, there are 42 attributes. The first 41 attributes describe the flow of the data and the 42nd attribute is the label calssifying the given record as being normal request or an attack type [11].

Table 1. Instances in the NSL-KDD dataset

	Normal	DoS	Probe	U2R	R2L	Total
KDDTrain+	67,343	45,927	11,656	52	995	125,973
KDDTest+	9,711	7,458	2,421	200	2,754	22,544
KDDTest−21	2,152	4,342	2,402	200	2,754	11,850

Table 1 gives a description of number of instances of each attack type in the NSL-KDD Dataset. There are 41 attributes in the dataset, 9 of which are categorized as basic features, 13 being content related features, and the rest 19 of them being traffic related. The 19 traffic related features are further categorized as 9 time related and 10 host based.

3 Evaluation Criteria

The performance of the model is analyzed with the help of a confusion matrix as shown in table 2[13]. TP stands for True Positives, it means the number of positive records correctly classified. FP stands for False Positives, the number of negative records wrongly classified. FN stands for False Negatives, the number of positive records wrongly classified. TN stands for True Negatives, the number of negative records correctly classified.

Table 2. Confusion matrix

		Predicted	
		Attack	Normal
Actual	Attack	TP	FN
	Normal	FP	TN

Based on the above matrix following measures calculated.

$$Accuracy = \frac{TP + TN}{TP + FP + FN + TN}$$

$$Recall = \frac{TP}{TP + FN}$$

It is interesting to note here that while accuracy is good measure for performance evaluation, recall is an equally important measure because of the highly unbalanced nature of the data. In other words catching all the attacks is more important than correctly classifying the genuine requests.

4 Methodology

Figure 3 provides the overall architechture for the proposed model involving the use of SVM and k-NN. The entire process is divided into two steps.

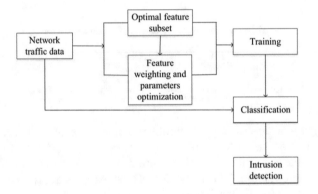

Fig. 3. Architecture of the proposed model

We may write our classifier using the w and b parameters as [1]

$$h_{w,b}(x) = g(w^x + b)$$

The following is the functional margin of the classifier (w, b) for the i_{th} training example

$$\gamma_i = y_i(w^T x + b)$$

Now the geometric margin may be defined for the classifier (w, b) as

$$\gamma_i = y_i \left(\left(\frac{w}{||w||} \right)^T x_i + \frac{b}{||w||} \right)$$

The larger the geometric margin, greater is the probability of our classification to be correct.

4.1   Step 1

In the first step we use Support Vector Machine to determine which class the input connection belongs to. In the given dataset there are five classes named Normal, Probe, DoS, U2R or R2L. The respective classes are mapped to 0, 1, 2, 3 and 4. After the classification via SVM of the given class, we find the certainty of our classification as the geometric margin of our given example from the hyperplane. If this value exceeds a given threshold, we call this classification as certain and otherwise as uncertain. Now the connections for which classification came out to be uncertain are forwarded to step 2 for further evaluation.

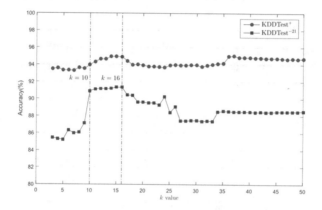

Fig. 4. Value of k vs accuracy for the test set

4.2   Step 2

In the second step we only receive those records as inputs which were classified as being uncertain along with their predicted class. We then apply the process of classification via k-Nearest Neighbor approach to classify the given record into the desired class [10]. After the process is complete we finally output the desired class which the connection must belong to, i.e., normal or anomaly.

5   Experiment and Analysis

To analyze the proposed model's performance, we conducted several experiments on the NSL-KDD dataset and compared the performance with some other recent works.

Figure 4 compares the accuracy for different values of k in k-NN classification. In both of the test sets, the KDDTest and the KDDTest21, we observe that the best value of accuracy is found from the values k = 10 to k = 16. Thus choosing a value of k in this range would yield best results [2].

Table 3. KDDTest classification - confusion matrix

	Predicted	Step 1		Step 2		Step 1 & 2	
		Attack	Normal	Attack	Normal	Attack	Normal
Actual	Attack	8,634	201	3,208	790	11,842	991
	Normal	41	2,138	113	7,419	154	9,557

Table 4. KDDTest21 classification - confusion matrix

	Predicted	Step 1		Step 2		Step 1 & 2	
		Attack	Normal	Attack	Normal	Attack	Normal
Actual	Attack	5,499	201	3,284	714	8,783	915
	Normal	35	540	75	1,502	110	2,042

Table 3 denotes the confusion matrix in case of classification on KDDTest dataset and shows a quantitative comparison between the approaches using only SVM, only k-NN and both.

Table 5. Confusion matrix for classification of all the given classes using the proposed approach

Predicted class / actual class	Normal	Dos	Probe	R2L	U2R
Normal	9107	372	217	11	4
Dos	1051	6206	96	103	2
Probe	332	91	1982	14	2
R2L	2005	53	93	597	6
U2R	136	13	16	9	26

While Table 4 shows the similar comparison for the KDDTest21 test set.

The Table 5 Confusion matrix stating True and false detections for all the 5 given classes.

Finally the Table 6 presents a quantitative comparison of the proposed model's performance with some general as well as some hybrid approaches used in some of the recent works.

Table 6. A comparison of performance of the proposed method with existing popular methods from recent works

Testing set	Method	Accuracy	Precision	DR	F1-score	FAR
KDDTest$^+$	C4.5	81.01%	96.67%	69.02%	80.54%	3.14%
	RF	77.06%	96.49%	61.96%	75.46%	2.98%
	k-NN	76.70%	97.21%	60.81%	74.82%	2.31%
	BPNN	75.34%	91.74%	62.29%	74.20%	7.41%
	NB	77.75%	93.36%	65.57%	77.04%	6.16%
	SVM + k-NN	**94.92%**	**98.72%**	**92.28%**	**95.39%**	**1.59%**
KDDTest^{-21}	C4.5	63.95%	95.08%	59.00%	72.82%	13.75%
	RF	55.34%	94.27%	48.37%	63.93%	13.24%
	k-NN	55.73%	95.54%	48.15%	64.03%	10.13%
	BPNN	57.46%	93.30%	51.73%	66.56%	16.73%
	NB	58.12%	90.20%	54.77%	68.16%	26.81%
	SVM + k-NN	**91.35%**	**98.76%**	**90.57%**	**94.49%**	**5.11%**

6   Conclusion and Future Works

In this paper we presented an effective two step approach for implementing an Intrusion Detection System. In step 1 we used Support Vector Machines to try to classify a given connection as a normal connection or an attack. The incoming connections which could not be classified into a desired class with certainty were forwarded to the second step in order to further attempt to clsasify with the help of k-Nearest Neighbor algorithm. We then with several analysis and experimentation analyzed the performance gained by the proposed model against the regular approaches as well as some hybrid approaches in some recent works. The proposed model showed to have better accuracy as well as precision.

In future steps can be taken in order to improve the model further. Different hybrid models may be simulated based on different classification techniques to improve the results. Also one may involve deep learning approaches and attempt to improve the performance of the system with multiple layers of neurons combining them with other binary classifiers for optimized results.

References

1. Wu, K., Chen, Z., Li, W.: A novel intrusion detection model for a massive network using convolutional neural networks. IEEE Access **6**, 50850–50859 (2018). https://doi.org/10.1109/ACCESS.2018.2868993
2. Naseer, S., et al.: Enhanced network anomaly detection based on deep neural networks. IEEE Access **6**, 48231–48246 (2018). https://doi.org/10.1109/ACCESS.2018.2863036
3. Chen, L., Gao, Y., Chen, G., Zhang, H.: Metric all-k-nearest-neighbor search. IEEE Trans. Knowl. Data Eng. **28**(1), 98–112 (2016). https://doi.org/10.1109/TKDE.2015.2453954
4. Lee, J., Lee, D.: An improved cluster labeling method for support vector clustering. IEEE Trans. Pattern Anal. Mach. Intell. **27**(3), 461–464 (2005)

5. Ahmad, I., Basheri, M., Iqbal, M.J., Rahim, A.: Performance comparison of support vector machine, random forest, and extreme learning machine for intrusion detection. IEEE Access **6**, 33789–33795 (2018)
6. Wang, X., Zhang, C., Zheng, K.: Intrusion detection algorithm based on density, cluster centers, and nearest neighbors. China Commun. **13**(7), 24–31 (2016)
7. Bryant, A., Cios, K.: RNN-DBSCAN: a density-based clustering algorithm using reverse nearest neighbor density estimates. IEEE Trans. Knowl. Data Eng. **30**(6), 1109–1121 (2018)
8. Khan, F.A., Gumaei, A., Derhab, A., Hussain, A.: A novel two-stage deep learning model for efficient network intrusion detection. IEEE Access **7**, 30373–30385 (2019)
9. Wang, Z.: Deep learning-based intrusion detection with adversaries. IEEE Access **6**, 38367–38384 (2018)
10. Angiulli, F., Astorino, A.: Scaling up support vector machines using nearest neighbor condensation. IEEE Trans. Neural Netw. **21**(2), 351–357 (2010)
11. Yu, D., Liu, G., Guo, M., Liu, X., Yao, S.: Density peaks clustering based on weighted local density sequence and nearest neighbor assignment. IEEE Access **7**, 34301–34317 (2019)
12. Tao, P., Sun, Z., Sun, Z.: An improved intrusion detection algorithm based on GA and SVM. IEEE Access **6**, 13624–13631 (2018)
13. Pan, S., Morris, T., Adhikari, U.: Developing a hybrid intrusion detection system using data mining for power systems. IEEE Trans. Smart Grid **6**(6), 3104–3113 (2015)
14. Yost, J.R.: The march of IDES: early history of intrusion-detection expert systems. IEEE Ann. Hist. Comput. **38**(4), 42–54 (2016)

Grammar Based Computing: A New Paradigm for Solving Real Life Problems

Krishn Kumar Mishra(✉)

Computer Science and Engineering Department,
Motilal Nehru National Institute of Technology Allahabad,
Prayagraj, India
kkm@mnnit.ac.in

Abstract. A computer is very useful device in solving real life problems. However due to limited storage and processing power, it can not handle problems operating on big data values. Many cryptographic problems involve operations on big data values and these operations are very important in deciding the strength of security algorithm. Some problems belong to bioinformatics and require processing of large genomes. In current scenario, no method can be designed for solving such problems in reasonable time. In fact for some problems no solution exists. In proposed work, we define a new strategy which can be applied in those problems where input structure has some repetitive pattern. To solve such problems, we will create training data by using a new method which is called as grammar based method. We will study input and output of related problems and use it to predict answer of complex problem. This is a completely new method and can be used in solving many problems. This method is very fast and easy. The performance of proposed method is very fast as compared to other methods. Currently, the work is in primary phase and has been used in many problems. Later on it will be applied on other problems.

Keywords: Grammar · Computation · New method for solving real life problems · Pattern

1 Introduction

One very important application of computers is to solve real life problems by transforming input data into output data. This transformation of data is done with the help of available data types and the available operators. For example numerical data can be processed by selecting suitable data type like integer, float, boolean and manipulating it with applicable arithmetic operators like (+, −, /, *) and logical operators. Similarly non numeric data can be processed by using suitable data type like char, string and their operators like (concatenation, union, intersection and others). In very complex real life problems, both numeric and non-numeric data types can be used and will be manipulated by

M. Tripathi and S. Upadhyaya (Eds.): ICDLAIR 2019, LNNS 175, pp. 127–136, 2021.
https://doi.org/10.1007/978-3-030-67187-7_14

corresponding operators. This method of solving the problem is known as procedural programming [1]. In procedural programming, a programmer has to use traditional operators to define complex functions for manipulating data values. Although this method is very useful when we want to manipulate those data values which can be stored in the computer memory efficiently yet such method will not be useful when we want to manipulate big/complex data. Let us elaborate this statement, for example if I want to multiply 9998 with 9997 (it can also be expressed as three 9s followed by one 8 and three 9s followed by one 7) it will be very easy for a machine because these numbers can be stored efficiently in a computer memory and can be multiplied by the supported hardware. However multiplication of 2000 9s followed by one 8 (9999...written for 2000 times followed by one 8) with 2000 9s followed by one 7 (9999...written for 2000 times followed by one 7) is not possible due to available machine limitations. No computer can calculate multiplication of such a big numbers due to limited hardware and software requirements. When we think about it many things will come to our mind. Can we calculate the output of such a big multiplication in reasonable time? If yes, then in which kind of situations will I find such problem? What kind of algorithms can be used to solve such problems? Can we design new algorithms which can handle such a big data? What is the usefulness of such problems? I will give you the answer to these questions one by one and show the applicability of these problems. However before discussing the procedure let us have a look on other important problems of similar type and their applicability.

In addition to these multiplication problems, big numbers involving other arithmetic operations like (/, + and −) also create similar problems. A big number in which we want to identify divisor of the number will be very useful in cryptography [2–4]. If we could find divisors of a big number in reasonable time, we can predict whether the number is prime or not. Moreover some non-mathematical problems involve a huge data, for example defining DNA properties and understanding its structure [5,6] is a very important problem. The whole genetic information is composed with the help of nucleotide and contains a lot of repetition. Therefore the method developed for previous problems can be used to solve these problems.

In this paper, a new type of method is discussed to solve programs with huge data inputs. As an example problem, we will design a new number multiplication algorithm for large numbers, that purports to mimic the operations that we, humans, would find ourselves drawn to do if the inputs happen to have some "structure" (say, 55555567*9998 rather than 7473426*5286). This method of multiplication will reduce the overall processing time. However in those cases, traditional multiplication techniques will not be useful. We have to design the solution of such problems by using some latest (machine learning) methods [7–9]. The whole paper is structured into five segments. First part explains problem. Second part explains proposed solution. Third section shows the applicability of the problem. Fourth part discusses some problems and their solutions. Finally in fifth section conclusions will be drawn.

About Example Problem

Number multiplication is very old and important problem. Although many number system exists but mostly binary and decimal system are used. Decimal number system is widely used in business and manual transactions whereas binary system is used in computer arithmetic therefore both number systems are important. Common operations on these systems are addition, multiplication, subtraction and division and are often interrelated. Traditional binary and decimal multiplication take $O(n^2)$ time where 'n' is the total number of digits in numbers. There are various other algorithms for large numbers multiplication. Karatsuba algorithm takes $O(n^{1.585})$ time complexity, 3-way Toom-Cook algorithm takes $O(n^{1.465})$ time complexity and Schonhage-Strassen [10,11] algorithm takes $O(n \log n \log logn)$ time complexity where 'n' is the total number of digits numbers. Since, almost all arithmetic computation multiplication is most common so any reduction of time may cause a big improvement in computation capability of the computer. Keeping limitation of machine in mind scientists suggested many hardware and software refinement to reduce the time complexity of multiplication operations.

2   Brief Overview of Existing Methods Used for Solving Real Life Problems and Proposed Approach

All real life problems can be categorized in to three categories. One class of real life problems can be easily solved by procedural programming whereas others which do not belong to previous class but can be stored and manipulated using available hardware and software structure can be solved by artificial neural network [12]. Artificial Neural Network mimic the biological behavior of human brain to solve those problems which are easy for a human being but are difficult to implement on computer using traditional deterministic algorithms (like Pattern Recognition, Function mapping and pattern association problems). A neural network consist of some set of nodes N divided into various layers, a weight vector W, activation function F(x) and a training set T. Training data is provided to neural network to solve problems. A learning method is used to update the weight vector until desired mapping is found. There are two types of learning available supervised and unsupervised learning. If we look into the mathematical model of neural network, we see that we are trying to implement an unknown function so that we can generate output for any given piece of input.

Apart from these two classes there is one more class of problem which cannot be solved either by procedural programming or by artificial neural network. These problems require handing of very big data and cannot be operated by existing hardware and software methods. However we can solve such problems by implementing a new learning method which we call as grammar based learning. This learning method is called as grammar based learning method because in such problems we first create a training set that consist of sub problems which are connected with the main problem by some grammatical rule [13–15] (i.e. Regular Grammar, context free grammar, etc.). After creating training data, each solution of sub problem is studied and some grammatical rules are established to check how output value can be predicted with the help of input data. This learning can be used to understand how the changes in input are affecting the output of the problem. Once such kind of training is done, it can be used to predict the output of actual problem. The beauty of this approach is its prediction accuracy. We can predict the output of any such problem without error. This learning method can predict correct output without updating weight vector and without forming a functional mapping between input and output. Here we do not need any weight vector and activation function.

Let us discuss what kind of problems can be treated with this approach. Our first example of number multiplication involves multiplying two big numbers 9999999999999998*6666666666666667. After checking both input values, we can define a grammatical rule to create sub-problems of the same type. For defining the grammatical rule, we will assume numbers as mathematical objects and will try to understand the structural relationship between two consecutive objects. To define the grammatical rule, we need to identify a pattern which defines connectivity in sub-problem and the problem. The value of this pattern is user defined. For example 9999999999999998 has repetitive occurrence of digit '9'. Similarly 6666666666666667 has repetitive occurrence of digit '6'. So a pattern of '6' and '9' is appearing in these numbers. These numbers belongs to a series in which new number can be created by appending digit 9 or 6 to previous sub-problem. For example, this problem can be understood as a part of series in which first pair will be (8, 7) second will be (98, 67) and third will be (998, 667). Similarly we can approach up to our pair (9999999999999998, 6666666666666667). So these inputs can be generated by a grammar rule which can be written as follows.

$$S \to aS|b, \text{ where a and b} \in [0, 1, 2, 3, 4, 5, 6, 7, 8, 9]$$

Here 'a' has the value '9' forming a pattern in first number (9999999999999998) and for the second problem the value of 'a' is '6' (6666666666666667).

In the above example we were able to generate sub pairs in which single digit '9' and '6' repeats in every number of the pair. Sometimes it is not possible to form a pattern of one digit. In such case we will see whether we can find pattern of two digits or three digits so that we can create training data. For example, if we want to multiply 12121212124 with 23232323237, pattern of one digit will not work. However pattern of two digits will give us training instances. Thus the multiplication of these numbers will be the part of series of pairs $\{(4, 7), (124, 237), (12124, 23237)...\}$. Here a context '12' is repeating in the first number and '23' is repeating in the second number. So by analyzing input we can identify previous sub-problems of the series and training data set can be created.

After generating a series of input numbers, we will try to identify how output of particular sub-problem is related to input of that sub-problem. Moreover we will also learn how the input and output of any sub-problem is related to input and output of previous sub-problem. This knowledge will be used to predict the correct answer of the problem.

3   Proposed Approach

Proposed mathematical model for grammar based learning consist of three tuples (I, O, R) where I represent the grammar for creating input set of training set, O is output grammar which shows the production rules which generate output training set, R is a set of production rules which defines how output is linked to input set. However for creating production rules for different sets, we have to start with some assumptions. In proposed learning method, there is no role of function. Instead, rules are learned with the help of structural relationship between input and output data/objects.

In our algorithm we treat numbers differently. We will assume that these numbers can be created from the basic object set D (0, 1, 2, 3, 4, 5, 6, 7, 8, and 9). For example a decimal number of digits 'n' can be constructed by appending the 'n' basic numbers (Objects) of the set D. If we carefully check the whole set of positive integers we can find that, not all numbers are independent, there are some numbers which are interrelated to each other with some rule. For example, consider following set of interrelated numbers $\{6, 86, 886, 8886 \ldots \ldots\}$. Here we can see that every consecutive pair of numbers has some relationship. For example in pair (6, 86), 86 can be created by appending '8' to previous number '6'. We can see here that any new number of the set can be created by appending '8' to previous number. So a pattern of '8' is appearing. We can find many sets of numbers in which new number can be created by appending the digits of pattern to previous number of that set. These sets will form the basic structure of grammar based learning. This method is applicable only for those numbers which can be connected with previous number by some pattern and belongs to such sets. Let us suppose we want to identify what will be the answer of multiplication 9999999998 and 9999999997. Since in such numbers a pattern is appearing, to identify the multiplication of numbers first we need to generate training set for prediction. These two numbers are connected to set S1 (8, 98, 998, 9998...) and S2 (7, 97, 997...) respectively. With the help of these input values, we will create input set $\{(8, 7), (98, 97), (998, 997) \ldots\}$. We can see here that next pair can be generated by appending some digits to the previous pair. Now we will check the output of these sets and will try to establish some structural relationship among the resultant of those pair of numbers.

Before discussing the actual process, we will define some rules and assumptions which are required to find out whether result can be predicted for the problem or not.

First rule is regarding basic numbers/objects. Any number which is generated by random composition of basic numbers will also belong to the category of basic number. For example 45678 is a basic number for us. Since for such numbers no training data can be produced, so multiplication between basic numbers can be done by simple multiplication method only. For example, if we want to multiply 6 and 7, we have to multiply it by simple approach and its resultant will be 42.

Second assumption is very necessary to relate results of sub-problems by defining production rules which follow the law of conservation of objects (basic numbers). We assume that if we are multiplying two numbers each of '3' digits, it will generate a '6' digit number as output. If the size of output is not equal to the required length we will append sufficient '0' as prefix to conserve the total number of digits in both input and output.

Also to define the grammar for output, we will assume that if we multiply numbers of same size, it will generate resultants in two context, first context and second context and each context will be of equal size. For example when we multiply (6, 7) it will generate '42' and this number can be separated into two context. First context will be '4' and second context will be '2'. Similarly (99, 97) will generate '9603' in which '96' is the first and '03' is the second context respectively. Now, (99, 9) can be multiplied only when we consider both the numbers to be of the same size i.e., (99, 09) and its multiplication will produce a result of '891'. But to proceed further, as per our rule, we will have to convert the result to '0891' in which first context will be '08' and second context will be '91'.

4   Proposed Algorithm

Our algorithm is completely based on different learning procedure which works only if we could produce grammatically related training instances. This process can be used to predict the output. In order to predict the multiplication of given pair, we present some example of multiplication of related sub pairs using our model. We will follow our rule and will try to form production rules to understand the relationship among different results. If we find some relationship, we will then check the result of new pair. If new result confirms our mapping we can then apply this mapping to find result of any related pair that belongs to the given set.

As we have already stated that this algorithm can be applied to only those numbers which are a part of connected series of pairs, so before multiplying any two numbers we generate its sub pairs. Following all rules and assumption stated above, we will start from first pair. Let us number this pair as 1 and check the multiplication of this pair. Let us assume that size of each basic number in this pair is same and it is equal to 'n' then we will calculate the value of result for this pair by using our traditional multiplication method. As we are multiplying two numbers each of size 'n' then we assume that our result will contain '2n' digits. If the size of resultant is not equal to '2n' then we append some '0' at the beginning of resultant number and make it of size '2n'. Then we will divide this result into two parts, first context will be of size 'n' and second context will also be of size 'n'. We will store the value of each context. Again for next consecutive pair we will generate the result using traditional method. As new pair is generated by applying the some repeating digits 'p' to each number of previous pair. The size of each number in new pair will be $(p + n)$, where 'p' is those digits which will repeat in every pair. Now we will repeat the same process and multiply these numbers by traditional approach. We again divide this result in two contexts first and second each of size (p + n). We will store first 'p' digits of each context as new generated digits due to application of repeated pattern on the numbers and match the last 'n' digits of each context with the previous context values. If the last 'n' digit of both context of second pair matches with the values of both context of first pair then we can generate a mapping for other pairs. If it does not match we will check another consecutive pair. We will repeat the process for 2 to 3 pairs and store the value of first 'p' digits of both the context. If these 'p' values are equal with the previous 'p' values, then our mapping will become fix and we will retrieve these values. In order to calculate any multiplication we simply repeat these digits appropriate times to the first and second context of starting pair to generate the resultant value. In other way, this is a simple approach and can be understood by following procedure. We assume that there will be no loss of objects during multiplication and let us multiply first pair (a, b) in which two digits are multiplied. We assume that after multiplying these two digits we get result 'cd' in which two new digits appears we assume that this result belongs to a pair in which first digit is first context and second digit belongs to second context so for this result our pair will be (c, d). Now we will take other related pair in which two particular digits appear on different numbers. Let 'm' appear before 'a' and 'n' appears before 'b' then the new pair will be (ma, nb). We will generate its result by traditional multiplication and let the result be 'pcqd'. Now, we will further divide this number into two different contexts 'pc' and 'qd'. We assume that digits 'p' and 'q' appear in the prefix of the pair (c, d) respectively because of the application of pattern (m, n) appearing in the prefix of pair (a, b) respectively. If we are right then other two digits must belong to previous results in our case it is (c, d) which belong to previous result. So we can assume that whenever we add one digit of 'm' and 'n' we get 'p' and 'q' prefixed in both the contexts obtained from the resultant product. To verify our mapping we repeat our process for (mma, nnb) and check its result. If the result is 'ppcqqd' then we will further divide this result into two contexts (ppc, qqd) and check for the new digits appended to the previous context which forms the result. We find it is (p, q) since this pair is same as previous one. So, we can infer that whenever digit of pair (m, n) appear on the input pattern, on multiplication it will generate pair (p, q) on the previous contexts of result. With this inference we can find any multiplication without doing any mathematical calculation. Sometimes this pairs is seen very clearly after some steps of the above process is carried out, until then we have to wait for the mapping to fix.

Step by step algorithm for this process

1. Generate sub pairs and start numbering these pair from N = 1 that are connected with given pair. Note the size 'p' of repeating pattern.
2. For any consecutive pair starting from the first pair (N = 1) and repeat the following with (N = N + 1) until mapping is found.
   - Calculate the resulting value for first pair having each number of size 'n'.
     • Divide this value into two contexts first and second.
     • Store the value of each context as First1 (n) and Second1 (n).

- Calculate the value of second consecutive pair (N = 2) having input numbers of size (p + n).
  - Divide this result also in two contexts first and second each of size (p + n).
  - Let the new generated contexts are First2 (p + n) and Second2 (p + n).
  - Check if First2 (n) = First1 (n) and Second2 (n) = Second1 (n).
    * If yes, then store the value of First2 (p) and Second2 (p).
3. If for two consecutive pair values of First2 (n) and Second2 (n) matches with previous pair.
  - Print the resulting pattern First2 (p) and Second2 (p).

Let us explain this algorithm with one example. Let us suppose we want to multiply (9999998, 9999997). Since these numbers have the repetition of same number on certain digits we can generate sub pairs for them. Let us start with {(8, 7), (98, 97), (998, 997), (9998, 9997) . . . . . .}. We can generate our number by simply repeating '9' six times before '8' and '7' respectively. We will start with first pair (8, 7) and multiply them and we will get '56'. We will calculate the first and second context for this number and the two contexts will be '5' and '6'. Now as we see all pairs are generated by appending the digit '9' to the prefix of each pair, so results must have some relationship. We will try to trace out this relationship. We will now multiply the next pair (98, 97) and get the result '9506'. We will again divide this number into two contexts, first and second. Its first context will be '95' and second context will be '06'. As we have append one '9' to every number of previous pair, so we guess that the new digit on each context is the result of this pattern. So '9' on first context and '0' on second context is appended due to application of pattern '9' on the pair of digit (8, 7). So we check the remaining digits of the resultant contexts and they are '5' and '6'. As they are equal to the value of previous contexts, we can assume that every time the digit '9' is appended on the pair of numbers to be multiplied, the pair of digit (9, 0) are prefixed to the first and second context of the previous contexts to generate the answers, but we have to verify it again. Then we will check next two pairs (98, 97) and (998, 997) again. For the first pair context will be '95' and '06' and for the second pair we get new contexts '995' and '006'. So newly appended digits will be (9, 0). As we check the remaining digits of each context after removing newly appended digits in the result of second pair of numbers, we get that they are '95' and '06' which is equal to the multiplication value of previous pair. So we can say for each repetition of digit '9' on initial numbers, every time we have to append (9, 0) to the first and second context of initial pair to generate any result. For example, our multiplication answer will be 99999950000006. Since six 9s were repeated on the basic pair (8, 7). Let we discuss one more example and suppose we want to multiply (999999994, 999999993). Its sub pairs will be {(4, 3), (94, 93), (994, 993), (9994, 9993) . . .} and their results will be {12, 8742, 987042, 99870042 . . .}. If we try to find mapping from first two pairs we will get no mapping. But if we see the next pairs we get a mapping of (9, 0) which will be added to initial context (87, 42) every time.

Similar mapping can be found in case of repeated patterns. Here, some results, which are generated by proposed algorithm, are being shown. These result are based on starting numbers (4, 3) and (9, 8) respectively (Fig. 1).

A. Result for 4*3 = 12
- Add pair (9, 3)

B. Result for 9*8 = 72
- Add pair (6, 9)

```
Enter two numbers: 4 3                          |Enter two numbers: 9 8                                        |
-----------------------------------------------|--------------------------------------------------------------|
Iteration number : 0                           |Iteration number : 0                                          |
Number 1 : 4  and Number 2 : 3                 |Number 1 : 9  and Number 2 : 8                                |
Product of the two numbers is : 12             |Product of the two numbers is : 72                            |
-----------------------------------------------|--------------------------------------------------------------|
Iteration number : 1                           |Iteration number : 1                                          |
Number 1 : 94  and Number 2 : 33               |Number 1 : 69  and Number 2 : 98                              |
Product of the two numbers is : 3102           |Product of the two numbers is : 6762                          |
Contexts are : 31 and 02                       |Contexts are : 67 and 62                                      |
-----------------------------------------------|--------------------------------------------------------------|
Iteration number : 2                           |Iteration number : 2                                          |
Number 1 : 994  and Number 2 : 333             |Number 1 : 669  and Number 2 : 998                            |
Product of the two numbers is : 331002         |Product of the two numbers is : 667662                        |
Contexts are : 331 and 002                     |Contexts are : 667 and 662                                    |
-----------------------------------------------|--------------------------------------------------------------|
Iteration number : 3                           |Iteration number : 3                                          |
Number 1 : 9994  and Number 2 : 3333           |Number 1 : 6669  and Number 2 : 9998                          |
Product of the two numbers is : 33310002       |Product of the two numbers is : 66676662                      |
Contexts are : 3331 and 0002                   |Contexts are : 6667 and 6662                                  |
-----------------------------------------------|--------------------------------------------------------------|
Iteration number : 4                           |Iteration number : 4                                          |
Number 1 : 99994  and Number 2 : 33333         |Number 1 : 66669  and Number 2 : 99998                        |
Product of the two numbers is : 3333100002     |Product of the two numbers is : 6666766662                    |
Contexts are : 33331 and 00002                 |Contexts are : 66667 and 66662                                |
-----------------------------------------------|--------------------------------------------------------------|
Iteration number : 5                           |Iteration number : 5                                          |
Number 1 : 999994  and Number 2 : 333333       |Number 1 : 666669  and Number 2 : 999998                      |
Product of the two numbers is : 333331000002   |Product of the two numbers is : 666667666662                  |
Contexts are : 333331 and 000002               |Contexts are : 666667 and 666662                              |
************************************************|**************************************************************|
Pattern exists and the mapping digits are : 3 and 0 |Pattern exists and the mapping digits are : 6 and 6      |
************************************************* *************************************************************
```

Fig. 1. Results

5   Conclusion and Future Work

Even though the algorithm is very interesting, yet there are some questions which are still to be answered. First question is that for what numbers we can find mapping through this algorithm. Till now we were able to find mapping for those numbers in which pattern size was one and one of the pattern digits belongs to the set 3, 6, 9. For example, we can find the answer of (999998, 888885), (9999956, 8888867) and (666667, 777772) but we did not obtain any mapping when we were trying other numbers like (777778, 888886). We are still trying to update our algorithm for other numbers. Also we are still trying to work with pattern of size more than one digit. Although it is very difficult to understand the relationship when pattern size becomes more than 1, yet, efforts can be made to understand the relationship. By this algorithm, new approach can be searched for reorganization of natural language. Also the algorithm can be extended to other operators as well.

References

1. Aho, A.V., Hopcroft, J.E.: The Design and Analysis of Computer Algorithms. Pearson Education, India (1974)
2. Forouzan, B.A.: Cryptography & Network Security. McGraw-Hill Inc., New York (2007)
3. Diffie, W., Hellman, M.: New directions in cryptography. IEEE Trans. Inf. Theory **22**(6), 644–654 (1976)
4. Katz, J., Lindell, Y.: Introduction to modern cryptography. Chapman and Hall/CRC, Boca Raton (2014)
5. Burge, C., Karlin, S.: Prediction of complete gene structures in human genomic DNA. J. Mol. Biol. **268**(1), 78–94 (1997)
6. Hingerty, B.E., Figueroa, S., Hayden, T.L., Broyde, S.: Prediction of DNA structure from sequence: a build-up technique. Biopolymers: Original Res. Biomol. **28**(7), 1195–1222 (1989)
7. Bishop, C.M.: Pattern Recognition and Machine Learning. Springer, Heidelberg (2006)
8. Domingos, P.M.: A few useful things to know about machine learning. Commun. ACM **55**(10), 78–87 (2012)
9. Dietterich, T.G.: Machine-learning research. AI Mag. **18**(4), 97–97 (1997)
10. Karatsuba, A.A., Ofman, Y.P.: Multiplication of many-digital numbers by automatic computers. In: Doklady Akademii Nauk, vol. 145, pp. 293–294. Russian Academy of Sciences (1962)
11. Bodrato, M.: Towards optimal toom-cook multiplication for univariate and multivariate polynomials in characteristic 2 and 0. In: International Workshop on the Arithmetic of Finite Fields, pp. 116–133. Springer (2007)
12. Zurada, J.M.: Introduction to Artificial Neural Systems, vol. 8. West publishing company St. Paul (1992)
13. Parekh, R., Honavar, V.: An incremental interactive algorithm for regular grammar inference. In: International Colloquium on Grammatical Inference, pp. 238–249. Springer (1996)
14. Watson, B.W.: A new regular grammar pattern matching algorithm. In: European Symposium on Algorithms, pp. 364–377. Springer (1996)
15. Sakoe, H.: Dp matching system for recognizing a string of words connected according to a regular grammar, November 26 1985. US Patent 4,555,796

Air Pollution Monitoring Using Blue Channel Texture Features of Image

Sukanta Roga[1](✉), Shawli Bardhan[2], and Dilip H. Lataye[1]

[1] Department of Mechanical Engineering and Civil Engineering, Visvesvaraya National Institute of Technology, Nagpur 440010, Maharashtra, India
rogasukanta@gmail.com, diliplataye@rediffmail.com
[2] Department of Computer Science and Engineering, Tripura University (A Central University), Suryamaninagar 799022, Tripura, India
shawli.cse@gmail.com

Abstract. Increasing level of air pollution is now threatening issues and results severe health problem . Predicting the air pollution level from visual image can be helpful for society for awareness and maintaining safety from this. Availability of high-resolution visual camera in smart phone will accelerate the aim of decreasing air pollution effect by developing simple system for pollution level detection using pixel information. Aiming at the requirement, in this study we propose a technique to detect the pollution level through image texture feature extraction and classification. Online available image dataset of Beijing has been used for experiment and particulate matter 2.5 (PM2.5) is used for validation of classification. Neighborhood pixel influence oriented five texture level features are computed from each image and used as the input of artificial neural network for classification. The outcome of classification shows 84.2% of correct pollution level detection comparing with PM2.5 value. Therefore, outcome of the method's accuracy and less complexity of processing makes it acceptable and easy for image base pollution level detection.

Keywords: Air pollution · Classification · Feature extraction · Image · Machine learning

1 Introduction

Air pollution is a high alarming environmental issue that today's world is facing due to rapid growth of airborne particulate matter (PM). In present scenario, rapid industrialization and urbanization increases the PM in air with diameter less than 2.5 μm (PM2.5). Higher the rate of PM in air generates harmful effects in human body by spreading chemicals in lung and blood which causes life taking disease like cardiac attack, lung failure, cerebrovascular diseases etc. Therefore, PM2.5 is considered as the standard air quality measuring matric for air pollution monitoring. The value of PM2.5 greater than 100 μg/m3 is considered as unhealthy for sensitive group of person and PM2.5 > 150 μg/m3 is considered as unhealthy for everyone [1]. Also PM2.5 > 200 μg/m3 is

© The Author(s), under exclusive license to Springer Nature Switzerland AG 2021
M. Tripathi and S. Upadhyaya (Eds.): ICDLAIR 2019, LNNS 175, pp. 137–143, 2021.
https://doi.org/10.1007/978-3-030-67187-7_15

very unhealthy and hazardous of human health [1]. As per the survey of World Health Organization (WHO), the harmful effect of air pollution causes over 7 million deaths in every year [2]. Industrial pollution, wild fires, fuel burning are the major causes of increasing pollution causing airborne particles and mostly found in China, Southeast Asia and India for years.

The popular method for air pollution measure is performed by monitoring stations, controlled by sophisticated sensors and complicated as well as costly setup. Therefore, particular region based air pollution monitoring using the mentioned technology is cost-effective and tough due to complexity of setup. In recent years, image analysis and machine learning approaches are used as the alternative strategy for air quality monitoring. The availability of high resolution and portable digital image acquisition sensors present in cameras and smartphones increases the prospect of image base computerized air quality analysis. Monitoring the level of PM2.5 can be predictable by analyzing digital images. Accurate measurement of air quality through image analysis will provide an affordable, easy and efficient approach of air pollution monitoring. The principle of air quality monitoring using image basically lies on the visibility of a scene. Presence of pollution in the air causes increase of airborne particles and sunlight scatters due to those and results poor visibility of a scene. In recent years, the concept of air quality monitoring using image getting growing attention. Due to the lack of visual image dataset, satellite images are popularly used in this purpose. Using smartphone camera images, Vahdatpour MS et al. [3] performed the air pollution level prediction using Gabor features and Convolutional Neural Network (CNN). The sky regions of the images are used in their analysis and achieved maximum accuracy of 59.38% using CNN. Spyromitros-Xioufis E et al. [4] also used the sky region of visual image for development of air quality estimation model. They performed the whole analysis into three-step: sky region detection, sky region localization, air quality estimation. The combine heuristic rules proposed by air quality experts and Fully Convolution Network, and achieved 80.3% of accuracy towards accurate air quality monitoring. Chakma, A et al. [5] also used deep convolutional network to classify visual images into three categories as no pollution, moderate pollution, and severe pollution. They achieved maximum 68.74% accuracy of correct categorization using holistic images as the input of the network. Liu, C. et al.[6] also extracted holistic image features for pollution estimation. Wang, H. et al. [7] found out the correlation between the PM2.5 and the degradation of the processed visual image which is 0.829.

Observation of the existing studies contains the complexity of sky image segmentation and convolutional network processing. In general, identification of pollution level based on PM2.5 (unhealthy, hazardous etc.) is important for taking required precaution. Focused on minimization of the mentioned complexity of existing research, the study aims to classify the polluted (PM2.5 > 100 μg/m3 and < 200 μg/m3) and highly polluted(PM2.5 > = 200 μg/m3) visual image by extracting Neighbor-hood Gray-Tone Difference Matrix (NGTDM) features of image. Identification of the pollution level will help in providing prior knowledge to the individual in maintaining the required safety. The rest of the manuscript mentions the detailed methodology of the classification.

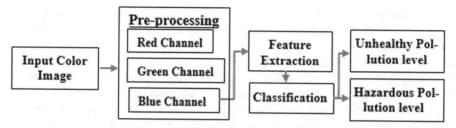

Fig. 1. Overall flow diagram of the methodology of image base pollution level classification.

2 Methodology

The methodology targets to classify the images into two groups aiming on pollution classification. One group contains unhealthy environmental images which represents PM2.5 range is from 100 μg/m3 to 200μg/m3 and another is hazardous environmental images representing PM2.5 greater than 200μg/m3. The method is performed by taking pixel intensity features of blue channel and Artificial Neural Network (ANN) for classification. The Fig. 1 shows the flow diagram of the overall method.

2.1 Dataset Description

In this study, we collected visual images of Beijing Television tower captured on the month January to May of the year 2014 by Yi Zou. All the images are captured at almost in the same time of morning. From the collected images, we selected those whose corresponding PM2.5 value is above or equal to 100 μg/m3 for classification. In this purpose, 55 images are selected following the PM2.5 range. Each of the captured images contains the date and time of acquisition. Reference to the information, we collected the PM2.5 value corresponding to each of the images from www.stateair.net which is under U.S Department of State Data Use Statement from Mission China Air Quality Monitoring Program. The Fig. 2 shows sample images of the collected dataset.

2.2 Image Pre-processing

Observation of the review work shows that, the sky region is the important part of a visual image for pollution analysis. The texture of sky region depending on the color of sky is considered as an important feature in this purpose. In our analysis also, we consider the blue channel for analysis as it gives maximum information regarding the texture of the sky and about image clarity related to level of pollution. With the increase of the pollution, the clarity, contrast, decreases along with the blue channel value of sky image [8]. Therefore, to extract maximum relevant information for pollution classification, in preprocessing step, the blue channel is extracted from all the images and used for further feature extraction.

2.3 Feature Extraction

In natural scene, texture information is used for analysis by human eyes and practically applicable for environmental change analysis [9]. In image processing, texture stands for interrelation representation between image pixels. Thus automatic analysis of texture pattern is performed in this study for environmental change detection due to air pollution. The Neighborhood Gray-Tone Difference Matrix (NGTDM) based five texture features are extracted in this study named as coarseness, contrast, busyness, complexity, and texture strength. Here the NGTDM generation is performed because the changes of the texture due to presence of pollution will effect a single pixels depending on neighborhood region also. Therefore, presence of huge pollution in a scene will make pixel intensities similar to each other and will change the texture pattern massively. Therefore amount of pollution will be correlated with texture feature values. The generation of NGTDM is performed for i[th] entry by taking summation of the differences between the gray level of all pixels with gray level i, and the average gray level of their surrounding neighbors [9]. Using the NGTDM, all the five features are extracted as follows [9].

Fig. 2. Samples images of collected dataset with corresponding PM2.5 value.

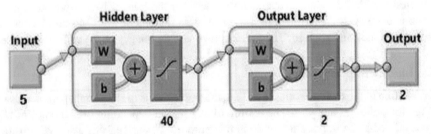

Fig. 3. Artificial Neural Network for classification with 5 extracted image features for each image, 40 hidden layers, and two class output.

Coarness

$$\left[\partial + \sum_{i-0}^{H_{GT}} P_i N(i) \right]^{-1} \tag{1}$$

Contrast

$$\left[\frac{1}{G_N(G_N-1)}\sum_{i=0}^{H_{GT}}\sum_{j=0}^{H_{GT}}P_iP_j(i-j)^2\right]\left[\frac{1}{n^2}\sum_{i=0}^{H_{GT}}N(i)\right] \qquad (2)$$

Business

$$\left[\sum_{i=0}^{H_{GT}}P_iN(i)\right]\Bigg/\sum_{i=0}^{H_{GT}}\sum_{j=0}^{H_{GT}}iP_i-jP_j,\ P_i\neq 0,\ P_j\neq 0 \qquad (3)$$

Complexity

$$\left[\sum_{j=0}^{H_{GT}}\sum_{j=0}^{H_{GT}}\left\{(i-j)\Big/\left(n^2(P_i+P_j)\right)\right\}\{P_iN(i)+P_jN(j)\}\right],$$
$$P_i\neq 0,\ P_j\neq 0 \qquad (4)$$

Texture Strength

$$\left[\sum_{i=0}^{H_{GT}}\sum_{j=0}^{H_{GT}}(P_i+P_j)(i-j)^2\right]\Bigg/\left[\partial+\sum_{i=0}^{H_{GT}}N(i)\right],\ P_i\neq 0, P_j\neq 0 \qquad (5)$$

The above equation represents the texture features extracted from each image of size. *MxM* where,

∂: a small number for preventing coarseness to become infinite.
H_{GT} : highest graytone value of a particular image.
i: graytone value.
P_i: probability of occurance of i P_j: probability of occurance of j.
N: corresponding i^{th} entry in NGTDM.
G_N: total number of different gray levels present in a image.
$n = M$ - 2d.
d: neighborhood size(1).

3   Classification

For classification, Artificial Neural Network(ANN) is used in this study with two- layer feed-forward channel. The sigmoidal activation function is applied to generate non-linearity in the network. Here we used 40 hidden neurons for optimal training of the network through trial and error mechanism. The network diagram is shown in Fig. 3 for clear understanding. 19 images are send for testing purpose of the system where each represented with 5 features. The output layer divides the input images into two classes: one is unhealthy pollution level and another is hazardous level of pollution.

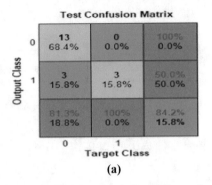
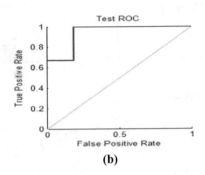

Fig. 4. (a) Confusion matrix depending on classifier output. (b) ROC curve of classification output.

4   Result and Discussion

The measurement of the ability of a system is performed depending on the quantitative outcome. Here also to analyze the system performance, we calculate the Sensitivity, Specificity, and Accuracy of classifier. The sensitivity of the system is 81.3% with specificity 100% and accuracy 84.2%. The Confusion Matrix of the corresponding classification is given in Fig. 4(a). The Receiver Operating Characteristic (ROC) curve shown in Fig. 4(b) indicates the relation in between true positive and false-positive classification rate. The Area Under the Curve(AUC) value of the given ROC in 0.86 which indicates acceptable performance of the classifier for level of pollution detection. Both the accuracy and AUC are higher for the classification. Also with the increase of input image as training and testing phase, the sensitivity and accuracy will also improve. On the other hand, the specificity of the system is very high and overall quantitative measurements of performance indicate the acceptable performance of the system in case of image-based pollution classification with less complexity.

5   Conclusion

Presence of PM2.5 generates pixel oriented special texture in visual image. In this study we extracted image texture level features for classification in between level of pollution through blue channel feature analysis of color image. Those features are coarseness, contrast, business, complexity, and texture strength. The coarseness of image increases when local uniformity of image is high and basic pattern making textures are large. If the visibility of an image is high with clear edge, it results high contrast. On the other hand, business indicates rapid change in intensity distribution among neighborhood. Complexity of an image increase with the increase of patches which results increased amount of sharp edges /lines. Higher texture strength indicates clearly visible and definable primitives/patches composing an image. The defined five texture features are used in this study for pollution level detection and generates 84.3% of correct classification rate. The advantage of the method is its less complexity of execution and processing time compare to existing studies. In the extended version of the work, we will collect more dataset and will try to implement more specific classification based on PM2.5 value.

Acknowledgment. The authors would like to thank Yi Zou, Beijing Kinto Investment Management Co., Ltd, Beijing, China for acquisition of the dataset which is used in this study. Authers also would like to thank Dr. M. K. Bhowmik, Assistant Professor, Department of Computer Science and Engineering, Tripura University for his support during the initial level of knowledge development in the area of image feature extraction and classification.

References

1. Eason, G., Noble, B., Sneddon, I.N.: On certain integrals of Lipschitz-Hankel type involving products of Bessel functions. Phil. Trans. Roy. Soc. London **A247**, 529–551 (1955)
2. Guidelines (2017). In: World Health Organization. https://www.who.int/airpollution/guidelines/en/. Accessed 21 Aug 2019
3. Vahdatpour, M.S., Sajedi, H., Ramezani, F.: Air pollution forecasting from sky images with shallow and deep classifiers. Earth Sci. Inform. **11**, 413–422 (2018). https://doi.org/10.1007/s12145-018-0334-x
4. Spyromitros-Xioufis, E., Moumtzidou, A., Papadopoulos, S., et al.: towards improved air quality monitoring using publicly available sky images. Multimed. Tools Appl. Environ. Biodivers. Inform. 67–92 (2018). https://doi.org/10.1007/978-3-319-76445-0_5
5. Chakma, A., Vizena, B., Cao, T., Lin, J., Zhang, J.: Image-based air quality analysis using deep convolutional neural network. In: Proceedings - International Conference on Image Processing, ICIP, Vol. 2017–September (2018). https://doi.org/10.1109/ICIP.2017.8297023
6. Liu, C., Tsow, F., Zou, Y., Tao, N.: Particle pollution estimation based on image analysis. PLoS ONE, **11**(2) (2016). https://doi.org/10.1371/journal.pone.0145955
7. Wang, H., Yuan, X., Wang, X., Zhang, Y., Dai, Q.: Real-time air quality estimation based on color image processing. In: 2014 IEEE Visual Communications and Image Processing Conference, VCIP 2014, pp. 326–329). Institute of Electrical and Electronics Engineers Inc. (2015). https://doi.org/https://doi.org/10.1109/VCIP.2014.7051572
8. Amritphale, A.N.: A digital image processing method for detecting pollution in the atmosphere from camera video. In: IP version (2013).
9. Amadasun, M., King, R.: Textural features corresponding to textural properties. IEEE Trans. Syst. Man Cybern. **19**(5):1264–1274 (1989). https://doi.org/10.1109/21.44046

Transmission Map Estimation Function to Prevent Over-Saturation in Single Image Dehazing

Teena Sharma[1(✉)], Isha Agrawal[2], and Nishchal K. Verma[1]

[1] Department of Electrical Engineering, IIT, Kanpur, Kanpur 208016, India
[2] Department of Computer Science and Engineering, PDPM IIITDM, Jabalpur, Jabalpur 482005, India
`tee.shar6@gmail.com`, `ishaagrawal2017@gmail.com`, `nishchal@iitk.ac.in`

Abstract. This paper proposes a transmission map estimation function with the objective to prevent over-saturation in fine haze and haze-free regions. The conventional transmission map estimation function often results in over-saturation and over-dehazing due to its lack of variance with the corresponding scene depth. To overcome this, the proposed transmission map estimation function considers dark channel along with saturation and value (or brightness) channels' information of the input hazy image. Furthermore, the proposed method is independent of any heuristics such as the scattering coefficient of the atmosphere, and depends only on the input image. The performance of the proposed method in this paper has been analyzed quantitatively and qualitatively on various standard images frequently used in literature. The experimental results demonstrate the effectiveness of the proposed method with respect to the state-of-the-art.

Keywords: Image dehazing · Transmission map estimation · Dark channel · Saturation channel · Value channel

1 Introduction

The presence of various particulate matters in the atmosphere, such as dust, water drops and aerosol, often obscure the clarity of the environment. One of the most common phenomena amongst such inclement weather conditions is haze. It results in the reflected light being absorbed and scattered which leads to low brightness and possible lack of color information in the image. A number of dehazing methodologies have been developed in the literature to deal with such conditions.

Literature Review: One of the pioneer models that contributed towards the development of dehazing methodologies is the atmospheric scattering model [1], mathematically represented as

$$I_O(m) = I_T(m)T_M(m) + A_G(1 - T_M(m)) \tag{1}$$

M. Tripathi and S. Upadhyaya (Eds.): ICDLAIR 2019, LNNS 175, pp. 144–152, 2021.
https://doi.org/10.1007/978-3-030-67187-7_16

where I_T is the true scene radiance, A_G is the global atmospheric light, T_M is the transmission map describing the portion of light that reaches to the camera, I_O is the observed hazy image, and m is the pixel location.

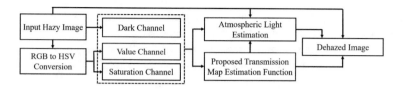

Fig. 1. Block diagram of the proposed methodology.

A number of traditional methods make use of (1) by calculating $T_M(m)$ and A_G to output a clear image. The work presented in [2] introduced an additional prior, called dark channel prior (DCP) based on the observation that some pixels often have a very low intensity in at least one of the color channels, and used this to estimate the transmission. However, it fails in the sky regions and on objects having an intrinsic colour similar to their background. It is also computationally expensive, and does not vary much with scene depth. In [3], some more haze relevant features were presented like maximum contrast, hue disparity, and max saturation. Some other initial attempts for dehazing were based on contrast restoration. However, these techniques tend to produce unrealistic images due to their underlying assumption of the intensity levels being uniform throughout, such as in histogram equalization (HE) [4]. Others are adaptive but computationally expensive, like contrast-limited adaptive histogram equalization (CLAHE) [5]. In [6], the local contrast has been maximized based on markov random field (MRF) which gives over-saturated results. Boundary constraint and contextual regularization (BCCR) [7] was proposed to efficiently remove haze. In [8], a refined image formation model was used to remove haze, but this is quite time-consuming and fails in the regions with dense haze. To reduce the computational time due to soft matting in DCP [9], various filters were introduced by [10]. Nowadays, deep learning based approaches have also gained tremendous popularity. Some of these models use the conventional atmospheric scattering model [1] to approximate a haze-free output, such as color attenuation prior (CAP) [11], multi-scale convolutional neural network (MSCNN) [12], DehazeNet [13], etc. In recent literature, deep learning based approaches directly map from hazy to haze-free images without producing any intermediate transmission map, such as atmospheric illumination prior network (AIPNet) [14]. However, all of these models take haze relevant features as input. The generic model-agnostic convolutional neural network (GMAN) [15], on the other hand, requires no such input features, and thus suggests the possibility of a global restoration model.

Contributions: The dehazing methods, especially those using dark channel [2], often overestimate the airlight in the atmospheric scattering model, leading to over- intensification of the colour channels. This is because the transmission maps generated using dark channel do not vary much with the corresponding scene depth, leading to an overestimation of haze thickness even in regions having fine haze. This paper proposes a transmission map estimation function for unsupervised single image dehazing to overcome the over-saturation problem. The proposed function uses saturation and value channels along with the dark channel of the input image. This estimation approximates a haze-free image without over-intensification of the color channels.

The remainder of the paper comprises of Sect. 2 for the proposed method, Sect. 3 for the experimental results, and Sect. 4 to highlight the conclusions.

2   Proposed Methodology

In this section, the proposed method of single image dehazing has been explained in detail. Figure 1 shows the block diagram of the proposed method. The following subsections include discussions of basic building blocks of Fig. 1.

2.1   Dark Channel

The proposed method extracts the dark channel from a input hazy image as

$$I_O^{\mathrm{dark}}(m) = \min_{n \in P(m)} \left(\min_{C_c \in \{\mathrm{R, G, B}\}} I_O^{C_c}(n) \right) \qquad (2)$$

where C_c represents R, G, and B color channels of the $I_O(m)$, and $P(m)$ is a local patch centered at pixel m. The size of the local patch considered in this paper is 3×3. Dark channel states that the intensity of dark or low pixels is mainly determined by the airlight. This directly produces haze transmissions, which sometimes fails to accurately measure the image depth. Accordingly, the proposed method considers saturation and value channels of HSV color domain for more accurate estimation of the haze.

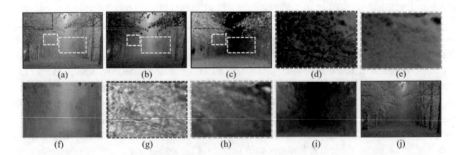

Fig. 2. Effect of over-saturation: **(a)** Observed hazy image, **(b)** Dark channel of (a), **(c)** Saturation channel of (a), Zoomed views of: **(d)** Red patch in (b), **(e)** White patch in (b), **(f)** Yellow patch in (b), **(g)** Red patch in (c), **(h)** White patch in (c), **(i)** Yellow patch in (c), and **(j)** Dehazed image using the proposed method. Here, left block is red, center block is white, and right block is yellow in (a), (b), and (c).

2.2   RGB to HSV Conversion

Consider an input RGB image $I_O(m)$ having $\{R, G, B\} \in [0, C_{\alpha_0}]$, where C_{α_0} is the maximum possible intensity. The value $(I_O^V(m))$ and saturation $(I_O^S(m))$ channels in the corresponding HSV color space [16] are given by

$$I_O^V(m) = \frac{C_{\alpha_1}(m)}{C_{\alpha_0}}, \quad \text{and} \quad I_O^S(m) = \begin{cases} \dfrac{C_r(m)}{C_{\alpha_1}(m)} & C_{\alpha_1}(m) > 0, \\ 0 & \text{otherwise} \end{cases} \tag{3}$$

where $C_{\alpha_1}(m) = \max(R, G, B)$ and $C_r(m) = C_{\alpha_1}(m) - C_{\alpha_2}(m)$ for $C_{\alpha_2}(m) = \min(R, G, B)$ at pixel location m. The value channel indicates the quantity of light reflected. On the other hand, saturation is the measure of purity of a color which gets reduced when another pigment is added to the base color. The presence of haze in the atmosphere adversely affects the purity of the color observed, and as a result, the saturation of the captured image decreases. For better understanding, the dark and saturation channels of a hazy image have been shown in Fig. 2. Since, the thickness of haze in an image increases with depth, it is minimum near the leaves (marked by red), and maximum towards the center (marked by yellow). It can be observed that as the haze is lesser in the region enclosed by red, the intensity of dark channel is close to zero (Refer Fig. 2(d)). This is because the minimum intensity value in regions having little to no haze will be zero. On the other hand, the intensity of saturation channel is close to one due to more purity in the absence of much haze (Refer Fig. 2(g)). Similarly, the dark channel in the region enclosed by yellow is close to white as minimum intensity will be high (Refer Fig. 2(f)) and the saturation channel, on the other hand, is close to zero due to lack of purity (Refer Fig. 2(i)). It can therefore be inferred that as haze increases, the intensity of dark channel moves from zero to one, whereas saturation decreases from one to zero. Therefore, the regions having intermediate haze, such as marked by white, both, dark and saturation channel, have an intermediate gray value due to the presence of moderate haze (Refer Figs. 2(e) and 2(h)). Accordingly, considering both the channels together can give better insight for estimation of the haze thickness.

2.3   Proposed Transmission Map Estimation Function

The transmission channel $T_M(m)$ according to the conventional atmospheric scattering model [1] is given by

$$T_M(m) = \exp\left(-\beta \times I_O^{\text{dark}}(m)\right) \tag{4}$$

where $I_O^{\text{dark}}(m)$ represents the pixel depth, and β is the scattering coefficient of the atmosphere. (4) considers only dark channel for haze estimation. It generates a transmission map that does not vary much with the corresponding scene depth. As a result, the haze thickness gets over-estimated. Furthermore, (4) also takes a heuristic parameter β which has to be manually adjusted for different images. As a result, additional information is required to estimate haze thickness. Inspired by DCP [2], AIPNet [14], and CAP [11], this paper considers saturation and value channels of $I_O(m)$, as they are the most affected by haze relative to the hue channel. Accordingly, the proposed transmission map estimation function is given by

$$T'_M(m) = \exp\left(-\frac{I_O^{\text{dark}}(m)}{\exp\left(\left(I_O^S(m)\right)^4 \times \left(I_O^V(m) + I_O^S(m)\right)^{0.01}\right)}\right) \tag{5}$$

where $T'_M(m)$ is the new estimated transmission map. It can be inferred from Sect. 2.2 that saturation decreases in the presence of haze. As a result, the saturation in the regions having thick haze will be closer to zero. Therefore, the denominator of (5) will be close to one and transmission value will mainly depend on the dark channel. On the other hand, the saturation in the regions having fine haze will be higher. Therefore, the transmission map will be more sensitive to it. This prevents over-enhancement of the image. To ensure that, along with preventing over-saturation, color fidelity and contrast information is also maintained by value channel.

2.4   Atmospheric Light Estimation and Image Dehazing

The transmission map generated using proposed function in (5) is further used to calculate atmospheric light (A_G) as

$$A_G = \max_{i \in \{m | T'_M(m) < t_0\}} (I_O(i)) \tag{6}$$

where t_0 is the threshold for medium transmission. In this paper, $t_0 = 0.1$. Finally, a haze-free output image $I_T(m)$ is obtained using (1) as

$$I_T(m) = \frac{I_O(m) - A_G}{T'_M(m)} + A_G \tag{7}$$

Another merit of using (5) is that it depends only on $I_O(m)$, and does not require any manual adjustment of parameters such as β.

3   Experimental Results

In this section, the qualitative and quantitative results[1] of the proposed method have been discussed and compared with some famous state-of-the-art methods[2], namely CLAHE, DCP, BCCR, CAP, DehazeNet, and GMAN.

---

[1] The results of the proposed method on images in Figs. 3 and 4 are available at https://drive.google.com/open?id=1vXc2SwzZVyiHkg9OQUsV5XNTtlSUjpyK.

[2] The results of existing literature have been obtained using publicly available source codes or libraries by their corresponding authors' support.

Fig. 3. Qualitative comparison on natural hazy images. **From left to right and top to bottom:** Input hazy image, dehazed output using CLAHE, DCP, BCCR, CAP, DehazeNet, GMAN, and the proposed method.

Fig. 4. Top row: Synthetic hazy images. **Bottom row:** Corresponding ground truths. **From left to right:** Image of "Computer", "Cones", "Laundry", "Moebius", and "Reindeer". These images are from Middlebury dataset [17,18].

Qualitative Analysis: The comparison of experimental results on some natural and synthetic hazy images commonly used in literature are shown in Figs. 3 and 5. The synthetic hazy images are a part of Middlebury dataset [17,18], and were generated using the conventional atmospheric scattering model [1] by considering different transmission values for different haze densities. These figures illustrate the efficacy of the proposed function (5) by producing visually appealing results without any over-saturation effects. For both, natural and synthetic hazy images, it can be observed that CLAHE fails to remove haze completely, and also loses color information while BCCR and GMAN are accompanied by significant artifacts. DCP too generates artifacts, which are noticeable in the sky regions of

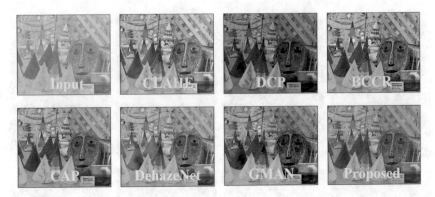

Fig. 5. Results on "Cones" image. (Please refer Fig. 4 for ground truth image).

Table 1. Performance comparison in terms of PSNR and SSIM on images of Fig. 4

Dataset	Metric	CLAHE	DCP	BCCR	CAP	DehazeNet	GMAN	Proposed
Computer	**PSNR**	27.428	27.648	27.928	27.548	27.565	28.232	**29.962**
	SSIM	0.907	0.922	0.793	0.962	0.948	0.835	**0.958**
Cones	**PSNR**	27.558	27.751	28.939	30.123	27.630	28.726	**30.817**
	SSIM	0.858	0.956	0.936	0.976	0.952	0.886	**0.964**
Laundry	**PSNR**	27.436	27.502	28.634	29.304	28.164	28.128	**29.215**
	SSIM	0.867	0.890	0.922	0.975	0.957	0.830	**0.957**
Moebius	**PSNR**	27.623	27.648	28.500	28.541	29.140	28.436	**29.221**
	SSIM	0.850	0.901	0.928	0.957	0.949	0.861	**0.952**
Reindeer	**PSNR**	27.545	28.673	27.986	27.725	27.335	29.909	**27.912**
	SSIM	0.652	0.884	0.764	0.831	0.790	0.863	**0.798**

Fig. 3. It also over-estimates the haze leading to over-intensification of the color channels. The proposed method presents no such distortions. The absence of any such artifacts suggests that the proposed transmission map estimation function successfully estimates the haze thickness.

Quantitative Analysis: The peak signal-to-noise ratio (PSNR) and structural similarity (SSIM) index values comparison on synthetic hazy images of Fig. 4 have been tabulated in Table 1 (Please refer Fig. 5 for result visualization on "Cones" image). It can be seen that the proposed method significantly outperforms nearly all the considered state-of-the-art methods in terms of PSNR, and gives comparable performance in terms of SSIM. The higher PSNR and SSIM values of the proposed method as compared to DCP evidence that the use of additional channels efficiently estimates haze thickness. Furthermore, the proposed method does not require any soft matting either, unlike in DCP. This implies that the proposed transmission map estimation function directly outputs a haze-free image without requiring any further operations. This also suggests that the proposed method will be much faster than DCP, as soft matting employed by DCP is often time consuming. Although the results of CAP are slightly higher than the proposed method, it should be noted that CAP is a deep learning based model which has been trained on millions of images. On the other hand, the proposed method is an unsupervised model which requires no training. Considering these parameters, the proposed transmission map estimation function gives remarkable results.

4   Conclusions

The paper proposed a transmission map estimation function for single image dehazing which considers value and saturation channels along with the dark channel of the input hazy image. It successfully removes haze without any over-saturation or over-intensification effects, and thereby, overcomes the limitations of the conventional transmission estimation function. Moreover, it does not require any additional operations such as filtering or soft matting. Although the results of deep-learning based models may slightly be better, the proposed method is unsupervised and does not require any training. These considerations evidence the suitability and efficiency of the proposed method for dehazing applications. This function can further be improved by considering more color domains, and other factors affecting haze.

References

1. Narasimhan, S.G., Nayar, S.K.: Vision and the atmosphere. Int. J. Comput. Vision **48**(3), 233–254 (2002). https://doi.org/10.1023/A:1016328200723
2. He, K., Sun, J., Tang, X.: Single image haze removal using dark channel prior. IEEE Trans. Pattern Anal. Mach. Intell. **33**(12), 2341–2353 (2011). https://doi.org/10.1109/TPAMI.2010.168
3. Tang, K., Yang, J., Wang, J.: Investigating haze-relevant features in a learning framework for image dehazing. In: 2014 IEEE Conference on Computer Vision and Pattern Recognition, Columbus, OH, USA, 23-28 June 2014. https://doi.org/10.1109/CVPR.2014.383
4. Gonzalez, R.C., Woods, R.E.: Digital Image Processing, 3rd edn. Prentice-Hall, Inc., Upper Saddle River (2006)
5. Zuiderveld, K.: Contrast limited adaptive histogram equalization. In: Graphic Gems IV, pp. 474–485. Academic Press Professional, San Diego, August 1994
6. Tan, R.T.: Visibility in bad weather from a single image. In: 2008 IEEE Conference on Computer Vision and Pattern Recognition, Anchorage, AK, USA, 23–28 June 2008. https://doi.org/10.1109/CVPR.2008.4587643
7. Meng, G., Wang, Y., Duan, J., Xiang, S., Pan, C.: Efficient image dehazing with boundary constraint and contextual regularization. In: 2013 IEEE International Conference on Computer Vision, Sydney, NSW, Australia, pp. 617–624, 1–8 December (2013). https://doi.org/10.1109/ICCV.2013.82
8. Fattal, R.: Single image dehazing. ACM Trans. Graphics (TOG) **27**(3) (2008). https://doi.org/10.1145/1360612.1360671
9. Levin, A., Lischinski, D., Weiss, Y.: A closed-form solution to natural image matting. IEEE Trans. Pattern Anal. Mach. Intell. **30**(2), 228–242 (2008). https://doi.org/10.1109/TPAMI.2007.1177

10. He, K., Sun, J., Tang, X.: Guided image filtering. IEEE Trans. Pattern Anal. Mach. Intell. **35**(6), 1397–1409 (2013). https://doi.org/10.1109/TPAMI.2012.213
11. Zhu, Q., Mai, J., Shao, L.: A fast single image haze removal algorithm using color attenuation prior. IEEE Trans. Image Process. **24**(11), 3522–3533 (2015). https://doi.org/10.1109/TIP.2015.2446191
12. Ren, W., Liu, S., Zhang, H., Pan, J., Cao, X., Yang, M.-H.: Single image dehazing via multi-scale convolutional neural networks. In: European Conference on Computer Vision, pp. 154–169, September 2016. https://doi.org/10.1007/978-3-319-46475-6_10
13. Cai, B., Xu, X., Jia, K., Qing, C., Tao, D.: DehazeNet: an end-to-end system for single image haze removal. IEEE Trans. Image Process. **25**(11), 5187–5198 (2016). https://doi.org/10.1109/TIP.2016.2598681
14. Wang, A., Wang, W., Liu, J., Gu, N.: AIPNet: image-to-image single image dehazing with atmospheric illumination prior. IEEE Trans. Image Process. **28**(1), 381–393 (2019). https://doi.org/10.1109/TIP.2018.2868567
15. Liu, Z., Xiao, B., Alrabeiah, M., Wang, K., Chen, J.: Single image dehazing with a generic model-agnostic convolutional neural network. IEEE Signal Process. Lett. **26**(6), 833–837 (2019). https://doi.org/10.1109/LSP.2019.2910403
16. Burger, W., Bruge, M.J.: Digital Image Processing: An Algorithmic Introduction Using java. Springer, London (2016)
17. Scharstein, D., Szeliski, R.: High-accuracy stereo depth maps using structured light. In: IEEE Computer Society Conference on Computer Vision and Pattern Recognition (CVPR 2003), Madison, WI, vol. 1, pp. 195–202, June 2003
18. Hirschmüller, H., Scharstein, D.: Evaluation of cost functions for stereo matching. In: IEEE Computer Society Conference on Computer Vision and Pattern Recognition (CVPR 2007), Minneapolis, MN, June 2007

Stock Price Prediction Using Recurrent Neural Network and Long Short-Term Memory

Avanish Kumar$^{(\boxtimes)}$, Kaustubh Purohit, and Krishan Kumar

Department of Computer Science and Engineering,
National Institute of Technology Uttarkhand, Srinagar Garhwal, Uttarakhand, India
avanishkr018@gmail.com

Abstract. The stock market or equity market has a significant impact on our world. An increment or the decrement in the share price determines whether an investor is going to be in gain or not. The existing prediction models with the help of both dynamic and definite algorithms focuses on foreseeing the stock value of a single organization using the daily opening and closing stock price while the recommended model uses a different approach, where instead of using the data for a specific model, latent dynamics of the data-set is identified with the help of deep learning algorithms. In the proposed model, the price of Google Stock is predicted using two deep learning models with the least possible error. We are applying the Recurrent Neural Network for predicting the price on a short term basis.

Keywords: RNN · Time stamp · LSTM · Stock prediction

1 Introduction

Predicting the future possibilities with the help of historical data and analyzing them is known as forecasting. Applications of forecasting can be seen in various fields form finance, business, mathematics, and science. There are basically two types of forecasting problems:

1) Long-term (prediction for more than 2 year)
2) Medium-term (prediction that varies form 1 to 2 year)
3) Short-term (prediction from a few seconds, minutes to months)

Most of the problems in the forecasting field involves the analysis of time [1]. A data stored in the time series is defined in the chronological order for the observations of a selected variable. In the problem we are trying to solve, stock price is that variable, it can either be uni-variate or multi-variate. Uni-variate data carries the status of only one stock while the multi-variate data has the information of more than one organization at various occasion in time. Time series data helps us in analyzing different types of data from predicting the trends,

M. Tripathi and S. Upadhyaya (Eds.): ICDLAIR 2019, LNNS 175, pp. 153–160, 2021.
https://doi.org/10.1007/978-3-030-67187-7_17

patterns and periods. Early knowledge of the market and the company helps in investing money profitably, also recognising the market leader can be easily done with the help of analysis of patterns, thus making the Stock Price Prediction an important area for research work. The existing methods for predicting the stock price can be categorized into three categories [2].

1) Time-Series Forecasting
2) Fundamental Analysis
3) Technical Analysis

Time-Series Forecasting is used when the analysis of time-series is necessary. Fundamental Analysis is used when the values of profit, sale and revenue is given and share value is estimated using all these factors and is best suited for long-term analysis while the Technical analysis uses previous values of the stock for prediction, it is basically the moving average using unweighted mean of 'x' data points [1] (Fig. 1).

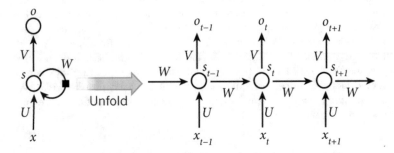

Fig. 1. Recurrent neural network

The stock price of an organisation in a market is highly volatile which makes it complicated for the financial indicators to predict the value, but with the advancement in the technology the chances of gaining the constant profit from the market is increasing and helping the experts in making a better predictions. Better prediction helps in decreasing the risk and maximizing the profit through the trade of stocks [3]. Recurrent neural networks (RNN) have proved one of the most powerful models for processing sequential data. Long Short-Term memory is one of the most successful RNNs architectures. LSTM introduces the memory cell, a unit of computation that replaces traditional artificial neurons in the hidden layer of the network. With these memory cells, networks are able to effectively associate memories and input remote in time, hence suit to grasp the structure of data dynamically over time with high prediction capacity [2].

The paper is structured as follows Sect. 2 explains the previous existing models and their Drawbacks, Sect. 3 explains the Methodology of the proposed model, Sect. 4 tells us about the Results and Discussions and Sect. 5 includes the conclusion.

2   Previous Models and Drawbacks

Previous models for the analysis of wall street include fundamental analysis, which takes the past performance of the company and the credibility of the company, and statistical analysis, which is plays with, how the stock price is varied. Algorithms which are generally used in predicting the trend are Genetic Algorithms (GA) and Artificial Neural Networks (ANN's), but both of them fails in determining the correlation between the LSTD and the stock prices [4]. Simple ANN's also lack in predicting the exploding gradient condition, i.e weights become either too high or less, thus slowing down the rate of convergence Which arises because of the two factors: Random initialization of weights, and because weights tend to change at a much faster rate at the end of the network.

3   Methodology

The data set consists of day-wise data for National Stock Exchange listed companies for the period of Jan 2012 to Dec 2016. Data Set includes information like Opening Rate, Closing Rate, Highest Day Rate, Lowest Day Rate, Volume of Transaction on a daily basis. For calculation, we choose a company 'X' data set and used its values from the past year to train our model and thus tested the trained model with a test data. The data for the company were extracted from the available data and was subjected to pre-processing to obtain the stock price. The work is based on the application of Recurrent Neural Network using Long Short Term Memory where 1257 values are used in determining the values of the future 20. Libraries used in the proposed model are [5].

NumPy: As it helps in doing the mathematical and scientific operation and ease off the work in multi-dimensional arrays and matrices.

Pandas: As it provides high-performance with the help of easy-to-use data structures and analysis data for Python.

Keras: As is can develop and evaluate deep learning models using Theano and TensorFlow and can train the neural network in a short line of codes.

SkLearn: As it can be used in the normalization of data set into a confined boundary.

The data of the company varies within a range of 250 to 900 for Google. To unify the data-range, the price cap was normalized to the range of 0 to 1 using the SkLearn library of Python. The normalized data is then as given as an input to the prescribed model for training. The model was repeatedly trained for 100, 250, 500, 1000 epochs for much fine tuning. If the current epoch loss (mean square error) is less than the loss of the previous epoch, then the weight matrices for that epoch is stored. After multiple tutelage processes, the model was tested with the test data-set and the one with least RMSE(Root Mean Squared Error) for a particular epoch is taken as the final model for prediction [6]. The model consists of two neural network architectures, Recurrent Neural Network and Long Short Term Memory. Class of neural networks that which makes a connection between the computational units from a directed circle is

Recurrent Neural Network. This is a special case of Neural Network where the output from the previous level is given as an input to the contemporary level. Traditionally, in all the neural networks input and output are independent of each other, but when we need to predict the future, previous values are required hence RNN's requirement is a must, as it solves this issue using the Hidden layer concept as these remember the information about the sequence [7].

The values of the current state, activation function, and the output state can be calculated using these formulas.

$$CurrentState$$

$$a(x) = b(a(x - 1), i)$$

$$where$$

$$a(x) = CurrentState$$

$$a(x - 1) = PreviousState$$

$$i = InputState$$

$$ActivationFunction$$

$$a(x) = tan(a)[Wa(x - 1) + Yi]$$

$$where$$

$$W = WeightofRecurrentNeural$$

$$Y = WeightofInputNeural$$

$$OutputFunction$$

$$Y = W(o) * a$$

$$where$$

$$W(o) = WeightofOutputLayer$$

$$Y = Output$$

The second type of neural network is Long Short Term Memory LTSM which is a special type of RNN as it helps by protecting the loss in the error that is back-propagated with time and layers of the proposed model. A constant error is maintained by the LSTM model, thus allowing the error not to exceed a defined value, thus allowing the model to train over 1000 steps, thus acts independently. This is one of the main provocations for artificial intelligence and neural network as algorithms are mostly resisted by the environment. LSTMs contain data in the form of gated cells. Information can be read or written into these cells, and the cell decides about what information to preserve and what to erase with the help of the gates. These are generally sigmoidal functions that are analog in nature, i.e either close or open. Analogs are differentiable in nature, hence gives a slight advantage over the traditional back-propagation [2]. LSTM takes input as [BatchSize, TimeStamp, Feature] in a 3- Dimensional format, where Batch Size determines after how many inputs sets the weight of the network is going to update, more the size less the time and vice-versa. Time Stamp is defined as how much previous data ones wish to view in order to predict the future value and Feature tells the number of attributes needed to calculate the time stamp [1].

The hyperbolic Tangent function is used as an activation function in this model. The derivative of the activation function is used in the update the weights with the help of error loss calculated (Fig. 2).

Hyperbolic Tangent Function

$$tanh(x) = (e^x - e^{-x})/(e^x + e{-}x) \tag{1}$$

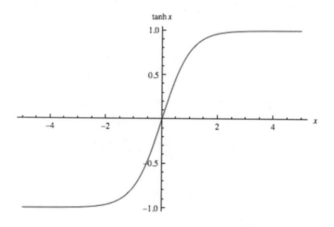

tanh x

Fig. 2. Range of hyperbolic tangent function

It gives an output in the range 0,1 and most importantly it is a continuous function, i.e gives the output for each and every value of x.

Hyperbolic tangent function and its derivative plays a vital role in determining the true value of the stock by reducing the error:

$$f(x) = (e^x - e{-}x)/(e^x + e{-}x) \tag{2}$$
$$d(f(x))/dx = 1 - (f(x))^2 \tag{3}$$

The training data-set was then normalized in the range of 0 to 1 using Min-MaxScalar function of the SkLearn library of the python. After each and every iteration, denormalization was done and the percentage error was calculated. The error was calculated using the RMS (Root Mean Square Value) of the predicted value using the formula:

$$error = sigma((TV^2 - PV^2)\cdot 5)/N \qquad (4)$$

$$where$$

$$TV = truevalue$$

$$PV = predictedvalue$$

$$N = TotalNoofValues$$

4   Result and Discussion

The model was trained on google data-set using the different values of Batch Size, Time Stamp and Epochs. Effect of Batch Size on loss, the time required per iteration for a particular epoch is calculated and shown in Table[1]. It is clear from table[1] that Batch Size gives the best result when its value is 10 (Figs. 3 and 4).

Epoch	Batch size	Loss	Time/epoch
100	5	0.0011	23 s 9 ms
100	10	0.0011	11 s 9 ms
100	32	0.0014	4 s 9 ms
100	50	0.0018	3 s 9 m

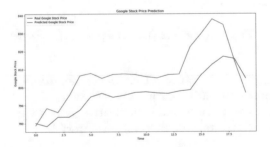

Fig. 3. Predicted Vs true value graph

When the value of epoch was changed keeping the value of Batch Size and Time Stamp constant the result came out to be in Table[2]. Its clear from Table[2] that model gives the best result when the epoch value is

Epoch	Batch size	Loss	Time/epoch
100	32	0.0015	4 s 3 ms
250	32	9.7833e−04	4 s 3 ms
500	32	7.8043e−04	4 s 3 ms
1000	32	7.5204e−04	4 s 3 ms

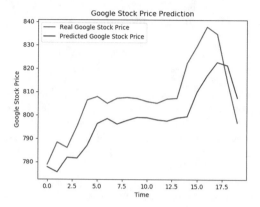

Fig. 4. Predicted Vs true value graph

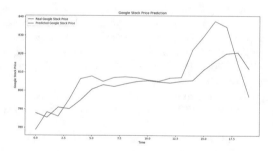

Fig. 5. Final output

From Table[1] and Table[2] it can be concluded that the most appropriate value of Batch Size, Time Stamp and Epoch came out to be 10,60,1000 and the error loss came out to be 7.3543e−04. Thus decreasing the batch size increases the time taken by the model to train but on the other hand decreases the loss per iteration, on the other hand as the number of epochs has increased the loss suffered by the model decreases, i.e the Model becomes better (Fig. 5).

5   Conclusion

We propose a Recurrent Neural Network and Long Short Term Memory based Stock Prediction model. These are used as both of them are good at handling the time-based problems and can make predictions about the future. We trained the model using Google stock price data-set from Jan 2012 till Dec 2016 and then tested it by predicting the future stocks of the same from 1st Jan 2017 to 31st Jan 2017. The study showed that the model is capable enough in predicting the near future with the help of the two. Also, it is clear that the proposed model can even predict the prices correctly whenever there is a sudden change in the market. Changes in the stock are not defined on a confined pattern, hence the actuality of the same will differ. This analysis will help the investors and people investing there riches will gain much more profit.

References

1. Selvin, S., Vinayakumar, R., Gopalakrishnan, E.A., Menon, V.K., Soman K.P.: Stock price prediction using LTSM, RNN and CNN and CNN-sliding window model. In: Proceddings of the 2017 International Conference on Advances in Computing, Communications and Informatics (ICACCI), Centre for Computational Engineering and Networking (CEN), Amrita School of Engineering, Coimbatore. https://doi.org/10.1109/ICACCI.2017.8126078
2. Jeenanunta, C., Chaysiri, R., Thong, L.: Stock price prediction with long short-term memory recurrent neural network. In: International Conference on Embedded Systems and Intelligent Technology & International Conference on Information and Communication Technology for Embedded Systems (ICESIT-ICICTES) (2018)
3. Hegazy, O., Soliman, O.S., Salam, M.A.: A machine learning model for stock market prediction, Faculty of Computers and Informatics, Cairo University, Egypt Higher Technological Institute (H.T.I), 10th of Ramadan City, Egypt
4. Rajput, V., Bobde, S.: Stock market forecasting techniques: literature survey. IJC-SMC **5**(6), 500–506 (2016). Department of Computer Engineering, Maharashtra Institute of Technology, Pune, India
5. Yu, L.Q., Rong, F.S.: Stock market forecasting research based on neural network and pattern matching, INSPEC Accession Number: 11562445 Date of Conference, 7–9 May 2010
6. Chen, K., Zhou, Y., Dai, F.: A LSTM-based method for stock returns prediction: a case study of China stock market. In: 2015 IEEE International Conference on Big Data (Big Data). IEEE (2015)
7. Karpathy, A.: The Unreasonable Effectiveness of Recurrent Neural Networks. http://karpathy.github.io/2015/05/21/rnn-effectiveness/

Artificial Neural Network Model for Path Loss Predictions in the VHF Band

Segun I. Popoola[1], Nasir Faruk[2], N. T. Surajudeen-Bakinde[3], Aderemi A. Atayero[1], and Sanjay Misra[1(✉)]

[1] Department of Electrical and Information Engineering, Covenant University, Ota, Nigeria
{segun.popoola,atayero,sanjay.misra}@covenantuniversity.edu.ng
[2] Department of Telecommunication Science, University of Ilorin, Ilorin, Nigeria
faruk.n@unilorin.edu.ng
[3] Department of Electrical and Electronics Engineering, University of Ilorin, Ilorin, Nigeria
deenmat@unilorin.edu.ng

Abstract. Artificial Neural Networks (ANNs) have been recently exploited to develop suitable models for path loss predictions . However, the ANN algorithm that provides the best results has not been well established neither has the models been characterized to limit their performances and applications in the various frequency bands. In this paper, we characterize the propagation path loss in the Very High Frequency Band (VHF) using different ANN learning algorithms and activation functions based on the measurement data collected at 203.25 MHz in an urban environment (Ilorin, Nigeria). The prediction results of Hata, COST 231, ECC-33, and Egli models at varying distances were fed into a feed-forward neural network and mapped to each corresponding measured path loss value. Statistical analysis shows that the ANN model that was trained with hyperbolic tangent activation function (HTAF), Levenberg-Marquardt (LM) algorithm, and 80 neurons in the hidden layer produced the most satisfactory results with Mean Error (ME), Root Mean Square Error (RMSE), Standard Deviation (SD), and coefficient of determination (R^2) values of 3.75 dB, 5.10 dB, 3.46 dB, and 0.95. However, the HTAF with Scale Conjugate Gradient (SCG) is more stable even though its prediction errors were slightly higher than that of LM.

Keywords: Path loss prediction · Radio propagation · ANN · Activation function · Learning algorithm

1 Introduction

Path loss prediction is very essential in the coverage design and optimization of wireless communication systems. The success of these systems depends largely on how adequately the system is planned to provide satisfactory coverage within the interference constraint. These models are either empirical or theoretical; the empirical models are developed using data gathered from the observations of an environment while the theoretical models requires the knowledge of the fundamental principles of radio wave

© The Author(s), under exclusive license to Springer Nature Switzerland AG 2021
M. Tripathi and S. Upadhyaya (Eds.): ICDLAIR 2019, LNNS 175, pp. 161–169, 2021.
https://doi.org/10.1007/978-3-030-67187-7_18

propagation [1]. Empirical models are widely used for the prediction of path loss usually owing to their simplicity but they are prone to errors when applied to diverse set of environments other than the one it was initially built for. Previous works [2–7] revealed how inconsistent the models can be without proper tuning.

However, heuristic methods such as the Artificial Neural Networks (ANNs) have been recently applied to predict path loss in various bands and locations [8, 9]. This was aimed at improving the accuracy by minimizing prediction errors. Erik *et al.* [10] developed an ANN model that generalizes well for measurement data taken in rural part of Western Australia. The proposed model was found to be efficient when compared with the predictions of ITU-R.P.1546 and Okumura-Hata models. Popescu *et al.* [11] proposed a Neural Network (NN) model for path loss and the model was found to yield better results when compared with the predictions of COST 231-Walfisch-Ikegami (CWI) model. Interestingly, even when hybrid approach (i.e. combination of CWI and NN) was employed, no significant improvement was observed when compared to the NN model. Sotirious *et al.* [12] also carried out an investigation on the optimal ANN for propagation path loss prediction using the Adaptive Differential Evolutionary (ADE) Algorithm. Among all the ADE algorithms investigated, the Composite DE (CoDE) proved to be the most effective, even when compared to a deterministic model i.e. the Ray-Tracing model. Angeles and Dadios [13] also showed that NN model outperformed the Free Space Loss (FSL) and Egli models. Similarly, Benmus *et al.* [14] proved that NN based path loss model provides better RMSE when compared to Hata model based on measurements conducted in Great Tripoli area at 900, 1800, and 2100 MHz Bands.

In recent times, different machine learning and geospatial approaches have been proposed for path loss prediction. Ayadi *et al.* [15] proposed a new model based on Artificial Neural Network (ANN) for path loss prediction in multi-bands heterogeneous wireless networks. In [16], some propagation parameters, such as distance between transmitting and receiving antennas, transmitting power and terrain elevation were used as inputs to develop an ANN based path loss model in the GSM frequencies. The Levenberg-Marquardt (LM) was used for training algorithm and the number of neurons in the hidden layer was varied from 31 to 39. The ANN model was found to be efficient in terms of RMSE when compared with the basic empirical path loss models such as: Hata; Egli; COST-231; Ericsson models. Sotiroudis *et al.* [17] proposed a model for path loss prediction based on ANN for an urban environment. The model showed that if the correct neural network size is chosen, it can help to increase its response speed and the overall performance of the model. In [18], a three stage approach was employed to develop an ANN model in the GSM band. The model usesd 33 neurons in the hidden layer and tansig activation, function was also used.

Although it is well established in the literature that the ANN model provides a better performance when compared to the empirical models, there are variant algorithms within the ANN model. In essence, the ANN algorithm that provides the best results has not been well established neither has the models been characterized to limit their performances and applications in the various frequency bands. This paper, therefore, characterized the propagation path loss in the VHF by using the received signal strength (RSS) data collected based on extensive measurement campaign conducted at 203.25 MHz in an urban environment (Ilorin, Nigeria). The prediction results of widely used empirical models (Hata, COST 231, ECC-33, and Egli models) were fed into the ANN as input data that map a corresponding path loss output. The ANN model was trained, tested, and verified based on LM and SCG algorithms with HTAF and logistic sigmoid activation function. The Mean Error (ME), Root Mean Square Error (RMSE), Standard Deviation (SD), and coefficient of determination (R^2), relative to real measurement data, were computed to appraise the performance of the ANN model under different configurations.

2   Materials and Methods

This section is divided into two, the first part describes the propagation environment and the data collection procedure while the second part explains the ANN model configurations.

2.1   Field Measurement Campaigns

Field measurements were conducted in Ilorin (Long 4°36′25″E, Lat 8°25′55″N) and its environs within Kwara State, Nigeria. Ilorin is a large city characterized by a complex terrain due to the presence of hills and valleys within the metropolis. Outside the metropolis, the routes are covered with thick vegetation. The altitude within the transmitter's coordinates is 403.7 m; which can be as low as 150 m when travelling within and outside the metropolis. The radio transmitter of the Nigerian Television Authority (NTA) Ilorin which was used transmits on channel 5 at frequency of 203.25 MHz. A dedicated Agilent spectrum analyzer was properly positioned in a vehicle was used for RSS measurements along a predefined route [19–22]. An average speed limit of 40 km/h was maintained throughout the campaign to reduce Doppler Effect. The RSS was measured continuously and stored in an external drive for subsequent analysis [23]. Measured data were properly filtered to minimize noise and preserve the shadowing effects. This reduces the number of data sets per route to 500 points.

2.2   ANN Model Development for Path Loss Predictions

In this work, the ANN was configured to characterize the complex non-linear relationship that exist between the propagation distance and the corresponding path loss in an urban environment. Firstly, path losses were computed at various distances within 10 km based on Hata, COST 231, ECC-33, and Egli models. The prediction results and the separation distances between the transmitter and the receiver were expressed in matrix form to serve as the ANN input variables.

The ANN has a single output variable that represents the corresponding path loss for each given distance. A single vector of path loss values acted as the ANN targets. Therefore, the ANN was a single layer feed-forward neural network of five input layers, one hidden layer, and a single output layer. The input matrix was transformed by the hidden layer based on the requirements of Levenberg-Marquardt (LM) and Scale Conjugate Gradient (SCG) training algorithms. Also, the ANN model that was trained with hyperbolic tangent activation function (HTAF) and the logistic sigmoid activation function were used for the function approximation performed by the artificial neurons because they are continuous and differentiable, thereby speeding up the learning rate relative to other activation functions. The HTAF and the logistic sigmoid activation function are both S-shaped and can be represented by Eqs. (1) and (2) respectively.

$$f(x) = \frac{1 - e^{-x}}{1 + e^{-x}} \tag{1}$$

$$f(x) = \frac{1}{1 + e^{-x}} \tag{2}$$

Series of ANN experimentations were performed with different training configurations. At each instance, the number of neurons in the hidden layer was varied between 10 and 100 at 10 intervals to determine the optimal solution. The entire dataset used for the ANN modelling was used for training, validation and testing in the proportion of 70%, 15%, and 15% respectively. The ANN development process, model validation, and simulations were done using MATLAB R2016a computation software produced by MathWorks Inc. Coefficient of determination (R^2) helps to understand the degree of relationship between the measured values and the values predicted by the ANN model under different scenarios understudied. The mathematical formula is given in Eq. (3).

$$R^2 = \frac{\sum_{i=1}^{n}\left(P_{i,measured} - P_{measured,mean}\right)^2 - \sum_{i=1}^{n}\left(P_{i,predicted} - P_{i,measured}\right)^2}{\sum_{i=1}^{n}\left(P_{i,measured} - P_{measured,mean}\right)^2} \tag{3}$$

The validity of the ANN model for path loss predictions was further affirmed with other statistical performance metrics: the ME, RMSE and the SD given by Eqs. (4)--(6).

$$ME = \frac{1}{n}\sum_{i=1}^{n}\left(P_{i,predicted} - P_{i,measured}\right) \tag{4}$$

$$RMSE = \sqrt{\frac{1}{n}\sum_{i=1}^{n}\left(P_{i,predicted} - P_{i,measured}\right)^2} \tag{5}$$

$$SD = \sqrt{\frac{1}{n}(|P_{i,predicted} - P_{i,measured}| - \mu)^2} \tag{6}$$

where, μ = the mean prediction error in decibel (dB).

3   Results and Discussion

Figure 1 shows the comparison of the path loss predictions obtained based on ANN model and the empirical path loss models. For the ANN Model, Hyperbolic Tangent, SCG Algorithm and 83 Neurons were used. It can observe that the ANN model prediction provides the best fit along the route. ECC-33 model performed optimally when compared with the other three contending models, i.e. Hata, COST 231 and Egli models which all underestimated the path loss over the distance. In Fig. 2, the Hyperbolic Tangent, LM Algorithm and 80 Neurons was used. Similar results were obtained as in the case of Fig. 1.

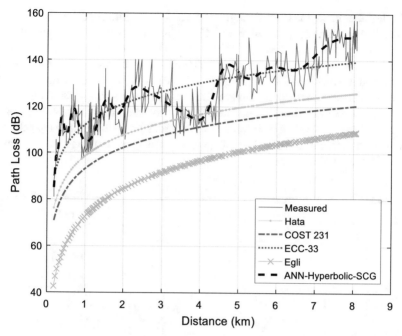

Fig. 1. ANN model with hyperbolic Tangent, SCG algorithm and 83 Neurons

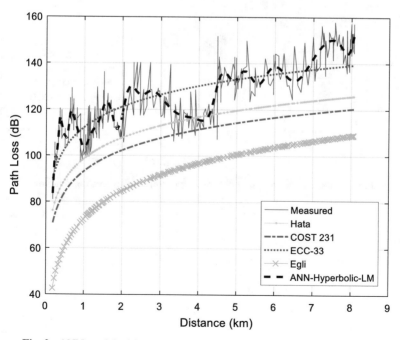

Fig. 2. ANN model with hyperbolic tangent, LM algorithm and 80 Neurons

On the other hand, the Logistic Sigmoid Scaled Conjugate Gradient and Levenberg-Marquardt algorithms provided higher RMSE values higher than the acceptable range. Although, the values of the RMSE obtained for the Logistic Sigmoid as shown in Table 2, is found to be higher, but, better than the empirical path loss predictions as side the ECC-33 model which has RMSE value of 9.1463 dB. Furthermore, we investigated the impact of epoch size on the prediction accuracy of the two contending ANN algorithms. The results are presented in Fig. 4. Interestingly, observed that the ANN Model with Hyperbolic Tangent, SCG Algorithm and 83 Neurons attained stability over the ANN Model with Hyperbolic Tangent, Levenberg-Marquardt (LM) training algorithm, and 80 neurons (Fig. 3 and Table 1) .

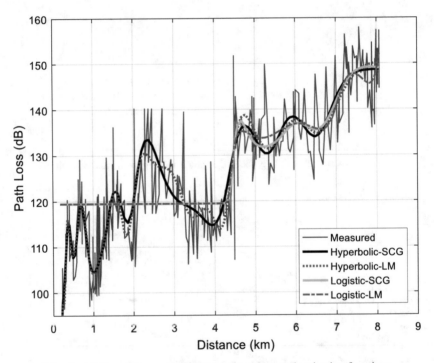

Fig. 3. ANN modeling using different algorithms and activation functions

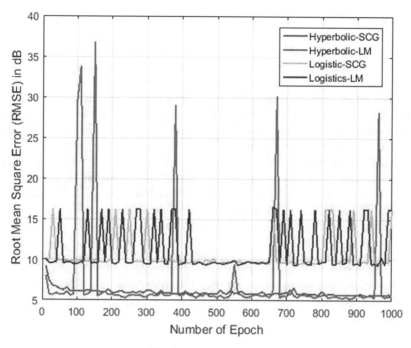

Fig. 4. Error distribution

Table 1. Statistical analysis of ANN models

Activation function	Training algorithm	No. of Hidden Neurons	MAE	RMSE	SD	R
Hyperbolic	SCG	83	4.24	5.65	3.74	0.93
Tangent sigmoid	LM	80	3.75	5.10	3.46	0.95
Logistic	SCG	71	6.98	9.50	6.44	0.86
Sigmoid	LM	64	6.86	9.36	6.37	0.86

Table 2. Statistical analysis of the predictions of existing path loss models

Empirical models	ME	RMSE	SD	R
Hata	15.11	17.29	8.39	0.82
COST 231	20.15	22.05	8.94	0.82
ECC-33	7.47	9.14	5.27	0.82
Egli	37.27	38.69	30.27	0.82
Optimal ANN	3.74	5.10	3.46	0.95

4 Conclusion

Empirical propagation models are prone to errors when applied to diverse set of environments. Heuristic methods have been gained considerable momentum in path loss prediction and have been found to be efficient. This paper have characterized The performance of the ANN model algorithms for path loss prediction in the Very High Frequency Band (VHF). Findings from this work showed that among the algorithms, the ANN model which employed hyperbolic tangent activation function (HTAF), Levenberg-Marquardt (LM) training algorithm, and 80 neurons in the hidden layer produced the most satisfactory results with least prediction errors. However, the hyperbolic tangent activation function with Scale Conjugate Gradient was found to be more stable. The prediction errors for all the ANN algorithms were better than that of the empirical models. In further work, we intended to study and characterized these algorithms for various routes across wide band that covers VHF and UHF bands.

Acknowledgment. The authors wish to appreciate the Center for Research, Innovation, and Discovery (CU-CRID) of Covenant University, Ota, Nigeria for funding this research. We, also, appreciate the University of Ilorin, Ilorin, Nigeria for the provision of dedicated spectrum analyzer used for this study.

References

1. Rappaport, T.S.: Wireless Communications: Principles and Practice, vol. 2. Prentice hall PTR, New Jersey (1996)
2. Oseni, O.F., et al.: Comparative analysis of received signal strength prediction models for radio network planning of GSM 900 MHz in Ilorin, Nigeria. Int. J. Innov. Technol. Explor. Eng. (IJITEE) **4**(3), 45–50 (2014)
3. Oseni, O.F., et al.: Radio frequency optimization of mobile networks in Abeokuta, Nigeria for improved quality of service. Int. J. Res. Eng. Technol. **3**(8), 174–180 (2014)
4. Popoola, S.I., Oseni, O.F.: Empirical path loss models for GSM network deployment in Makurdi, Nigeria. Int. Refereed J. Eng. Sci. **3**(6), 85–94 (2014)
5. Popoola, S.I., Oseni, O.F.: Performance evaluation of radio propagation models on GSM network in urban area of Lagos. Nigeria. Int. J. Sci. Eng. Res. **5**(6), 1212–1217 (2014)
6. Popoola, S.I., et al.: Calibrating the standard path loss model for urban environments using field measurements and geospatial data. In: Lecture Notes in Engineering and Computer Science: Proceedings of The World Congress on Engineering 2017, London, U.K., 5–7 July 2017, pp. 513–518 (2017)
7. Popoola, S.I., et al. (eds.): Standard propagation model tuning for path loss predictions in built-up environments. In: International Conference on Computational Science and its Applications, pp. 363–375. Springer (2017)
8. Popoola, S.I., Misra, S., Atayero, A.A.: Outdoor path loss predictions based on extreme learning machine. Wirel. Pers. Commun. **99**, 1–20 (2017)
9. Popoola, S.I., et al.: Optimal model for path loss predictions using feed-forward neural networks. Cogent Eng. **5**(1), 1444345 (2018)
10. Ostlin, E., Zepernick, H.-J., Suzuki, H.: Macrocell path-loss prediction using artificial neural networks. IEEE Trans. Veh. Technol. **59**(6), 2735–2747 (2010)

11. Popescu, I., Nafornita, I., Constantinou, P.: Comparison of neural network models for path loss prediction. In: IEEE International Conference on in Wireless and Mobile Computing, Networking and Communications, 2005 (WiMob'2005). IEEE (2005)
12. Sotiroudis, S.P., et al.: Optimal artificial neural network design for propagation path-loss prediction using adaptive evolutionary algorithms. In: 2013 7th European Conference on Antennas and Propagation (EuCAP). IEEE (2013)
13. Angeles, J.C.D., Dadios, E.P.: Neural network-based path loss prediction for digital TV macro-cells. In: International Conference on Humanoid, Nanotechnology, Information Technology, Communication and Control, Environment and Management (HNICEM), 2015. IEEE (2015)
14. Benmus, T.A., Abboud, R., Shatter, M.K.: Neural network approach to model the propagation path loss for great Tripoli area at 900, 1800, and 2100 MHz bands. In: 2015 16th International Conference on Sciences and Techniques of Automatic Control and Computer Engineering (STA). IEEE (2015)
15. Ayadi, M., Zineb, A.B., Tabbane, S.: A UHF path loss model using learning machine for heterogeneous networks. IEEE Trans. Antennas Propag. **65**(7), 3675–3683 (2017)
16. Eichie, J.O., et al.: Comparative analysis of basic models and artificial neural network based model for path loss prediction. progress. In: Electromagnetics Research M, vol. 61, pp. 133–146 (2017)
17. Sotiroudis, S.P., et al.: Optimal artificial neural network design for propagation path-loss prediction using adaptive evolutionary algorithms. In: 2013 7th European Conference on Antennas and Propagation (Eucap), pp. 3795–3799 (2013)
18. Eichie, J.O., et al.: Artificial neural network model for the determination of GSM Rxlevel from atmospheric parameters. Eng. Sci. Technol. Int. J. **20**(2), 795–804 (2017)
19. Ayeni, A., et al.: Comparative assessments of some selected existing radio propagation models: a study of Kano City, Nigeria. Eur. J. Sci. Res. **70**(1), 120–127 (2012)
20. Bakinde, N.S., et al.: Comparison of propagation models for GSM 1800 and WCDMA systems in selected urban areas of Nigeria. Int. J. Appl. Inf. Syst. (IJAIS) **2**(7), 6–13 (2012)
21. Faruk, N., Adediran, Y.A., Ayeni, A.A.: Error bounds of empirical path loss models at VHF/UHF bands in Kwara State, Nigeria. In: 2013 IEEE Eurocon, pp. 602–607 (2013)
22. Faruk, N., Ayeni, A., Adediran, Y.A.: On the study of empirical path loss models for accurate prediction of TV signal for secondary users. Progress Electromagn. Res. B **49**, 155–176 (2013)
23. Popoola, S.I., Atayero, A.A., Faruk, N.: Received signal strength and local terrain profile data for radio network planning and optimization at GSM frequency bands. Data Brief **16**, 972–981 (2018)

ShallowFake-Detection of Fake Videos Using Deep Learning

Aadya Singh[✉], Abey Alex George[✉], Pankaj Gupta[✉],
and Lakshmi Gadhikar[✉]

Fr. Conceicao Rodrigues Institute of Technology, Vashi, India
aadyas2409@gmail.com, abeyalex17@gmail.com, pankajgupta.pg877@gmail.com,
lmgadhikar@gmail.com
www.fcrit.ac.in

Abstract. In recent years, we have come across a vast range of software tools like "Photoshop" and techniques like DeepFake that have made it easier to create unrealistic and believable face swaps in videos that end up leaving very few traces of manipulation. The realistic nature of DeepFake videos are exploited for carrying out unethical and false practices such as generation of pedopornographic materials, Fake News, Fake Surveillance footage, Fake Hoaxes, videos for blackmailing amongst many more. Many AI based tools have been developed to detect DeepFake based video manipulation by extracting tampered face features. Through our approach, we aim to provide a forensic tool for investigators to detect these DeepFake videos which analyses specific facial artifacts like Eye Blink and Pulse. Each artifact will give their own probabilistic output via there classifier model and then combine them to give probability of fakeness. Thus instead of allowing resource intensive algorithms to detect facial artifacts for DeepFake detection, we aim to analyze and interpret immutable facial artifacts for detecting a DeepFake video.

Keywords: DeepFake · Fake videos · Eye blinking · Pulse detection · LRCN

1 Introduction

In the recent years, with rapid advancement in digital and camera technology along with the growth of social networks have made the editing and dissemination of digital media more easier and convenient than ever before. As an after effect, it has also opened doors and fostered the probability of digital tampering and manipulation of videos as an effective way to propagate falsified information. It thus paves its way in creating an overall negative impact in the society. Hence, a surge of such fake videos has disoriented the concept of identity and privacy all across the globe.

2   Background

Drawing a distinct demarcation in between a fake and a real image or video has proven to be challenging for the forensic investigators. Some methods detect fakeness at signal level (JPEG compression) or physical level like lighting conditions. Most of these techniques exploit the regularities in statistical features from the wavelet domain [1,2]. In recent years, non-statistical approaches that trace the pattern of facial expressions have also come into existence [3].

A Reddit User in December 2017 used AI methods to create fake pornographic videos. This is how the concept of DeepFake began. Since then, DeepFake videos are both constructively and destructively used in many applications. For instance DeepFake is used in Film industries to reduce costs that incur to add CGI texture.

DeepFake primarily involves AutoEncoders for detection of facial features of the target and victim involved. They are trained with the warped models of their faces with different orientations for accurate detection of facial points. The encoder of victim extracts feature points based on features extracted by target's encoder. The target video is then reconstructed using victim's decoder. The model is given the input of the target system which will generate fake video by applying decoder to each frame of the target video to replace the feature points of the target with victims's features. Thus, the decoder will generate the victim's face with respect to the target's environment.

3   Related Work

'Yuezun Li, Ming-Ching Chang and Siwei Lyu' in their paper 'In Ictu Oculi: Exposing AI Generated Fake Face Videos by Detecting Eye Blinking' [6] have led to a conclusion that developments in neural networks have certainly led to a significant increase in realistic face swapping videos. Their approach includes a deep learning model that is a combination of CNN (Convolutional Neural Network) for feature extraction of eye thereby getting the eye state and use of LSTM (Long Short Term Memory) and RNN (Recurrent Neural Network) to capture the temporal regularities in eye blink patterns to distinguish open and closed eye state with the consideration of previous temporal knowledge. The mean resting blinking rate is 17 blinks/min or 0.283 blinks per second (during conversation this rate decreases to 4.5 blinks/s and increases to 26 blinks/min). The major disadvantage of this system is that upcoming sophisticated forgers can still create realistic blinking effects with post-processing and more training data [7]. Thus eye blinking alone in that case won't stay as an efficient parameter to detect tampered videos. This approach also would not compute efficient results if the face is not aligned with the camera. The paper does not comment on the video being original or fake either. Our application would study other physiological signals like pulse which would add an additional level of analysis to arrive at a more accurate result. This would be helpful in cases where the face is not aligned with the camera and the eye blink rates are highly realistic to be used as a parameter to detect fakeness.

David Guera and Edward J. Delp in their paper 'DeepFake Video Detection Using Recurrent Neural Networks' [8] proposed a temporal aware pipeline to detect fakeness in DeepFake videos. The frame level extraction takes place with the help of CNN. The extracted features are then used to train RNN (Recursive Neural Network) that is further used for classification. They use a convolutional LSTM for frame sequence processing. The implementation involves training the model with intensive videos that would require a lot of resources and would end up being computationally expensive for common users. We aim to target specific features like eye and pulse that have a higher probability of getting tampered while creation of DeepFake videos which thus would help in improval of performance.

Conotter, E. Bodnari, G. Boato, H. Farid in their paper 'Physiologically-Based Detection Of Computer Generated Faces In Video' [9] took advantage of the rich temporal data specific to a video to discriminate computer generated images from the real one. It measures differences in facial color that result from human pulse that would stay absent in the CG videos. They use Eulerian Algorithm to magnify the pulses at a certain frequency which is further used in computation of average luminance across a skin patch. The drawback of the system is its vulnerability to situations when a modeler would induce artificial pulses to a computer generated character. Our application overcomes the same by an addition of the Eye Blinking Detection model. This is used to attain more confidence and an added security on the output regarding the video being tampered. This works in cases where the Pulse detection model fails due to introduction of artificial pulses.

Umur Aybars Ciftci et al. in their paper "FakeCatcher: Detection of Synthetic Portrait Videos using Biological Signals" [10] present a detector using biological signals. They have enlightened how human signals work in synthetic AI videos. The drawback of the paper is the latent representation of biological signals which has been used for authenticity classification. This can be learnt by an AutoEncoder over a period of time which would result in the failure of their proposed model. Our system will overcome this drawback by the deployed Eye Detection model that would function in cases where the Pulse Detection model fails.

4   Proposed System

Our system aims to detect the fake videos by exploiting the artifacts that the Fake video generators can't probably manipulate or manipulate with comparatively lesser accuracy. These include detecting pulse variations in fake videos using techniques such as remote photoplethysmography (rPPG) with Kalman Bound Filters [12] at 4 independent facial regions of interest namely forehead, mouth and the region around both the cheeks. We also aim to detect the eye blink pattern using Convolutional neural network and interpreting this eye blink sequence with respect to other facial artifacts like illumination, warping, geometrical misalignments [11] etc. The output of these models is then provided to an

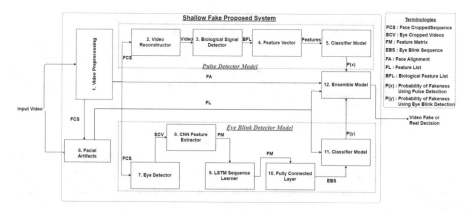

Fig. 1. The system architecture of ShallowFake-Detection of fake videos using deep learning

ensemble model which aggregates and gives a decision regarding fakeness of the entire video (Fig. 1).

We provide the input video simultaneously to two different modules namely Facial Artifacts and The Video PreProcessor [Fig. 2]. The Video PreProcessor breakdowns the video into frames. The dlib face detector is used to detect the faces in the video and are used for obtaining landmarks. The faces found in the entire video are then encoded using a encoder and stored in a matrix. This will help us in analysing any face specifically present in a video, a drawback of many existing systems. The output is a list consisting of multiple lists where each inner list has all the face frames belonging to a particular individual in the video. We align these faces by performing affine warp and then crop out the face from the video. We also calculate a Face Alignment coefficient that acts as an input for our ensemble model to decide which model should be given more importance. We run algorithms in the Facial Artifact model to obtain the amount of blurriness, illumination and prediction of geometric and eyeball misalignment along with the quality detection of the video. This output is given as input parameter to the final Ensemble model that helps us determine whether the video is fake or real. This output is also sent to two different modules simultaneously, namely the Eye Detection model for eye blink analysis and Video Reconstructor to reconstruct the video specific to individuals in the Pulse Detection Model. The Eye Detector module consists of using a dlib facial point detector to crop out the eye region as a preprocessing step for the eye blink detector model. The left and right eye frames of a particular individual are stored separately. Feature extraction from each eye image is done by the Convolution layer which generates a Feature matrix respective to that image. This CNN is implemented with the help of VGG16 framework. This output is fed into a LSTM (Long Short Term Memory) powered Recurrent Neural Network (RNN) that learns and retains the feature matrices of previous eye frames that are already read. This is also useful when there's ambiguity and one of the eye is not properly read by model. The

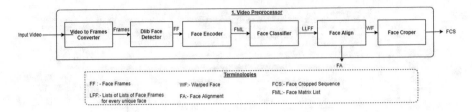

Fig. 2. The Video Preprocessor module

LSTM can then predict the eye state based on the previous eye states and the read state of one eye. In general, LSTM helps in temporal eye blink sequence learning. Now we avoid using the conventional CNN with LSTM as CNNs don't have a memory retention mechanism as provided by RNNs. Thus we get the Eye Blink State as output of our LSTM-RNN model. This output is sent to a Multidimensional Support Vector Machine (SVM). This Multidimensional SVM is trained along with labelled dataset to classify fake and real videos. The model receives the input (blurriness, illumination and eyeball misalignment) from the Facial Artifact model. These physiological parameters along with the eye blink rate helps in classifying the video as real or fake.

The Pulse Detector module has the Face Reconstructor that reconstructs the Face Cropped sequence of a particular person to video. The formed video is then sent to the Biological Sequence Detector that extract pulse using remote photo-plethysmography (rPPG or iPPG) [4,5] with Bounded Kalman Filters (effective for videos with head movement as their noise). We work on 12 signals which are a combination of G-channel based, chrominance based PPG and kalman filters on four facial ROIs namely the forehead, mouth and left and right cheek regions $(G_F, G_M, G_L, G_R, C_F, C_M, C_{LV}, C_R, K_F, K_M, K_L, K_R)$. The biological signals are transformed in frequency as well as time domain towards the following features (Table 1).

We aim to apply the feature extractor (mentioned in Table 2) on the transformed signals to obtain the final feature set. We decide to use a combination of features to improve the accuracy and hence conclude on custom feature set for classification of authentication. The output is given to a feature vector model which extracts signal based on features as mentioned below.

$$f = F_1(log(D_C) \cup F_2((log(S) \cup A_P(D_C)) \cup$$
$$F_3((log(S) \cup A_P(D_C)) \cup$$
$$F_2(log(S) \cup A_P(D_K) \cup F_2(log(S) \cup A_P(D_K)$$

These features are then used to train a multidimensional SVM classifier which classifies video as real or fake.

Each of the detectors has their own drawbacks. So, instead of considering their outputs individually we compute a combined output that can help us interpret the video authenticity even when one model might fail under certain

Table 1. Signals and functions for transformation

Symbol	Signal
S	G_F, G_M, G_L, G_R, C_F, C_M, C_L, C_R, K_F, K_M, K_L, K_R
D	Difference of Chrominance and G-based region respective to every region
A(S)	Autocorrelation
S_c	C_F, C_M, C_L, C_R
S_K	K_F, K_M, K_L, K_R
D_C	C_L-C_M, C_L-C_R, C_L-C_F, C_M-C_R, C_M-C_F, C_R-C_F
D_K	K_L-K_M, K_L-K_R, K_L-K_F, K_M-K_R, K_M-K_F, K_R-K_F
$A_P(S_C)$	pairwise cross spectral densities
$A_P(S_K)$	pairwise cross spectral densities based on kalman

Table 2. Feature set generation

Denoting symbol	Explanation
F_1	Mean and maximum of cross spectral density [10]
F_2	Count of narrow pulses in the spectral autocorrelation, count of spectral lines in the spectral autocorrelation, average energy of narrow pulses, maximum value of the spectral autocorrelation [14]
F_3	Standard deviation, standard deviation of mean values of 1 s windows, root mean square of 1 s differences, mean standard deviation of differences, standard deviation of differences, mean of autocorrelation, Shannon entropy [15]

circumstances. The outputs are combined using a simple weighted Ensemble model. The output features from the Facial Artifacts is fed as an input to the ensemble model which is then used for weight decision process. The probabilities from both the SVM models are also fed as an input to this model. The weights are calculated with facial artifacts and assigned to the SVM models accordingly. Weight calculation will be either done statically, by specifying standard values of those artifacts mentioned in various research work or dynamically by training a simple model with the same dataset which we are using for our SVM models. The output of the system is a binary value that tells whether the video is fake or real.

5   Expected Result and Analysis

The dataset and the expected result of the proposed system can be summarized as follows.

Dataset: Any classifiers works best on a dataset with wide variety of variation so that it can successfully classify any type of new input data. So we aim to train our dataset with 4 kinds of Data: Real Videos with High Quality, Real videos with Low Quality, Fake videos with High Quality and Fake videos with Low Quality. We randomize the amount of High and low quality of data but for Real and Fake videos in equal amounts. We randomly take about 500 videos from the FaceForensics++ dataset [13] for considering Real videos and For real videos we use Beautiful-Soup package of python for web-scraping and take about 500 real videos from various website with duration of 10 min. This will be mainly form news channels as these videos have high face times. With both these datasets, we train the SVM models for Pulse and Eye Blink detector. However for Ensemble model, we want only parameters from true videos to affect final combinative decision. Hence, we train using real videos only along with the weight assignments. The Video PreProcessor module does the processing of dataset as required by our model.

Expected Result: The First Crucial output is expected from Facial Artifact module where we expect 5 Feature Coefficients appended in Feature List. These Coefficients will be Geometrical Misalignment, Blurriness, Illumination, Quality and eyeball misalignment respectively. The SVM Model of the Eye Blink Detector will be trained with the dataset of open and closed eye. The output of facial Artifact model namely Blurriness, Eyeball misalignment, illumination and the eye blink rate from the LRCN model will be fed as an input. Secondly for the training of the Multidimensional SVM of Pulse Detector model we use the feature list as mentioned in Sect. 4. The outputs of both these classifiers along with the output of the Facial Artifact containing the Feature coefficients will be used to train the final classifier model to determine the authenticity of the input video.

6   Conclusion

The new development in generative networks has led to the unwanted development, propagation and creation of DeepFake videos using means that are not visible via a naked eye. We present ShallowFake, an AI fake video detector system through which we try and expose the DeepFake videos using a simple architecture that comprises of two different modules namely Eye Blink Detector and Pulse Detector. We aim to experimentally validate the irregular eye blink rates and pulse signals within the frames that are not well presented in these synthesized fake videos. To our knowledge, biological signals in these generated DeepFake videos is not explored much and we aim to exploit the same for fake detection. The proposed system thus detects visual tampering in videos to avoid miscreants taking advantage of technology to potentially ruin relationships and societal reputation. We believe that this system offers a powerful first line of defense to spot fake videos created using the AI tools.

There are several directions in which we would like to expand our vision for the proposed system. In the current model, we aim to consider only the irregular eye blink rates as a factor for the Eye Blink Detector module. However, the dynamic blink patterns-too frequent or too slow blinking can also be taken into consideration. The sophisticated forgers can still train data and use intensive models to create realistic eye blink rates. The Pulse Detector module can face hurdles provided the DeepFakes in future are enriched with biological signals, thereby being more realistic in nature. So in the long run, we aim to explore other physiological signals to detect fakeness.

References

1. Lyu, S., Farid, H.: How realistic is photorealistic? IEEE Trans. Signal Process. **53**(2), 845–850 (2005)
2. Wang, Y., Moulin, P.: On discrimination between photorealistic and photographic images. In: IEEE International Conference on Acoustics, Speech and Signal Processing, pp. II.161–II.164 (2006)
3. Dang-Nguyen, D.-T., Boato, G., De Natale, F.G.B.: Identify computer generated characters by analysing facial expressions variation. In: IEEE International Workshop on Information Forensics and Security, pp. 252–257 (2012)
4. Rouast, P.V., Adam, M.T.P., Chiong, R., Cornforth, D., Lux, E.: Remote heart rate measurement using low-cost RGB face video: a technical literature review. Front. Comput. Sci. **12**(5), 858–872 (2018)
5. Poh, M.-Z., McDu, D.J., Picard, R.W.: Non-contact, automated cardiac pulse measurements using video imaging and blind source separation. Opt. Express **18**(10), 10762–10774 (2010)
6. Li, Y., Chang, M.-C., Lyu, S.: In Ictu Oculi: exposing AI generated fake face videos by detecting eye blinking. arXiv:1806.02877v2. Accessed 11 June 2018
7. Donahue, J., Anne Hendricks, L., Guadarrama, S., Rohrbach, M., Venugopalan, S., Saenko, K., Darrell, T.: Long-term recurrent convolutional networks for visual recognition and description. In: CVPR, pp. 2625–2634 (2015)
8. Guera, D., Delp, E.J.: Deepfake video detection using recurrent neural networks. In: Video and Image Processing Laboratory (VIPER), pp. 1–6. Purdue University (November 2018). https://doi.org/10.1109/AVSS.2018.8639163
9. Conotter, V., Bodnari, E., Boato, G., Farid, H.: Physiologically-based detection of computer generated faces in video. In: 2014 IEEE International Conference on Image Processing (ICIP) (January 2015)
10. Aybars Ciftci, U., et al.: FakeCatcher: detection of synthetic portrait videos using biological signals. arXiv:1901.02212v2. Accessed Aug 2019
11. Matern, F., Riess, C., Stamminger, M.: Exploiting visual artifacts to expose Deep-Fakes and face manipulations. In: IEEE Winter Applications of Computer Vision Workshops, pp. 83–92. IEEE (2019)
12. Prakash, S.K.A.: Bounded Kalman filter method for motion-robust, non-contact heart rate estimation. Biomed. Opt. Express **9**(2), 873 (2018)

13. Rössler, A., Cozzolino, D., Verdoliva, L., Riess, C., Thies, J., Niener, M.: FaceForensics++: learning to detect manipulated facial images. In: International Conference on Computer Vision (ICCV) (2019)
14. Soleymani, M., Lichtenauer, J., Pun, T., Pantic, M.: A multimodal database for affect recognition and implicit tagging. IEEE Trans. Affect. Comput. **3**(1), 42–55 (2012)
15. Hu, H., Wang, Y., Song, J.: Signal classification based on spectral correlation analysis and SVM in cognitive radio. In: 22nd International Conference on Advanced Information Networking and Applications (AINA 2008), pp. 883–887 (March 2008)

Accessibility Analysis of Indian Government Websites

Nishtha Kesswani[(✉)]

Central University of Rajasthan, Ajmer, India
nishtha@curaj.ac.in

Abstract. With the emergence and popularity of the Internet, more and more users are using the Internet to accomplish varied tasks. While the websites are designed for the convenience of the users, accessibility is one such factor which is overlooked during the website design. In this paper, I have explored the accessibility of the Indian Government websites using automated tools TAW and AChecker. The results indicate that more than 30% websites of the Indian Government are poor in terms of accessibility.

Keywords: Website accessibility · Government websites · Accessibility

1 Introduction

More and more users are using the Internet for a variety of uses. Right from accessing the emails through Internet to online shopping, websites have played a major role. While the websites are designed to be catchy, one aspect that is overlooked while designing the websites is accessibility. Physical accessibility allows the users to access the physical infrastructure without barriers. And Web Accessibility includes empowering the persons with disability with the capability to access the websites and web tools. Making the web accessible is in the benefit of individuals, institutions and the society.

The Web Accessibility Initiative (WAI) [1] of the World Wide Web Consortium develops standards to implement accessibility. One such standard is Web Content Accessibility Guidelines (WCAG), which provides guidelines for web developers. WCAG 2.0 is ISO/IEC 40500:2012 standard. The components that contribute towards web accessibility are Web content, which includes any part of the website; user agents, software that people use to access the web content; and Authoring tools that people use to create web content. Broadly WCAG guidelines are divided into four principles:

- Perceivable: The information on the website is presented in such a manner that users can easily perceive it.
- Operable: The components of the website are easily operable.
- Understandable: The user interface should be understandable.

M. Tripathi and S. Upadhyaya (Eds.): ICDLAIR 2019, LNNS 175, pp. 179–189, 2021.
https://doi.org/10.1007/978-3-030-67187-7_20

– Robust: The website should support all kinds of user agents including assistive technologies.

Further these principles specify the Guidelines according to which accessible websites can be created. The WCAG guidelines have three Priorities:

– Priority 1: which includes the criteria that must be fulfilled.
– Priority 2: which includes the criteria that should be fulfilled.
– Priority 3: which includes the criteria that may be fulfilled.

Along-with this there are three Conformance levels A (lowest), AA and AAA (highest).

In this paper, I have done accessibility analysis of Indian Government websites. The reason why the Government websites have been used for this research is that due to physical accessibility barriers, visiting the Government Offices personally may not be possible. But, if the websites are designed with better accessibility then it would definitely help the differently abled. Major contributions of the research are (1) I have analyzed the accessibility of Indian Government websites using automated tools. (2) Major flaws in the existing websites have been identified. (3) Suggestions for improvement have been given.

Rest of the paper is organized as follows. Section 2 presents the related works. The results of accessibility analysis have been given in Sect. 3. Section 4 discusses the major flaws in the existing websites. Suggestions for improvement are given in Sect. 5. Section 6 concludes the paper.

2   Related Works

Since the awareness towards website accessibility has increased in the last few years, several researchers have tried to analyze the websites for accessibility. Different researchers have focused on different types of websites. In my previous work [2], we had done accessibility analysis of websites of top educational institutions of different countries. The study conducted on government websites of Saudi Arabia and Oman and found that government websites in these two countries are not accessible to all. There are guidelines and resources to make them accessible, but people are less aware about them. The government also does not know the importance of making websites accessible to all [3].

The comparative study conducted on government websites in Korean and USA country found that accessibility errors in the Korean government websites were approximately two times higher than those of the US government websites. It was also found that errors detected from manual evaluation of experts were less than the errors computed by automated software [14].

A study conducted on usability and accessibility of Malaysia e government websites found problems related to speed and broken links in the state websites were more as compare to the federal websites [4].

In Britain and Northern Ireland, there is legal requirement under disability discrimination act that website to be easily accessible. A survey found that although most of the facilities and transactions are conducted at local level, but local council websites were unusable by people with disability. Designing website for all may have business benefits but it is legal obligation too. Application of new principles and guidelines will improve websites and they will become usable by all [8].

On the basis of Automated analyses of WCAG 2.0 Level, top 100 government website in United States and United Kingdom were found to be violating a high percentage of Criteria. But these sites have shown extraordinary improvement over the years on the basis of number of accessibility criteria [9]. Alabama state level websites in USA have only 20% of the websites those meet the requirement of section 508 and only 19% of the websites meet the WAI priority I standards. An automated as well as manual review conducted on the state's website after adoption of ITS-530S2 found that compliance has not improved. Mere implementation of standards does not lead to compliance [7].

Websites of Higher education institutions in Malaysia also needs improvement to satisfy usability and accessibility criteria, but the study was based using WCAG 1.0 guideline but not latest version WCAG 2 [15]. In 1998 in USA, section 508 of the rehabilitation act of 1973 was changed by their parliament congress. It said that any agency getting funds from federal government should ensure that their websites are usable and accessible by public at large. A study conducted on 100 largest municipalities in USA revealed that although most of the municipalities have their accessibility statements and guidelines, but they are not fully compliant with the section. But it also says that compliance as per section 508 is improving over the period of time with the staff becoming more tech savvy and accessibility compliance aware [16]. Different sectors have different accessibility compliance. Twenty-three percent of the federal websites were found to be accessible as per compliance under section 508 and only 28% of the federal websites were accessible as per the WAI standards, but accessibility compliance was poor in case of NPO and corporate sector. Only 11% of NPO websites (14% as per WAI guidelines) and 6% of corporate websites (18% as per WAI guidelines) met section 508 Web accessibility requirements [17].

Effect of advances in technology was analyzed on accessibility and usability of general websites as well as federal websites in USA. It was found that with the increase of the complexity and advancement of technology, accessibility of general websites have decreased, while the accessibility of US government websites remained same over the period of time. Analyses of Variance, Tukey's HSD, Pearson's correlation coefficient were computed to conduct the study [18].

One accessibility criteria may be more important for one user while another may be more important for another user. The Australian government enforced WCAGv2 as accessibility criteria for their websites. Many websites failed on criterion one with only one or two errors. The reason may be that content editor may have made a mistake in process. These types of issues need to be resolved by proper training of the staff. Canada has its method for increasing accessibility of websites by fixing responsibility with respect to website management and legal requirement [19]. A study on five Asian countries government websites found that Asian e-Government websites require more efforts to meet performance and quality criteria in their website design. The website developers should use standard guidelines while designing websites [20]. A study conducted to measure e- government service accessibility and usability of top Saudi Arabia e- government services and their compliance to WCAG 2.0 and it was found that none of the evaluated service were in fully compliance with the four main guidelines for analysis of websites. According to it, website should be perceivable, operable, understandable and robust. But the experts were of the view that e-government websites in Saudi Arabia are well designed and user friendly [21]. Twenty five websites of turkey government do not address the issue of disability-accessibility, and they do not meet minimum accessibility requirements. As the automotive tool of evaluation was used, every tool detected different type of error. So it was difficult to find out that which website has better accessibility and which one has worst accessibility. One of the most detected accessibility issue was non-presence of text equivalent as well as static equivalent for dynamic content when the dynamic content changed. The study recommended for use of w3c guidelines for website design [22].

According to section 508 of the rehabilitation act in USA, every website should be user friendly for a person with disability. The study examined the complexities of accessibility and reasons for continued inaccessibility of federal e-government websites. The study used policy analysis, automated expert testing tools to provide reasons recommendations to curb inaccessibility of federal websites [23].

A study on citizens' perceptions of e-government website of Jordan found many variables significant for accessibility issues of websites. The study was first of its kind which used instrument in Arabic language [24]. The World Wide Web Consortium (W3C) has designed international standards for website accessibility which ensure that every one is on equal footage to publish material on World Wide Web. The study conducted on the accessibility of public library websites in Western Australia demonstrated the level of compliance of public library websites in Western Australia with Australian and International standards. The study also contained the willingness of public library websites to comply with standards, barriers to compliance and benefits of an accessible website [25].

A study of five websites of New Zealand government entities found that their websites provided orientation information about website, informed conditions for re-use of information, addressed privacy concerns, adapted print materials with the web environment properly, kept current & updated materials, made available contact details, made effective use of metadata, provided appropriate external links, provided accessibility to users with disabilities and provided help information on search engines to users [26].

The Australian Government has employed the Web Accessibility National Transition Strategy (NTS) which advices compliance with WCAG 2.0 for website accessibility. Apart from this, those websites which not covered by the NTS comes under the purview of the Australian Human Rights guidelines, which recommend WCAG 2.0 AA as a minimum standard [27].

Some researchers have also tried to focus on the accessibility of government websites of different countries. Few such contributions are listed in Table 1.

Table 1. Existing work on Government website accessibility

Reference	Domain	Demography
Abanumy [3]	E-government websites	Saudi Arabia and Oman
Isa [4]	E-government websites	Malaysia
Olalere [5]	Federal Government websites	United States
Kuzma [6]	E-government websites	United Kingdom
Youngblood [7]	Municipal Government websites	Alabama
Paris [8]	E-government websites and legislation	Ireland
Hanson [9]	Traffic and government websites	US and UK
Ismailova [10]	Government websites	Kyrgyz Republic
Karaim [28]	Government website usability and accessibility	Libya

To the best of my knowledge, none of the existing works has done analysis of the websites of Indian Government. Thus in this piece of work, major websites of the Government of India have been analyzed.

3   Experiments and Results

This section provides the analysis of the websites of government of India. For the purpose of the analysis the websites have been taken from the portal of the Government of India [11]. The websites were analyzed for the level AA using the automated tool TAW [12]. Following websites were analyzed for accessibility:

- President of India (https://presidentofindia.gov.in/)
- Vice President of India (https://vicepresidentofindia.nic.in/)
- Parliament of India (https://parliamentofindia.nic.in/)
- Rajyasabha (https://rajyasabha.nic.in/)
- Loksabha (https://loksabha.nic.in/) Apart from this following Ministries were analyzed:
- Ministry of Finance (https://www.finmin.nic.in/)
- Ministry of Home Affairs (https://mha.gov.in)
- Ministry of Electronics and IT (https://meity.gov.in/)
- Ministry of Social Justice and Empowerment (https://socialjustice.nic.in/)
- Ministry of Personnel, Public Grievances and Pension (https://www.persmin.nic.in/)
- Ministry of Railways (https://www.indianrailways.gov.in)
- Department of Science and Technology (https://www.dst.gov.in/)
- Ministry of Road, Transport and Highways (https://morth.nic.in/)
- Ministry of Tourism (https://tourism.gov.in/)

- Ministry of Women and Child Development (https://www.wcd.nic.in/)
- Prime Minister of India (https://www.pmindia.gov.in/en/)
- Cabinet Secretariat (https://cabsec.gov.in/)
- Election Commission of India (https://eci.gov.in/)
- Union Public Service Commission (https://upsc.gov.in/)
- National Human Rights Commission (http://www.nhrc.nic.in/)
- Comptroller and Auditor General (https://cag.gov.in/)
- NITI Aayog (http://niti.gov.in/)
- National Commission for women (http://ncw.nic.in/)
- National Commission for Scheduled tribes (https://ncst.nic.in/)
- Fifteenth Finance Commission of India (https://fincomindia.nic.in/)
- National Commission for minorities (http://ncm.nic.in/)
- Insurance Regulatory and Development Authority (https://www.irdai.gov.in/Defaulthome.aspx?page=H1)
- Office of Principal Scientific Advisor (http://psa.gov.in/).

The results are shown in Fig. 1, 2 and 3.

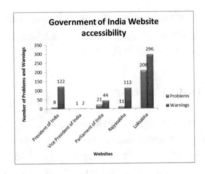

Fig. 1. Accessibility of Government of India websites

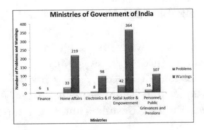

Fig. 2. Accessibility of Government of India websites

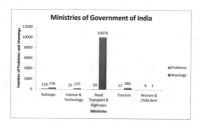

Fig. 3. Accessibility of Government of India websites

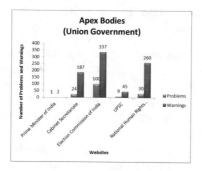

Fig. 4. Accessibility of websites of Apex Bodies of Government of India

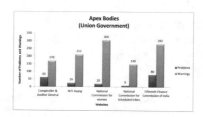

Fig. 5. Accessibility of websites of Apex Bodies of Government of India

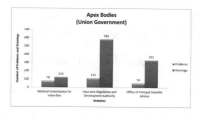

Fig. 6. Accessibility of websites of Apex Bodies of Government of India

Apart from this, the Apex bodies' websites of the Government of India were analyzed on TAW and the results are shown in Fig. 4, 5 and 6.

The results indicate that some of the websites are very well designed with Problems as low as only 1 in the websites of Vice President of India and Prime Minister of India. While the websites with poor accessibility with 100 or more problems include that of the Election Commission of India and Insurance Regulatory and Development Authority. Major flaws identified during the analysis are discussed in the next Section.

4   Major Flaws in Existing Websites

During the analysis, major flaws in the websites were identified. Some of the flaws in the websites are as follows:

- Text alternatives for non-text content was missing in most of the websites and the websites failed to be perceivable. For instance if it is a control that accepts user input, it should clearly mention the purpose of the control. Or it is a CAPTCHA, which clearly distinguishes a human from a computer, it should provide alternative ways to access the CAPTCHA to accommodate the persons with disabilities.
- Info and relationships were missing in many websites which made them less adaptable. This ensures that the information, structure and relationships are clearly indicated in the text.
- Link purpose was missing in many websites which made the websites difficult to operate. This makes sure that the link text makes sense to the user about the purpose of the link.
- In some case the content was less Readable.
- Other minor flaws that were found include that the pages were not titled and the user did not get enough time to respond, which is very important for the persons with disabilities.

The suggestions for improvement in the existing scenario are discussed in the next section.

5   Suggestions for Improvement

The detailed analysis was done using AChecker [13]. For instance the alt attribute was found missing in most of the image tags as shown below:

$$\langle imgsrc = ``/Common/images/dot1.gif" align = ``right" border = ``0"\rangle$$

Instead a better design could include alternative text as follows:

$$\langle imgsrc = ``/Common/images/dot1.gif" align = ``right" border = ``0" alt = "Image indicating the Office of the official\rangle$$

Similar such content indicating the CAPTCHA without alternative content is shown below:

$$\langle imgsrc = \text{``}CaptchaImage.axd?guid = efa78ba0 - 27f7 - 4dc2 - 83c4 - \\ 05ff0e1115f6\text{''} border =' 0'width = 160height = 40...\rangle$$

Another major flaw was that the content needs to be adaptable so that the information can be presented in different ways without losing the structure. For example, in the following tags, the input element 'text' and 'password' do not have an associated label which is desirable in a good design.

$$\langle inputname = \text{``}TxtSearch\text{''}type = \text{``}text\text{''}id = \text{``}TxtSearch\text{''}onkeypress = \\ \text{``}returnValidateChar(event);\text{''}maxlen...\text{``}\rangle$$

$$\langle inputname = \text{``}TxtPassword\text{''}type = \text{``}password\text{''}maxlength = \text{``}50\text{''} \\ id = \text{``}TxtPassword\text{''} \quad tabindex = \text{``}2\text{''} \quad onkeypress = ...\text{''}\rangle$$

Another important feature is that the website should be Keyboard Accessible. For instance, in the following tags, 'onMouseOut/onBlur' could have been mentioned in case the user is unable to use the mouse.

$$\langle tdid = \text{``}Menu1 - menuItem000 - subMenu - menuItem003 - subMenu - \\ menuItem000\text{''}onclick = \text{``}javascript : skmcloseSubM\text{''}\rangle$$

Also, the design should be such that it gives enough time to the users to read the content. For example in the following case, 'marquee' tag can be replaced with 'strong' or 'em'.

$$\langle marqueedirection = \text{``}up\text{''}scrollamount = \text{``}3\text{''}onmouseover = \text{``}javascript : \\ stop();\text{''}onmouseout = \text{``}javascript : sta\text{''}\rangle$$

Under the guideline 'Readable' the language of the page should be specified by adding the lang attribute and the ISO-639-1, two letter language code to the following content:

$$\langle html\rangle\langle head\rangle\langle title\rangle IRDAIWelcomesYou\langle/title\rangle$$

If some important design features are taken into consideration while designing the websites, more and more websites would be accessible for all.

6   Conclusion

In this paper, I analyzed the websites of the Government of India using accessibility tools TAW and AChecker. The reason why these tools have been used for the study is that they are widely used for accessibility analysis. Out of the websites that were analyzed, more than 30% websites had 50 or more problems. The results were analyzed in detail and major flaws in the design were also mentioned. I also tried to give a few suggestions for improvement so that the websites are designed to be more accessible.

References

1. W3C Web Accessibility Initiative. https://www.w3.org/WAI/fundamentals/accessibility-principles/. Accessed 22 Feb 2019
2. Kesswani, N., Kumar, S.: Accessibility analysis of websites of educational institutions. Perspect. Sci. **8**, 210–212 (2016)
3. Abanumy, A., Al-Badi, A., Mayhew, P.: e-Government website accessibility: in-depth evaluation of Saudi Arabia and Oman. Electron. J. e-Gov. **3**(3), 99–106 (2005)
4. Isa, W.A.R.W.M., et al.: Assessing the usability and accessibility of Malaysia e-government website. Am. J. Econ. Bus. Adm. **3**(1), 40–46 (2011)
5. Olalere, A., Lazar, J.: Accessibility of US federal government home pages: section 508 compliance and site accessibility statements. Gov. Inf. Q. **28**(3), 303–309 (2011)
6. Kuzma, J.M.: Accessibility design issues with UK e-government sites. Gov. Inf. Q. **27**(2), 141–146 (2010)
7. Youngblood, N.E., Mackiewicz, J.: A usability analysis of municipal government website home pages in Alabama. Gov. In. Q. **29**(4), 582–588 (2012)
8. Paris, M.: Website accessibility: a survey of local e-government websites and legislation in Northern Ireland. Univ. Access Inf. Soc. **4**(4), 292–299 (2006)
9. Hanson, V.L., Richards, J.T.: Progress on website accessibility? ACM Trans. Web (TWEB) **7**(1), 2 (2013)
10. Ismailova, R.: Web site accessibility, usability and security: a survey of government web sites in Kyrgyz Republic. Univ. Access Inf. Soc. **16**(1), 257–264 (2017)
11. Government of India, Web Directory. http://goidirectory.nic.in/index.php. Accessed 11 Oct 2018
12. The websites were analyzed for the level AA using the automated tool TAW
13. AChecker Accessibility. https://achecker.ca/checker/index.php. Accessed 19 Mar 2019
14. Hong, S., Katerattanakul, P., Lee, D.: Evaluating government website accessibility: software tool vs human experts. Manag. Res. News **31**(1), 27–40 (2007)
15. Aziz, M.A., Isa, W.A.R.W.M., Nordin, N.: Assessing the accessibility and usability of Malaysia higher education website. In: 2010 International Conference on User Science and Engineering (i-USEr). IEEE (2010)
16. Evans-Cowley, J.S.: The accessibility of municipal government websites. J. e-Gov. **2**(2), 75–90 (2006)
17. Loiacono, E.T., McCoy, S.: Website accessibility: a cross-sector comparison. Univ. Access Inf. Soc. **4**(4), 393–399 (2006)
18. Hackett, S., Parmanto, B., Zeng, X.: A retrospective look at website accessibility over time. Behav. Inf. Technol. **24**(6), 407–417 (2005)
19. Grantham, J., Grantham, E., Powers, D.: Website accessibility: an Australian view. In: Proceedings of the Thirteenth Australasian User Interface Conference, vol. 126. Australian Computer Society, Inc. (2012)
20. Jati, H., Dominic, D.D.: Quality evaluation of e-government website using web diagnostic tools: Asian case. In: 2009 International Conference on Information Management and Engineering. IEEE (2009)
21. Al-Faries, A., et al.: Evaluating the accessibility and usability of top Saudi e-government services. In: Proceedings of the 7th International Conference on Theory and Practice of Electronic Governance. ACM (2013)
22. Akgül, Y., Vatansever, K.: Web accessibility evaluation of government websites for people with disabilities in Turkey. J. Adv. Manag. Sci. **4**(3), 201–210 (2016)

23. Jaeger, P.T.: Assessing section 508 compliance on federal e-government Web sites: A multi-method, user-centered evaluation of accessibility for persons with disabilities. Gov. Inf. Q. **23**(2), 169–190 (2006)

24. Abu-Shanab, E.A., Baker, A.A.N.A.: Evaluating Jordan's e-government website: a case study. Electron. Gov. Int. J. **8**(4), 271–289 (2011)

25. Conway, V.: Website accessibility in Western Australian public libraries. Aust. Libr. J. **60**(2), 103–112 (2011)

26. Smith, A.G.: Applying evaluation criteria to New Zealand government websites. Int. J. Inf. Manag. **21**(2), 137–149 (2001)

27. Conway, V.L.: Website accessibility in Australia and the Australian government's national transition strategy. In: Proceedings of the International Cross-Disciplinary Conference on Web Accessibility. ACM (2011)

28. Karaim, N.A., Inal, Y.: Usability and accessibility evaluation of Libyan government websites. Univ. Access Inf. Soc. **18**(1), 207–216 (2019)

Preprocessing HTTP Requests and Dimension Reduction Technique for SQLI Detection

Nilesh Yadav[✉] and Narendra Shekokar

Department of Computer Engineering, D. J. Sanghvi College of Engineering,
Vile Parle, Mumbai, India
`nileshyadav2004@gmail.com, narendra.shekokar@djsce.ac.in`

Abstract. The rapid development of web applications leads to the security problems related to the web attacks. The detection of these attacks is a critical task specially SQL injection which is the top most web vulnerability. The existing signature based detection approaches lacks the functionality to cope up with the new signatures. Machine Learning (ML) becomes an alternative concept to existing solutions. However readymade labeled dataset or corpus with SQLI patterns is unavailable and also the current existing models have not yet used the feature reduction technique on most dangerous vulnerability logs. These are the well known issues in SQL Injection research. This paper contains an approach in which SQLI logs will be collected & further processed based on 'TFIDF-Ngram with Singular value decomposition' approach for dataset preparation and feature engineering. The paper explores the generation and classification of feature reduced SQLI data set from HTTP traffic logs like CSIC-2010 [1] and ECML/PKDD-2007 [2, 3]. The experiments carried out on supervised ML techniques with observed evaluations presented in Confusion Matrix (CM).

Keywords: Web application · Attack · SQLIA · Feature extraction · Term frequency · Dimension reduction · CSIC · ECML

1 Introduction

The web applications are made up of web technologies and widely used in various organizations. Significant financial losses to organizations because of the unnoticed web vulnerabilities. As per OWASP latest report [4] SQL injection is the top most web vulnerability in the world and hence it is important to detect it. This detection can be done mostly using two main approaches. One is by using pattern-matching to identify traffic of SQLI attacks and another one is an anomaly detection approach currently using machine learning (ML) techniques in which the labeled datasets train the model and then this predicts the testing data [5].

To classify the HTTP requests as malicious/non-malicious with ML, first we need to construct the dataset from the raw data and label this based on these featured values. A mathematical model will be developed to map the relationship between features and labels. ML has an ability to cope with new SQLI patterns but unavailability of the dataset for training the classifier is the biggest issue in SQLI research. In this paper, we explore the generation of data set containing extraction of Sql injection request from known HTTP attack logs using SQL keywords list. These requests are further processed for features construction and labeling using the TFIDF-Ngram concept. After this, dimension reduction method i.e. singular value decomposition (SVD) technique is applied for generation of low dimensional sets and further classified using supervised learning classifiers. Dimension reduction is one of vital process in ML while classification of the text. This characteristic helps in compact data representation and performance improvement.

After the Introductory section the paper representation is as follows. The Sect. 2 talks about related work. The TFIDF-Ngram with SVD combined methodology and its architecture is explained in Sect. 3. The experiments and results are presented in the Sect. 4. Finally Sect. 5 concludes the paper.

2   Related Work

We studied various ML techniques those have been used before for the extraction of features and classification of data.

D. das, U. Sharma & D.K. Bhattacharyya presented the new approach [6] for SQL injection detection. A web profile based on edit distance is prepared to clssify the dynamic SQL query as a normal or malicious. R. Dzisevic & D. sesok captured the text features [7] using 2 dimensionality reduction techniques LSA (Latent Semantic Analysis) & LDA (Linear Discriminant Analysis) combined with the TF-IDF method. This analysis is done on one of the advertisement website logs for text classification. A. Chandra, S. Khatri uses the CfsSubset evaluation technique [8] as a filter based attribute selection method for dimensionality reduction. K_Means classification technique is used here for detection of the attacks. K. Wrobel, M. Wielgosz [9] focused on text processing in which they analyzed the precision reduction concept along with dimensionality reduction while text classification. R. Paul, McWhirter, K. Kifayat, Qi Shi, B. Askwith [10] uses the a gap weighted string subsequence kernel scheme for feature vector creation where the similarity value gets generated from the original query & runtime query. The kernel matrix is developed & this will be acted as a input to classifier for detection of SQLI. Authors B. Asaad and M. Erascu [11] executed various techniques for features creation and dataset preparation. Extracted the features from data log using different text representation techniques like Bag-of-Words, TF-IDF and Bi-gram frequency. Tested these datasets with classifiers namely Naive Bayes and Support Vector Machine (SVM) for the detection of Fake News.

S. O. Uwagbole, W. J. Buchanan and L. Fan [12] have done the SQLI detection & prevention in big data. They used the predictive analytics where the automata states walk became the data extraction technique for corpus creation. Two Class SVM & Two Class Logistic Regression ML algorithms are used to mitigate SQLIA. S. Uwagbole, W. J. Buchanan and L. Fan [13] classify the big data for SQLI detection using ML technique. The main contribution of author is to pre-processing and generation of dataset using unique SQL word list from Microsoft SQL websites. S. Althubiti, X. Yuan [14] applied the various feature engineering & ML techniques to classify the web attacks for CSIC-2010 HTTP data-set. They selected best five important features from features created in paper (Nguyen [17]) using weka tool by attribute evaluator methods and then experimental results compared with the results reported by Pham [16] and Nguyen [17]. D. Kar, A. K. Sahoo, s. Panigrahi and M. Das [15] uses the high ranking nodes to classify the SQLI data. SQL query tokens became acts as actors and their interaction is captured as a graph here. Centrality of nodes is calculated and then the dimensionality reduction is done using information gain technique.

However, all these approaches used different techniques for dimensionality reduction and classification. Some author worked on the vulnerability classification of CSIC-2010 & ECML/PKDD-2007 traffic but, as of now no one is analyzed separately the SQL injection logs present in these data with dimensionality reduction. Here, we experimented the combine approach (TFIDF-Ngram with SVD) for processing, construction and reduction of features from these filtered SQLI logs. These filtered SQLI logs are obtained using 432 MySQL keywords. Finally we are also analyzing these datasets for the prediction of SQLI attack using ML concepts.

3   The Combine Approach

This section explains procedure of the new combined method. From the given CSIC & ECML logs, the SQLI HTTP requests are separated and further ML classifiers are applied on this reduced datasets for prediction of SQLI attack.

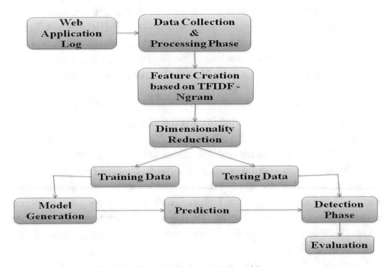

Fig. 1. The combine method architecture.

Figures 1 describe the combined method architecture. This technique is comprised of 4 steps: - Data Collection and Pre-Processing, Feature construction based on TF-IDF_Ngram phase, dimensionality reduction using SVD technique and Classification using supervised ML classifiers.

1) Data Collection and Pre-Processing Phase: - The experiments were conducted on the CSIC-2010 & ECML/PKDD-2007 logs. The CSIC-2010 dataset was constructed at

CSIC lab in Information Security Institute. This dataset contains more than 25000 malicious & 36000 normal requests. The ECML-2007 dataset will be composed of 20,000 normal samples. Testing samples contain 15,000 normal and abnormal datasets each. From the above unlabelled dataset, first the SQLI traffic is filtered using 432 unique SQL tokens. These tokens are nothing but the reserved key-words present at Mysql website [18]. A python utility is developed for SQLI filtration and also used to clean it from distorted and useless information. Finally we got only SQLI Http requests. Our final balanced CSIC (SQLI) dataset consisted of 4549 injected SQLI requests and 4549 genuine requests and final balanced ECML dataset consisted of 2274 injected SQLI requests and 2274 genuine requests.

2) Feature Construction Phase: - Once required SQLI requests have been received, it can't feed directly to ML Model. It is necessary to build proper labeled dataset. For construction of proper dataset matrix, we are using mix approach i.e. TF-IDF with Ngram. The term frequency (TF) can be defined as the number of times a word occurs in a document. Inverse document frequency (IDF) is inverse probability of finding a word in a document. The equation is a classic TF-IDF equation used to calculate weight:-

$$Wij = (TF)ij * \log(N/DFi).$$

In this equation, Wij is the weight of the word i in the document j, N is the total number of documents, (TF)ij is the frequency of the word i in the document j and DFi is the number of documents with the term i.

Ngram (n-grams) can be defined as the adjoining succession of n items from a given text/speech sample data e.g. size 3 means trigram. Here Ngram (3, 3) means while extracting the different n_grams, 3 is the lower and upper boundary of values. It helps to group the words together to form as an entity. Finally, the dataset items are labeled as malicious (0) and normal (1) based on the request URL values.

3) Dimensionality Reduction: - As explained above, we vectorize the text corpus. Here the matrices has large number of dimensions i.e. features, which may over-fit the classification model. This issue can be resolved by applying dimensionality reduction techniques i.e. PCA. Generally for square matrices, we can use eigendecomposition but here the generated data has the dimensionality too high than no. of data samples. In such a situation, we can use the SVD to find PCA analytically as a reduction technique. Also the calculation and storage of covariance matrix (in eigendecomposition) is very complex issue and this can be overcome by SVD. SVD is faster for decomposition as well as for computation. SVD is nothing but 3 simpler transformations (a rotation, a scaling and another rotation) from one complex transformation, and these 3 steps correspond to the three matrices U, S, V. So SVD of the data matrix H is given by: -

$$H = USV^{T}$$

Where, U is an m * m unitary matrix over K. S is a diagonal m * n matrix with non-negative real numbers on the diagonal, V is an n * n unitary matrix over K and V^{T} is the conjugate transpose of V. Where H is an m * n matrix whose entries come from field K. Unitary matrices are orthogonal matrices and focus on the changeability of the data in a small number of dimensions. Here the columns of US are the principal components and columns of V are principal directions. By fixing a boundary of the no. of components which used to represent the data, SVD reduces the dimensionality and provides ranking.

4) Classification Phase: - The combined new approach is used in the generation of reduced matrix from the given raw data. This will act as an input to ML classifiers. We are going to use the supervised ML classifiers for detection of SQLIA. The Support Vector Machine (SVM) and Random Forest (RF) are the supervised & robust algorithms present for document classification. SVM creates a hyperplane which separates the data into classes. As per SVM, the points nearest to the line from both the classes are called support vectors. To maximize the margin is our goal. The RF is made up of many decision trees from randomly chosen sub-set of training set. The final class of the test sample is decided by aggregation of the votes from diverse decision trees. We are examining these classifiers with our new scheme.

4   Experiments and Result Evaluation

The biggest problem in SQLI detection research is unavailability of dataset. As explained, we created the filtered datasets. Now the CSIC-2010 and ECML-2007 data contains only SQLI HTTP traffic. We used the 80% for training and 20% for testing matrix values (80:20 datasets) in all the experiments. Experiments were setup using Python language and Scikit library [19] in jupyter notebook [20]. To evaluate our ML model, we compare TF_IDF, TF_IDF-Ngram feature creation techniques with or without using SVD. We did experiments using two renowned classifiers i.e. linear SVM & RF. The classifier is used to predict the binary outcomes whether the given request is SQLI positive or negative. First convert a collection of raw data requests to a matrix of TF_IDF features using sklearn TfidfVectorizer then perform dimensionality reduction on this matrix using TruncatedSVD (from sklearn.decomposition) and produce the well reduced matrix. Samples further splits to 80:20 ratios for training and testing of the mentioned classifiers (LinearSVC from sklearn.svm & RandomForestClassifier from sklearn.ensemble). We use the standard measures. As Follows: -

   FP = false positive: the total number of attacks that do not detect.
   FN = false negative: the total no. of normal requests that are classified as anomalous.
   TP = True positive: the total number of attacks that are truly detected.
   TN = True negative: the total number of normal requests that are classified as normal.
   P = TP + FN: The No. of positive observations [possibly misclassified].
   N = TN + FP: The No. of negative observations [possibly misclassified].
   Accuracy = (TP + TN)/(TP + FP + FN + TN) × 100.
   Detection rate (DR) = Recall = True positive rate (TPR) = TP/(TP + FN) × 100.
   Precision (PR) = TP/(TP + FP) × 100.
   F1-score = (2 × PR × DR)/(PR + DR) × 100.
   Our method is evaluated using more favorable metrics of Accuracy and time. In our experiments, the http requests are examined. We have two datasets. Initially the Table 1, 2 and 3 are showing results of the models in which execution of both the mentioned classifiers for filtered CSIC dataset. The Table 1 and 2 shows the performance of scenarios

in which classifiers are executed using only TFIDF and TFIDF-Ngram technique to detect the SQLIA. The output of the combined scenario is consolidated in Table 3, where models are executed using TFIDF-Ngram with SVD vectorization method.

Table 1. CM on the SQLI-CSIC dataset using TFIDF

	ML		
	Algorithms	RF	Linear SVM
Measure	TP (True Positive)	905	908
	TN (True Negative)	881	895
	FP (False Positive)	15	12
	FN (False Negative)	19	05
	Accuracy (%)	98.13	99.07
	Precision (%)	98.37	98.70
	Recall (%)	97.94	99.45
	F1-score (%)	98.15	99.07
	Train time(s)	0.849	0.101
	Test time(s)	0.029	0.001

The Table 1 shows the Confusion Metrics (CM) for scenario where only TFIDF technique is used for vectorization. In Table 1 the execution time of training as well as testing phase is very less for SVM. Also this algorithm has better accuracy (99.07%) than RF algo. (98.13%). Table 2 shows the CM for scenario where TFIDF with Ngram technique is used for vectorization. Execution of this model is same as above but now we are using the TFIDF with Ngram for feature engineering. We observed the increase in no. of features due to usage of Ngram (3, 3). The value of n-gram is decided manually. The execution time of training & testing is very less for linear SVM than RF. Also here, accuracy of SVM is (99.89%) more than RF (96.48%).

The Table 3 shows the CM for our new combined scenario where, we are using the TFIDF-Ngram with SVD for dimensionality reduction. We selected the best no. of components in Truncated SVD using the explained _variance_ratio_. A utility is created for calculating no. of components required to pass the threshold. This threshold is decided manually. The execution time of training, testing is less for linear SVM than RF. Also here accuracy of SVM (99.90%) is more than RF (98.13%).

If we compare above scenarios or tables related to CSIC dataset. There is decrease in no. of features in combine scenario due to SVD dimensionality reduction method. It is observed here that both linear SVM & RF takes slightly more time for training & testing but we got slightly increase in accuracy. We acquired highest accuracy for SVM algorithm.

Table 2. CM on SQLI-CSIC dataset using TFIDF & Ngram

	ML		
	Algorithms	RF	Linear SVM
Measure	TP (True Positive)	884	920
	TN (True Negative)	872	898
	FP (False Positive)	36	0
	FN (False Negative)	28	2
	Accuracy (%)	96.48	99.89
	Precision (%)	96.09	100
	Recall (%)	96.92	99.78
	F1-score (%)	96.50	99.89
	Train time(s)	0.77	0.23
	Test time(s)	0.020	0.002

Table 3. CM on SQLI-CSIC dataset using TFIDF-Ngram with SVD

	ML		
	Algorithms	RF	Linear SVM
Measure	TP (True Positive)	904	920
	TN (True Negative)	882	898
	FP (False Positive)	16	0
	FN (False Negative)	18	2
	Accuracy (%)	98.13	99.90
	Precision (%)	98.23	100
	Recall (%)	98.05	99.78
	F1-score (%)	98.14	99.89
	Train time(s)	4.79	1.63
	Test time(s)	0.030	0.01

The Tables 4, 5 and 6 are showing results of the models in which execution of both the classifiers are done for the filtered SQLI-ECML dataset. The Table 4 shows performance of simple scenario in which only TFIDF technique is used. In this table the execution time of training and testing for the ECML SQLI dataset is very less for linear SVM algorithm as compared to RF. SVM has better accuracy (83.08%) than RF algorithm (79.45%). The Table 4 shows the entire CM for this scenario.

The Table 5 shows the CM for the scenario in which TFIDF & Ngram scheme is executed on this log. This time we are using the TFIDF with Ngram for vectorization and hence we observed the increase in no. of features due to usage of n-gram (3, 3). The execution time of training, testing is very less for linear SVM than RF. The accuracy of the SVM is still (83.08%) more than that of the RF algorithm (82.86%) here.

Table 4. CM on SQLI-ECML/PKDD dataset using TFIDF

	ML		
	Algorithms	RF	Linear SVM
Measure	TP (True Positive)	426	427
	TN (True Negative)	297	329
	FP (False Positive)	28	27
	FN (False Negative)	159	127
	Accuracy (%)	79.45	83.08
	Precision (%)	93.83	94.05
	Recall (%)	72.82	77.08
	F1-score (%)	82.01	84.70
	Train time(s)	5.46	0.061
	Test time(s)	0.018	0.001

Table 5. CM on SQLI-ECML/PKDD dataset using TFIDF & Ngram

	ML		
	Algorithms	RF	Linear SVM
Measure	TP (True Positive)	424	425
	TN (True Negative)	330	331
	FP (False Positive)	30	29
	FN (False Negative)	126	125
	Accuracy (%)	82.86	83.08
	Precision (%)	93.39	93.61
	Recall (%)	77.09	77.27
	F1-score (%)	84.46	84.70
	Train time(s)	1.64	0.25
	Test time(s)	0.027	0.002

Table 6. CM on SQLI-ECML/PKDD dataset using TFIDF-Ngram & SVD

	ML		
	Algorithms	RF	Linear SVM
Measure	TP (True Positive)	393	436
	TN (True Negative)	362	334
	FP (False Positive)	61	18
	FN (False Negative)	94	122
	Accuracy (%)	82.97	84.61
	Precision (%)	86.57	96.04
	Recall (%)	80.69	78.13
	F1-score (%)	83.53	86.16
	Train time(s)	18.53	0.15
	Test time(s)	0.085	0.001

The Table 6 shows the CM for our combined scenario which is again executed on this dataset. We are using the TFIDF-Ngram with SVD for feature reduction and executed this model same as above. We observed the decrease in no. of features due to usage of dimensionality reduction technique. Here the execution time of training, testing is less for linear SVM than RF. The accuracy of the SVM is still (84.61%) more than the RF algorithm (82.97%).

If we compare above scenarios or tables related to ECML dataset, there is decrease in no. of features in new scenario due to dimensionality reduction. For this combine technique, it is observed that RF takes slightly more time for training and testing. Overall in, we got the highest accuracy & less training, testing time for SVM.

We have used the combined (TFIDF_N-gram with SVD) method on both these SQLI-datasets. Especially in comparison with the second scenario, for CSIC dataset, it is observed that there is increase in training, testing time for SVM & RF algorithm. With respect to ECML dataset, there is increase in training and testing time for RF algorithm but there is decrease in training, testing time for SVM. For combined new approach, the accuracy of SVM is increased and it is more than RF algorithm.

5   Conclusion

In this paper we have applied combine ML scheme on SQLI CSIC-2010 and ECML/PKDD 2007 logs. First extracted the SQLI HTTP requests from these logs using Mysql keywords. Then applied the new combine technique with classifiers for detection. We have executed the TFIDF with Ngram technique for feature construction from SQLI logs and then applied the SVD as a dimensional reduction technique to reduce the feature space. This reduced dataset is classified and analyzed using two supervised

ML algorithms i.e. linear SVM & RF. The Obtained results are empirically assessed in the confusion matrices. It shows that SVM is very well suited for text categorization. It also depict that SVM is robust and its performance is very good in document data classification. Further experimentation will concentrate on various dynamic values of SVD components with used as well as different datasets using cross-validation if applicable. As per knowledge, the previous work was done on whole datalogs for web intrusion detection but SQLI attack analysis with dimensional reduction technique for these logs was missing before this paper.

References

1. Giménez, C.T., Villegas, A.P., Marañón, G.A.: HTTP Dataset CSIC 2010, CSIC (Spanish Research National Council) (2012)
2. Emanuel, K., Ute, S.: Knowledge discovery in databases: ECML/PKDD 2007. In: 18th European Conference on Machine Learning and 11th European Conference on Principles and Practice of Knowledge Discovery in Databases, Poland (2007)
3. Gallagher, B., Eliassi-Rad, T.: Classification of http attacks: a study on the ECML/PKDD 2007 discovery challenge. Technical Report No. LLNL-TR-414570. Lawrence Livermore National Laboratory, Livermore, CA (2009)
4. OWASP Group: Top 10 Most Critical Web Application Security Vulnerabilities. https://www. Owasp.org/index.php. Accessed 19 Sept 2019
5. Singh, J., Nene, M.J.: A survey on machine learning techniques for intrusion detection systems. IJARCCE J. **2**(11) (2013)
6. Das, D., Sharma, U., Bhattacharyya, D.K.: Defeating SQL injection attack in authentication security: an experimental study. Int. J. Inf. Secur. **18**, 1–22 (2019)
7. Dzisevic, R., Sesok, D.: Text classification using different feature extraction approaches. In: 2019 Open Conference of Electrical, Electronic and Information Sciences (eStream), Vilnius, Lithuania, pp. 1–4 (2019)
8. Chandra, A., Khatri, S.K., Simon, R.: Filter-based attribute selection approach for intrusion detection using k-means clustering and sequential minimal optimization techniq. In: AICAI Conference, Dubai, United Arab Emirates, pp. 740–745 (2019)
9. Wrobel, K., Wielgosz, M., Pietron, M., Karwatowski, M., Smywinski-Pohl, A.: Improving text classification with vectors of reduced precision. In: Proceedings of the 10th International Conference (ICAART), vol. 2, pp. 531–538 (2018)
10. McWhirter, P.R., Kifayat, K., Shi, Q., Askwith, B.: SQL injection attack classification through the feature extraction of SQL query strings using a gap-weighted string subsequence kernel. J. Inf. Secur. Appl. **40**, 199–216 (2018). ISSN 2214-2126
11. Al Asaad, B., Erascu, M.: A tool for fake news detection. In: 20th International Symposium on Symbolic and Numeric Algorithms for Scientific Computing (SYNASC), Timisoara, Romania, pp. 379–386 (2018)
12. Uwagbole, S.O., Buchanan, W., Fan, L.: An applied pattern-driven corpus to predictive analytics in mitigating SQL injection attack. In: IEEE Seventh International Conference on Emerging Security Technologies (EST), Canterbury, pp. 12–17 (2017)
13. Uwagbole, S.O., Buchanan, W.J., Fan, L.: Applied machine learning predictive analytics to SQL injection attack detection and prevention. In: IFIP/IEEE Symposium on Integrated Network and Service Management (IM), Lisbon, pp. 1087–1090 (2017)
14. Althubiti, S., Yuan, X., Esterline, A.: Analyzing HTTP requests for web intrusion detection. In: Conference on Cyber Security Education, KSU (2017)

15. Kar, D., Sahoo, A.K., Agarwal, K., Panigrahi, S., Das, M.: Learning to detect SQLIA using node centrality with feature selection. In: International Conference on Computing, Analytics and Security Trends (CAST), Pune, India, pp. 18–23 (2016)
16. Pham, T.S., Hoang, T.H., Vu, V.C.: Machine learning techniques for web intrusion detection—a comparison. In: Eighth International Conference on Knowledge and Systems Engineering (KSE), Hanoi, pp. 291–297 (2016)
17. Nguyen, H.T., et al.: Application of the generic feature selection measure in detection of web attacks. In: Computational Intelligence in Security for Information Systems, pp. 25–32. Springer, Berlin (2011)
18. Mysql.Com: MySQL 8.0 Reference Manual. https://dev.mysql.com/doc/refman/8.0/en/key words.html. Accessed 05 Dec 2019
19. Scikit-learn.org: scikit-learn: Machine Learning in Python. https://scikit-learn.org/stable/. Accessed 05 Dec 2019
20. Anaconda Distribution. https://www.anaconda.com/distribution. Accessed 5 Dec 2019

Swift Controller: A Computer Vision Based Mouse Controller

Pankaj Pundir[(✉)], Ayushi Agarwal, and Krishan Kumar

National Institute of Technology, Uttarakhand, Srinagar, India
{pankaj369.cse16,ayushi16.cse,kkberwal}@nituk.ac.in

Abstract. As technology advances and better virtual interfaces emerge, there is a demand for new kinds of interaction devices. The currently being used devices like keyboard, mouse, pens, etc. have been the most popular among these virtual interfaces. With the advancement of technology where speech and gestures have become the key components to interact with smart devices a replacement for currently used devices is needed. The development of user interfaces influences the changes in Human-Computer Interaction (HCI). This paper focuses to design an input device using computer vision and gesture recognition techniques. This device interacts with a computer using hand gestures, providing an intuitive cost-effective way of performing mouse controls. Various algorithms and techniques are used to make the product user-friendly. We have designed a colored glove that can work in various environments and control the navigation along with functions of the mouse such as left-click, right-click, dragging, and scrolling swiftly. This real-time application allows for practical interaction between users and the system.

Keywords: Hand gesture · Color segmentation · Human-computer interaction · Cursor control · Computer vision

1 Introduction

Computers are one of the most popular devices used around the globe. With various input/output devices the most commonly used input devices remain keyboard and mouse. There is a gamut of research going on in Human-Computer Interaction (HCI) [1]. This primarily involves interaction between humans and machines eliminating the use of any mechanical device.

Computer vision aids computers to gain comprehensive understanding from images and videos. It helps the computer to look at, understand and process images similar to human vision. HCI is a wide-ranging field that uses computer vision [2]. These two technologies when go hand in hand can help the system perceive and apprehend their environments on one side and have interactions with a human on the other side. Since these days computer vision offers promising results with motion capturing without limiting users in terms of speed, the possibility to create better natural gesture interactive systems has increased.

M. Tripathi and S. Upadhyaya (Eds.): ICDLAIR 2019, LNNS 175, pp. 201–209, 2021.
https://doi.org/10.1007/978-3-030-67187-7_22

Gesture recognition [3] can be used as a tool for communication between computers and humans. It is greatly different from traditional hardware-based methods and can accomplish human-computer interaction. It determines the user's intent through the recognition of the gesture or movement of the body parts. In the past decades, many researchers have strived to improve the hand gesture recognition technology. Hand gesture recognition [4] has great value in many applications such as sign language recognition, augmented reality (virtual reality) and robot control.

Intelligent HCI that uses visual survey of hand gestures has gained popularity and a huge amount of progress has been made to date. Using colors on fingertips for fingertip localization and gesture recognition has been used in various models. These color models mostly use RGB models [5]. But due to the difficulty in determination of features of colors, for example, variation in intensities, HSV color models tend to be more accurate. This is because this model can be related to how humans percept color, saturation, and fluorescence.

The most important step of color segmentation is preprocessing. The diversity of background conditions and noise captured in images tend to give poor results. Thus, processes such as removing noise using gaussian filters [6], erosion, dilation [7] are used to preprocess the image. But even after such processing, the results tend to deviate. In paper [8] Grif has implemented his work in a constant background.

We have designed a glove that implements mouse controls using hand gesture recognition. To process the above problem of diversity of backgrounds, we have designed an architecture that works even in backgrounds of similar colors. Thus making our application more robust. We have used an architecture that calculates distance from various contours formed and chooses the most appropriate distance. We have discussed the architecture in Sect. 3. Due to its unique architecture, our model works in various backgrounds. Since no deep learning is involved in implementing our model, the time to process each frame is very less and can be used in real-time applications.

2   Background Work

Finding an intuitive and handy source of communicating with the technology, is a creative challenge. We have advanced through various continuous input devices including trackball, mouse, touch-screen and sensor-based gesture input.

The most basic input devices are mouse and keyboard which are being used in our daily life, are being replaced with touch screen systems. In the near future virtual and augmented reality will cover all entertainment and graphical technologies. Thus there is a demand for advanced gesture-based techniques that can provide 3D input to such systems. Such devices witnessed eminent attention in the field of Human-Computer Interaction [9] which can be used to achieve more sophisticated interaction.

Currently more Deep learning-based techniques are implemented, such as hand segmentation, RGB-D videos with 3D hand pose estimation [10] to train a deep learning model for robust situations. In the paper [11] color glove having a specific pattern is trained corresponding with its 3D hand pose extracted from the sensors, using a machine learning model which in turn helps in estimating hand pose [11]. One of the intuitive technique uses a two-stage model to perform hand segmentation and gesture recognition using neural networks [12], this method achieves state of the art accuracy with the trade-off between computational time and device cost. Various techniques use Kinect sensors to track hand movement, in the paper [13] Hidden Markov Model is used to model the dynamics of the gestures. This technique leads to higher accuracy but uses external sensors similar to leap motion devices [14]. The depth camera is widely used, as it provides crucial feature information regarding the hand and its 3D position [15]. Moderate computation and cost-effective method use color segmentation techniques but it leads to various challenges related to background noise.

3   System Architecture

Our proposed method used a specially designed glove which enables us to track the hand and finger movement with minimum computation time and more robustness. We have used color stripes in a specific pattern to find the exact location of fingertips. This unique color-coding is implemented to maintain distinguishable features form the background. The color-coding of the glove is shown in Fig. 1.

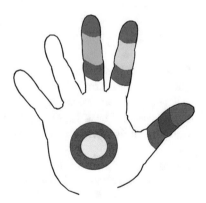

Fig. 1. Architecture of glove

To control the mouse and all its functionality we have optimally reduced the number of engaged fingers to 3. At center concentric circle with different colors is used to detect the presence of the hand. All the above centers are used to estimate the location of the fingertips and determine the performed gesture.

The distance between two points $A(x_1, y_1)$ and $B(x_2, y_2)$ is calculated using Manhattan formula given in Eq. 1. We have used Manhattan distance in place of Euclidean distance to simplify the calculations and speed up the process.

$$d(A, B) = |x_1 - x_2| + |y_1 - y_2| \tag{1}$$

To determine the presence of the hand, two concentric circles are painted over the glove. For increasing the robustness and accuracy contour detection is used. We selected the minimum of overlapping centers formed by two contours within a dynamic threshold i.e. threshold varies from hand's distance from the camera. If the distance is less than or equal to the threshold, we say that the hand is present as shown in Eq. 2 where t is the threshold, A is the set of centers of red contour and B is the set of centers of yellow contour.

$$min(distance(a, b)) \leqslant t \ \ \forall \ \ a \in A \ \ b \in B \tag{2}$$

For fingers, 3 stripes technique is used to find the correct centers as shown in Fig. 1. Top and bottom stripes are of the same color and the middle stripe act as a unique identifier for every fingertip. Calculation of the center coordinates of different contours is performed and the center X and Y coordinates of the desired contours are found. Keeping the track of these coordinates helps us to estimate the coordinate of the fingers. To find the correct fingertip we find the contour of every color which is surrounded by red color contour on both sides. If the sum of the minimum distance between all 3 contours is within the threshold then the unique color code will act as the fingertip position. This is shown in Eq. 3 where t is the threshold and A is the set of centers of red contour, B is the set of centers of unique identifier color contour.

$$min(distance(a, b) + distance(b, c)) \leqslant t \ \ \forall \ \ a, c \in A \ \ b \in B \ \& \ \ a \neq c \tag{3}$$

The threshold in both cases is determined by actual measured values and varied dynamically by using a scaling factor. This design is selected to trace the finger movement easily. All the basic functionality of the mouse can be easily performed with predefined gestures. These functionalities use relative features such as relative distance and the angle between two fingers to provide rotational and scale invariance.

We have selected thumb as the cursor pointer. Selecting the thumb over the fingers is due to biological reason compared to the rest of the fingers. Fingers tend to have more flexible motion, whereas thumb performs the supporting action while performing various tasks, this stability is achieved since the motion of the thumb is determined by muscles and not by joints [16]. Thus, assigning thumb as a mouse pointer will provide more accurate results and will be easy to use. Coordinate rescaling and validation are performed to avoid unnecessary jumping and detection of the false finger features due to background noise.

4   Proposed Algorithm

We have structured a glove to supplant the controls of a mouse using computer vision. The flowchart for the steps involved in the proposed algorithm is given in Fig. 2.

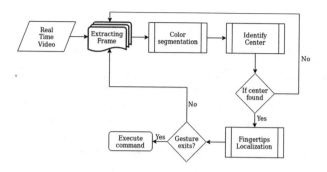

Fig. 2. Flowchart of Proposed method

The real-time video feed is captured using a webcam. Each frame of the video extracted is passed to the next phase, which is color segmentation along with the preprocessing.

4.1   Color Segmentation

The first step involved in our proposed algorithm is color segmentation [17]. This involves separating a particular color in the image. A Gaussian filter is applied first to each frame extracted This is used to increase the consistency of frame and remove noise from it. Its main purpose is to enhance the image structure within different scales. The second step involves converting an image from RGB to HSV color format. This helps to identify appropriate colors in diverse lightning conditions. Thus, increasing the robustness.

With the help of a range of colors specified while building the architecture, the mask of each color is separated. The process of erosion followed by dilation is then applied to each mask. Erosion is the process of removal of the outermost layer of pixels while dilation is the addition of the outermost layer of pixels. The use of both of these operations simultaneously increases the contour of the desired region by eliminating small noises from the frame. The contour of the segmented colors is found and their centers are stored.

4.2   Center Identification

After getting the center of contours, the center of the hand is identified. The procedure involved to find the hand center is given in Sect. 3. If the hand center is found it is assumed that the hand lies in the range of the camera and further processing is done.

4.3   Fingertips Localization

If the hand is found the next step is fingertips localization. Each finger is assigned a particular color as given in Fig. 1. The thumb is colored dark blue, index finger is colored light blue and middle finger is colored green. The procedure involved to find a fingertip is given in Sect. 3. The fingertips are labeled and gesture identification is done.

4.4   Gestures Identification

The next step followed to fingertip localization is gesture identification. The thumb acts as the cursor, wherever the thumb moves the cursor moves. If the thumb gets out of the frame the cursor stops wherever it is. The scaling factor is used to accomplish a complete degree of movement over the screen. The equation for x coordinate is given in Eq. 4 where windowX is width of the screen, frameX is width of image, S denotes sensitivity of cursor > 0 that controls the factor of movement of cursor with respect to hand, thumbX denotes the x coordinates of thumb, shiftX denotes the frame amount to be shifted to keep the center of hand in the frame of webcam. Similarly, the equation for y coordinate is given in Eq. 5

$$X = \frac{thumbX \times (windowX + S)}{frameX} - shiftX \qquad (4)$$

$$Y = \frac{thumbY \times (windowY + S)}{frameY} - shiftY \qquad (5)$$

The second gesture is the left click. The bending of the index finger results in a left-click. This is accomplished by comparing the lengths between index fingertip, center (A) and middle fingertip, center (B). If the $length(A)$ is less than $2/3length(B)$ then left button down operation is performed and as the length increases the left button up is performed resulting in a left-click. Scrolling and dragging can be achieved by keeping the index finger down and moving the thumb. The gesture of the left click is shown in Fig. 3 a.

The third gesture is right-click. This is obtained by bringing the middle finger in between thumb and index finger as shown in Fig. 3 b. When the fingertip of middle finger lies between fingertip of thumb and index finger the right button goes down and when it comes back, the right click is released. The comparison is done by keeping track of relative angles between the thumb-index and thumb-middle.

5   Experimentation and Results

This algorithm focused on controlling the mouse using a hand by wearing a colored glove. The experimentation was done on Intel(R) Core(TM) i5 − 7200U CPU @ 2.50 GHz and 4.0 GB RAM on Linux operating system. The first step

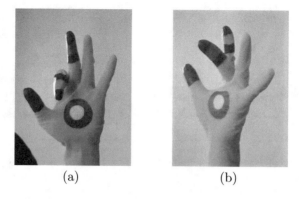

Fig. 3. Gesture: (a) Left click, (b) Right click

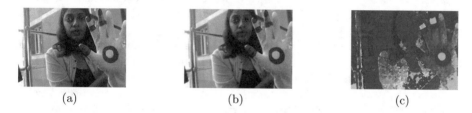

Fig. 4. Preprocessing: (a) Original image, (b) Denoised image, (c) HSV image

Fig. 5. Detection: (a) Contours, (b) Center and Fingertips

consisted of denoising the image using the Gaussian filter as shown in Fig. 4 b where the original image is shown in Fig. 4 a and then converting image into HSV format. HSV format is visualized by defining the *saturation* and *value* to be 230 as shown in Fig. 4 c, this displays the color specificity in HSV color format.

Then the masks of different colors were prepared followed by erosion and dilation. It is found that small noise is removed using this method, but portions containing large parts of the masked color remains as it is. In the next step the contours and center of contours are marked as shown in Fig. 5 a.

To refine the contours and to identify the center, the procedure involved in Sect. 2 is performed using Eq. 2. If the center is found then the fingertips are

Fig. 6. Center hidden: (a) Center hidden image, (b) Hand not found gesture

localized using Eq. 3 as shown in Fig. 5 b. If the center is not found as in image Fig. 6 a the result is as shown in Fig. 6 b, where the fingertip localization step is omitted and the next frame is processed. It was found that under good lighting conditions, center and fingertips were easily found.

The thumb acted as the mouse cursor. As long as thumb stayed in the frame the cursor moved. If the thumb went out of the frame the cursor stopped. The left click was performed using gesture given in Fig. 3 a with resulting left click as shown in Fig. 7 a. Similarly, the right click operation was performed using gesture given in Fig. 3 b with resulting right click operation as shown in Fig. 7 b. The time taken to process one frame on an average was **9 ms** with frame rate of **30 fps**. Thus, our model worked swiftly for controlling mouse without lags in timings or high computations.

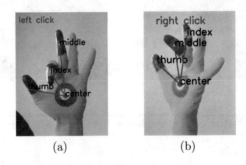

Fig. 7. Gesture identification: (a) Left click, (b) Right click

6   Conclusion and Future Scope

Our proposed model provided a smart method to remove background noise by using specific color patterns within the glove. This method aims to run on devices having moderate computational resources including smart T.V, netbooks, etc.

This technique can be used to define various custom gestures for different applications. This can be used, extensively in classrooms of schools, office meetings, and presentations. Using a color glove and a specific pattern made it easily traceable and cost-effective.

In future, we would also like to extend the functionality of the glove by introducing features such as swipe, scroll, and zoom. The main problem involved in this approach is light sensitivity. This glove will not give results in dark conditions. In the future we would like to extend our approach to meet the limitations we are facing now using radium color paints.

References

1. Jenny, P., et al.: Human-Computer Interaction. Addison-Wesley Longman Ltd. (1994)
2. Lenman, et al.: Computer vision based hand gesture interfaces for human-computer interaction. RIT, Sweden (2002)
3. Mitra, et al.: Gesture recognition: a survey. IEEE Trans. Syst. Man Cybern. Part C (Appl. Rev.) **37**(3), 311–324 (2007)
4. Smith, et al.: Hand gesture recognition system and method. U.S. Patent No. 6,128,003, 3 October 2000
5. Vincze, et al.: Hand gestures mouse cursor control. Sci. Bull. "Petru Maior" Univ. Tg. Mures **11**, 46–49 (2014)
6. Aurich, et al.: Non-linear Gaussian filters performing edge preserving diffusion. In: Mustererkennung 1995, pp. 538–545. Springer, Heidelberg (1995)
7. Gil, et al.: Efficient dilation, erosion, opening, and closing algorithms. IEEE Trans. PAMI **24**(12), 1606–1617 (2002)
8. Grif, et al.: Mouse cursor control system based on hand gesture. Procedia Technol. **22**, 657–661 (2016)
9. Meena, et al.: A study on hand gesture recognition technique. Dissertation (2011)
10. Hernando, G., et al.: First-person hand action benchmark with RGB-D videos and 3D hand pose annotations. In: Proceedings of the IEEE Conference on CVPR (2018)
11. Wang, et al.: Real-time hand-tracking with a color glove. ACM Trans. Graph. (TOG) **28**(3), 63 (2009)
12. Dadashzadeh, et al.: HGR-Net: a fusion network for hand gesture segmentation and recognition. arXiv preprint arXiv:1806.05653 (2018)
13. Gu, Y., et al.: Human gesture recognition through a Kinect sensor. In: 2012 IEEE International Conference on Robotics and Biomimetics (ROBIO). IEEE (2012)
14. Shao, L.: Hand movement and gesture recognition using Leap Motion Controller. Virtual Reality, Course Report (2016)
15. Sharp, et al.: Accurate, robust, and flexible real-time hand tracking. In: Proceedings of the 33rd Annual ACM ACI. ACM (2015)
16. https://en.wikipedia.org/wiki/Muscles_of_the_thumb
17. Meyer, et al.: Color image segmentation. In: 1992 International Conference on Image Processing and its Applications. IET (1992)

Simultaneous Vehicle Steering and Localization Using EKF

Ankur Jain$^{(\boxtimes)}$ and B. K. Roy

NIT Silchar, Silchar, Assam, India
ankurjainjob@gmail.com

Abstract. For successful operation in self-driving vehicle, we need to know its position on the map to navigate through the environment while keeping itself on the path. Here, we simulated vehicle steering, localization, and path tracking simultaneously. We have used a generalised workflow so that it can be used in other type of road curvature, map and control algorithms. For vehicle steering control, we designed pure-pursuit controller with given 2D map of path curvature and landmarks. For the purpose of estimation, extended kalman filter has been used. The results demonstrate the robustness of the proposed algorithm.

Keywords: Cyber physical system · Vehicle steering · Vehicle localization · Extended Kalman filter

1 Introduction

In the past decade, autonomous vehicles, advanced driver assistance systems [1], and cooperative vehicles [2] have been pursued across the globe. To cater huge demand for ever growing population of our country, we need to optimize their roadways and waterways infrastructure. Government of Indian planning to expand their roadways through Bharatmala project [3] and waterways through Sagarmala project [4] projects. Because of limited land resources and investment, no government could satisfy huge demand. Growing traffic also leads to vehicle accidents. One of the main objective of this research is to prevent the fatalities due to human prone accidents [5]. Navigation of the self-driving vehicle through the road networks require exact localization and steering to follow the given path in presence of noisy environment and sensors measurements [6]. Our aim is to achieve complete automation from low level vehicle dynamics control to high level connected vehicle management. For that, we need highly accurate robust localization system to exactly position our vehicle to follow the road track. In vehicle navigation problem, we guide our vehicle through steering input decided by the control algorithm to follow the pre-specified path on the given map[7]. We assumed that map consists of landmarks location so that we can calculate estimation confidence.

In the present discussion, our objective is to localise ego vehicle with respect to landmarks, design a vehicle steering control algorithm to track a given path.

There are very significant work has been carried out which is as follows. In [8], Kuutti et al. presents a survey of localisation techniques on the basis of map, sensor and cooperation. Fethi et al. addresses localization, mapping and path tracking in optimization point of view in [9]. Application of computer vision and history can be found in [10]. In most of the work, measurements are only produced by GPS system or GPS data is fused with other on-board odometry sensors. Morales et al. design an estimation algorithm using extended kalman filter in [11]. The algorithms fails if GPS data is unavailable.

For long-term sustainable operation and autonomous vehicle penetration into the current traffic, we need a generic structure in wide range of operation designs domains such as range of environmental conditions and weather throughout the year with/without some customization. GPS solutions are inefficient to keep vehicle within the lane markings due to radio frequency noise, multi-path fading and bad weather conditions. However, GPS/INS approaches, even fused with wheel odometry gives 2σ deviation more than 1m [12] which is not sufficient to keep vehicles on urban roads. Blockage and multipath appear as failure modes for GPS dependent solutions, for example, the anomalies in urban and heavily forested or mountaneous environments. LIDAR point cloud map based solutions fails in lack of optical features due to snow, dust, dirt etc.

Autonomous passenger car navigation in dynamic environments requires more accuracy than available from GPS. The above limitations can be solved using on-board RADAR mounting with given 2D map with landmarks information based vehicle localization. This method complement the LIDAR and camera based methods where optical obstruction present. Research into self-driving passenger vehicles has been ongoing since last few decades, but most of the work is focused on specific environments. The core of this work to represent the generic procedure to achieve the path tracking while estimating the position in given map under noisy measurement of RADAR. We illustrate through the suitability of extended kalman filter (EKF) algorithms to localize a vehicle, equipped with a 2D RADAR, within a previously-built reference metric map. The core idea is estimate the best position through the possible range by fusing the information received from the on-board RADAR output and the map data. The data flow architecture of the algorithm shown in Fig. 1a.

The rest of the paper organised as follows. We defined problem formulation in Sect. 2. The Sect. 3 describes the methodology used to solve the problem defined.

Table 1. Symbolic notations

Symbols		Symbols	
X	State vector	A	Fundamental matrix
u	Control input	B	Input matrix
x	longitudinal position	C	Measurement matrix
z	sensor measurements	P	State uncertainty covariance matrix
σ	Standard deviation	Q	Process uncertainty covariance matrix
Δt	Sampling time	R	Measurement uncertainty covariance matrix

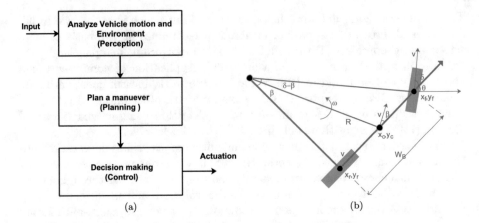

Fig. 1. (a) Data flow architecuture, (b) Bicycle model

Numerical values of parameters and simulation results are presented in the Sect. 4. The Sect. 5 concludes the research study.

Notation. Symbols used in the paper defined in Table 1.

2   Problem Formulation

2.1   Bicycle Model [13]

We can describe our generalised model as

$$\dot{x} = F(x, u) + q$$
$$y = H(x) + r \tag{1}$$

where $F(x, u)$ describes the nonlinear system dynamics, $H(x)$ describes the measurement function, q and r are random noise consideration in the dynamics and the measurement respectively. Here, we have utilized the bicycle model shown in Fig. 1b as process which is described as follows

– By triangle identity

$$tan\delta = \frac{W_B}{R} \tag{2}$$

– Applying Instantaneous center of rotation

$$\dot{\theta} = \omega = \frac{V}{R} = \frac{V\,tan\delta}{W_B} \tag{3}$$

– Final model

$$\begin{cases} \dot{x}_r = vcos\theta \\ \dot{y}_r = vsin\theta \quad + \mathcal{N}(0, Q) \\ \dot{\theta} = \frac{V tan\delta}{W_B} \end{cases} \tag{4}$$

2.2   Measurement Model

The vehicle's on-board RADAR provides noisy range and bearing measurements, coming from the multiple known landmarks locations mathematically represented by Eq. 5. It is used to convert the range and bearing into position and pose (x, y, θ) with respect to the landmarks.

– Measurement function

$$z = H(X, P) \qquad\qquad + \mathcal{N}(0, R)$$

$$\begin{bmatrix} range \\ bearing \end{bmatrix} = \begin{bmatrix} \sqrt{\triangle x^2 + \triangle y^2} \\ \tan^{-1}\frac{\triangle y}{\triangle x} - \theta \end{bmatrix} \quad + \mathcal{N}(0, R) \tag{5}$$

Now our objectives problems are listed out as follows

Problem 1. For a given vehicle model $\dot{x} = F(x, u)$, a series of controls u_k and sensor observations z_k over discrete time steps k , compute an estimate of the vehicle's location x_k in 2D map with respect to landmarks ld_k. All the states and measurements are distributed in Gaussian probabilistic manner. Find location posteriors $P(x_k|z_{1:k}, u_{1:k}, ld_k)$ using algorithm defined in Sect. 3.1.

Problem 2. Given a system $\dot{x} = F(x, u)$, reference path $x_{ref} : R \to R^n$, and velocity $v_{ref} : R \to R$, find a feedback law, $u(x)$, such that solutions to $\dot{x} = f(x, u(x))$ satisfy the following: $\forall \varepsilon > 0$ and $t1 < t2$, there exists a $\delta > 0$ and a differentiable $s : R \to R$ such that,

$$\|x(t_1) - x_{\text{ref}}(s(t_1))\| \le \delta \Rightarrow \|x(t_2) - x_{\text{ref}}(s(t_2))\| \le \varepsilon \tag{6}$$

$$\lim_{t \to \infty} \|x(t) - x_{\text{ref}}(s(t))\| = 0 \tag{7}$$

$$\lim_{t \to \infty} \dot{s}(t) = v_{\text{ref}}(s(t)) \tag{8}$$

3   Methodology

Accurate localization is a key component of any self-driving car software stack. If we want to drive autonomously, we certainly need to know where are we? This is the process of computing a physical quantity like position from a set of measurements. Since real-world measurements are imprecise, we developed a technique that try to find the best or optimal value about our sensors around the external world under given assumptions. The software stack for position and landmarks detection is shown in Fig. 2.

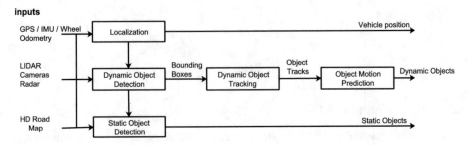

Fig. 2. Architecture of the data flow

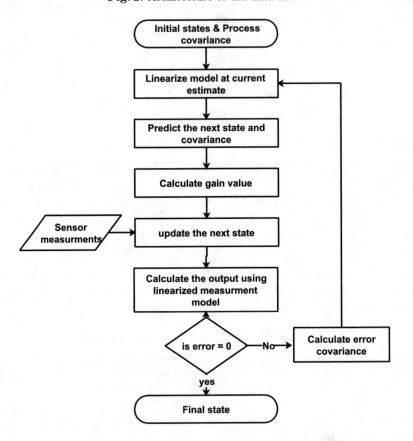

Fig. 3. Flow chart for EKF

3.1 Extended Kalman Filter [14]

It is the nonlinear filtering technique where we linearise the nonlinear system model at the current estimated value. The flow chart of the EKF algorithm could be seen in Fig. 3. Let's suppose, we have the system described in Eq. 1.

We do linearise this system the current estimated value at each iteration. Linearised equations are as

$$\dot{x} = Ax + Bu$$
$$y = Cx$$

Where, A, B, C is a fundamental matrix, input matrix, and output matrix respectively. Consider a Gaussian random variable $x \sim N(\hat{x}, P)$ and a nonlinear function $y = F(x, u)$. We linearise $F(x, u)$, and $H(\mathbf{x})$ by taking the partial derivatives of each to evaluate A and C. This gives us the the discrete state transition matrix and measurement model matrix:

$$A = \left. \frac{\partial F(x, u)}{\partial x} \right|_{\hat{x}_{k-1|k-1}, u_{k-1}}$$
$$C = \left. \frac{\partial H(x)}{\partial x} \right|_{\hat{x}_{k|k-1}} \tag{9}$$

A first order Taylor expansion of $F(x, u)$ around $\hat{x}$ gives

$$y \sim F(\hat{x}) + F\prime(\hat{x})(x - \hat{x}) \tag{10}$$

Then we get the posterior parameter as

$$E(y) = E(F(\hat{x}) + F'(\hat{x})(x - \hat{x})) \tag{11}$$
$$= E(F(\hat{x})) + E(F''(\hat{x})(x - \hat{x})) \tag{12}$$
$$= F(\hat{x}) \tag{13}$$

$$cov(y) = cov(F(\hat{x}) + F^{'}(\hat{x})(x - \hat{x})) \tag{14}$$
$$= cov(F(\hat{x})) + cov(F'(\hat{x})(x - \hat{x})) \tag{15}$$
$$= F'(\hat{x})cov((x - \hat{x}))F'(\hat{x})^T \tag{16}$$
$$= F'(\hat{x})PF'(\hat{x})^T \tag{17}$$
$$= APA^T \tag{18}$$

$$= APA^T + Q$$

Prediction
The EKF does not alter the Kalman filter's linear equations. Instead, it linearises the nonlinear equations at the point of the current estimate, and uses this linearisation in the linear Kalman filter. we have given as,

$$x_{k-1}|y_{1:k-1} \sim N(\hat{x}_{k-1|k-1}, P_{k-1|k-1})$$

Then,

$$\hat{x}_{k|k-1} = F(\hat{x}_{k-1|k-1}) \tag{19}$$

$$P_{k|k-1} = F'(\hat{x}_{k-1|k-1})P_{k-1|k-1}F'(\hat{x}_{k-1|k-1})^T + Q_{k-1} \tag{20}$$

Update

$$y = H(x_k) + r_k$$
$$\sim H(\hat{x}_{k|k-1}) + H'(\hat{x}_{k|k-1})(x - \hat{x}_{k|k-1}) + r_{k-1} \tag{21}$$

Then,

– Innovation:

$$v_k = z_k^T - H(\hat{x}_{k|k-1}) \tag{22}$$

– Innovation covariance:

$$S_k = H'(\hat{x}_{k|k-1})P_{k|k-1}H'(\hat{x}_{k|k-1})^T + R_k \tag{23}$$

– Extended Kalman gain

$$K_k = P_{k|k-1}H'(\hat{x}_{k|k-1})^T S_k^{-1} \tag{24}$$

– posterior mean:

$$\hat{x}_{k|k} = \hat{x}_{k|k-1} + K_k v_k \tag{25}$$

– posterior covariance:

$$P_{k|k} = (I - K_k H'(\hat{x}_{k|k-1}))P_{k|k-1} \tag{26}$$
$$= P_{k|k-1} - K_k S_k K_k^T$$

3.2   Controller Design

3.2.1 Steering Control

$$sin\alpha R = \frac{L/2}{R} \Rightarrow R = \frac{L}{2sin\alpha}$$

$$\kappa = \frac{2\sin(\alpha)}{L} = \frac{1}{R}$$

For a vehicle speed v_r, the commanded heading rate is,

$$\omega = \frac{2v_r sin(\alpha)}{L} = \frac{v_r tan\delta}{W_B} \tag{27}$$

$$\delta = tan^{-1}(\frac{2W_B sin(\alpha)}{L}) \tag{28}$$

Then α is given by

$$\alpha = \arctan\left(\frac{y_{\text{ref}} - y_r}{x_{\text{ref}} - x_r}\right) - \theta \tag{29}$$

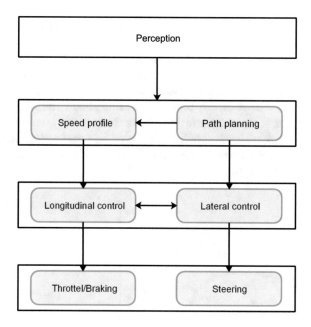

Fig. 4. Architecture of vehicle control strategy

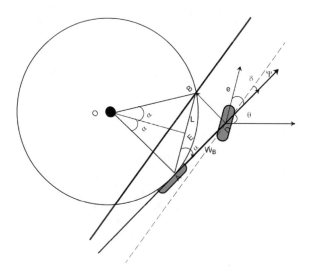

Fig. 5. Geometry pure pursuit control

4   Results

Here, we have python framwork to simulate the problem. For mathamatical operatations we have used numpy package and for ploting we have used matplotlib library. Figure 6a, shows the performance of our simulation, where red solid line represents the pre-defined path to be followed, blue squared block represents the landmarks, blue covariance ellipses represent the confidence of position estimate where boundary set at 6σ, blue dashed line shows the estimated position throughout the target course. Figure 6b shows the steering command being given by controller to track which is satisfactory for vehicle navigation. Initially we can that to move on track large steering input is required than small steering input is given at 50 s and 70 s due to high curvature.

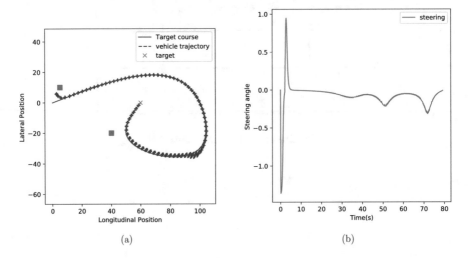

Fig. 6. (a) Path tracking with noisy sensors, (b) Steering angle input

5   Conclusions

This paper proposes a strategy for vehicle localization, guiding control and track following utilizing EKF and designed pure pursuit directing controller. While our methodology isn't boundlessly versatile and can be prevented by adequately serious changes in climate or natural hindrances, besides it is a critical advancement towards enabling vehicles to explore themselves in even the trickiest of genuine circumstances. simulation-based investigations have demonstrated the viability of the proposed strategy in accomplishing high-exactness (precision of centimetre-level).

This point holds incredible potential for future advancement. In spite of the fact that we used the (x,y) plane in our maps. For robustness, road slope inclusion in the algorithm is similarly helpful. While radio waves reflectivity holds some information apart from that an algorithm could be designed including road elevation. We can likewise meld information with different sensors and helpful information through the vehicular system. Other than this, the future work is to execute the proposed vehicle confinement technique in true situations. The detailed map information is still missing will be built up. At the point when the map information is accessible, exploratory outcomes on genuine information can be gotten and will be accounted for in future.

Acknowledgement. The authors would like to acknowledge the financial support is given by TEQIP-III, NIT Silchar, Silchar - 788010, Assam, India.

References

1. Kumar, A.M., Simon, P.: Review of lane detection and tracking algorithms in advanced driver assistance system. Int. J. Comput. Sci. Inf. Technol. **7**(4), 65–78 (2015). http://www.airccse.org/journal/jcsit/7415ijcsit06.pdf
2. Li, H., Nashashibi, F.: Cooperative multi-vehicle localization using split covariance intersection filter. IEEE Intell. Transp. Syst. Mag. **5**(2), 33–44 (2013)
3. Bharatmala project (2018). https://en.wikipedia.org/wiki/Bharatmala. Accessed 07 Oct 2018
4. Sagarmala project October 2018. https://en.wikipedia.org/wiki/Sagar_Mala_ project. Accessed 07 Oct 2018
5. Dey, K.C., Yan, L., Wang, X., Wang, Y., Shen, H., Chowdhury, M., Yu, L., Qiu, C., Soundararaj, V.: A review of communication, driver characteristics, and controls aspects of Cooperative Adaptive Cruise Control (CACC). IEEE Trans. Intell. Transp. Syst. **17**(2), 491–509 (2016)
6. Li, H., Nashashibi, F., Toulminet, G.: Localization for intelligent vehicle by fusing mono-camera, low-cost GPS and map data. In: 13th International IEEE Conference on Intelligent Transportation Systems, pp. 1657–1662. IEEE (2010)
7. Levinson, J., Thrun, S.: Robust vehicle localization in urban environments using probabilistic maps. In: 2010 IEEE International Conference on Robotics and Automation, pp. 4372–4378. IEEE (2010)
8. Kuutti, S., Fallah, S., Katsaros, K., Dianati, M., Mccullough, F., Mouzakitis, A.: A survey of the state-of-the-art localization techniques and their potentials for autonomous vehicle applications. IEEE Internet Things J. **5**(2), 829–846 (2018)
9. Fethi, D., Nemra, A., Louadj, K., Hamerlain, M.: Simultaneous localization, mapping, and path planning for unmanned vehicle using optimal control. Adv. Mech. Eng. **10**(1), 1687814017736653 (2018)
10. Dickmanns, E.D.: Vision for ground vehicles: history and prospects. Int. J. Veh. Auton. Syst. **1**(1), 1–44 (2002)
11. Morales, Y., Takeuchi, E., Tsubouchi, T.: Vehicle localization in outdoor woodland environments with sensor fault detection. In: 2008 IEEE International Conference on Robotics and Automation, pp. 449–454. IEEE (2008)
12. Kennedy, S., Rossi, J.: Performance of a deeply coupled commercial grade GPS, INS system from KVH and NovAtel INC. In: IEEE/ION Position, Location and Navigation Symposium 2008. pp. 17–24. IEEE (2008)
13. Rajamani, R.: Vehicle Dynamics and Control. Springer (2011)
14. Simon, D.: Optimal State Estimation: Kalman, H Infinity, and Nonlinear Approaches. John Wiley & Sons (2006)

Cloud-Based Clinical Decision Support System

Solomon Olalekan Oyenuga[1], Lalit Garg[1,2(✉)], Amit Kumar Bhardwaj[3],
and Divya Prakash Shrivastava[4]

[1] The University of Liverpool, Liverpool, UK
solomon.oyenuga@online.liverpool.ac.uk
[2] The University of Malta, Msida, Malta
lalit.garg@um.edu.mt
[3] Thapar Institute of Engineering and Technology, Patiala, India
akbhardwaj@thapar.edu
[4] Higher Colleges of Technology, Abu Dhabi, United Arab Emirates
dp_shrivastava@yahoo.com

Abstract. Cloud computing is spreading its scope to the different industries, where sensitive data are being maintained. The healthcare industry is facing the challenge of data integrity, confidentiality and huge operational cost of the data. Hence, there is a need to adopt the cloud-based clinical decision support system (CCDSS) in order to maintain the Electronic Medical Records (EMRs), where patients and healthcare professional can easily have access to their legitimate records irrespective of the geographical location, time and cost-effectively through a decision support system. The main objective of this research work is to build a clinical decision support system (DSS) through the deployment of a private Cloud-based System Architecture. This research work explored the pros & cons, challenges that could be faced during implementing a (CCDSS) in a Healthcare for optimization of the resources with time, money, and efforts along with the patient data security. A quantitative survey was conducted to gather the opinions of healthcare stakeholders. Based on the findings of the survey, a rule-based DSS was developed using a Data Flow Modelling tool that facilitated the Cloud deployment of applications and in-House private cloud deployment to ensure the security of data, protection of data, and the regulatory compliance policies in the health organizations. In order to evaluate the proposed system, the User Acceptance Testing was carried out to determine how well the proposed CCDSS meets the requirements.

Keywords: Clinical cloud-based decision support system (CCDSS) · Decision support system · Cloud computing · Medical records · Health-records

1 Introduction

It is painful when many people die due to not having the right treatment on-time because of unavailability of health-records. The right treatment in life-saving time can save a life. Wickramasinghe and Geisler (2008) described healthcare as an activity that involves

© The Author(s), under exclusive license to Springer Nature Switzerland AG 2021
M. Tripathi and S. Upadhyaya (Eds.): ICDLAIR 2019, LNNS 175, pp. 220–234, 2021.
https://doi.org/10.1007/978-3-030-67187-7_24

managing, preventing, and treating sickness and illness with the sole purpose of providing services that are effective and clean, which result in the protection of the physical and mental well-being of both humans and animals. The healthcare industry (HI) is generating voluminous and rapid data, and their trends show that there is a massive demand for using ICT solutions in healthcare safety, particularly regarding the safety, security, and protection of patient's data. To achieve the best results for the safety, security, and protection of patient data there is a need for transformation of healthcare processes and incorporating cloud-based IT solutions. Hence, to avoid the probability of any life-threatening situations from allergies and drug reactions because of wrong diagnosis and medical errors (Makary and Daniel 2016), there is a need to set up or develop an IT system known as decision support system (DSS), which can help the healthcare practitioners to resolve the problems that may arise during medical diagnosis of patients.

The main objective of this research work is to build a clinical decision support system (DSS) through the deployment of a private Cloud-based System Architecture. This research work explored the pros & cons, challenges that could be faced during implementing a (CCDSS) in a Healthcare for optimization of the resources with time, money, and efforts along with the patient data security.

In conclusion, this research proposes a framework of developing a DSS with a series of actions and steps that will be selected at the different stages during the development of the system.

The overview structure of the paper is as follows: the next section provides the background and the review of the literature. The following section discusses the research gap and problem formulation. Section 4 deals with the CCDSS analysis & design and data collection and analysis. Section 5 deals with the implementation of open source-based CDS system and its evaluation, which is followed by the section presenting the conclusion of this research work.

2 Background

According to Kohn et al. (2000) and Makary and Daniel (2016), there are more deaths due to medical errors or mistakes as compared to the deaths caused by Road Accidents and AIDS together. The non-availability of essential health records of an accidental patient causes a delay in treatment, which might be life-threatening. According to Jackson (2009) this problem amplifies thanks to the ever-increasing expenditure of healthcare organizations. The better growth and efficiency of clinical information support systems (CISS) help the physicians in ensuring the timely delivery of healthcare services.

Different users utilize different piecemeal of electronic health-care techniques to retrieve, display, analyze, transfer, and share an extensive collection of online and offline. Liang et al. (2006) emphasized the requirement of the web-based DSS which only helps the enterprises with smart business-level decisions but also helps its clients and customers in decision making about products and services. Even in healthcare industry DSS will help in minimizing wrong diagnosis and medical errors.

3   Literature Review

Peter et al. (2009), describe that the cloud computing enables on-demand full access of the shared collection of arranged computer resources, including hardware and software services, servers, applications, storage facilities, and networks, to ensure that there is minimal effort from the management or the service providers. Security issues have been raised by different researchers, and according to Edwards (2009), security risk and reliance on third party serve as a backdrop to the successful implementation of a cloud-based system.

Bonczek et al. (1980) describes a Decision support system described (DSS) as a group of three diverse interacting components; the language system media between the components and users of the DSS; a knowledge system - a storage facility of problem domain knowledge; and a system that links the processing of problems with the other two components (language and knowledge) with the ability to manipulate the common problems that are necessary for supporting decisions. Perrealult and Metzger (1999) describe a software-based clinical DSS which can assist in decision making in clinical domain with the features of matching the patients' data with clinical knowledge base system.

Turban and Aronson (2001), discussed that a DSS provision the relevant and timely information. Gibson et al. (2004) advocate the usage of business intelligence in clinical decision support system (CDSS) to improve the efficiency of DSS. Osheroff et al. (2005, 2007), described ae CDSS as a process of providing patient-specific information and relevant knowledge for professionals and stakeholders intelligently. He et al. (2010) and Kanagaraj (2011) advocate the use of cloud-based systems in the healthcare industry for rapid exchange of information about the patients' report images. Fernandez-Cardenosa et al. (2012) proposed a cloud-based hybrid Electronic Health Records (EHRs) system for hosting clinical/medical images in the cloud with the sole aim of sharing among the medical practitioners. Anooj (2012) highlights the challenges related technical, clinical, high cost of maintenance and issues related to implementation and evaluation. Hsieh and Hsu (2012), advocate the development of web-service for teleconsultation and telemonitoring service for ECG analysis.

Zafar and Koehler (2013) proposed that the adoption of a CDSS will improve the services rendered in a healthcare facility. Hussain et al. (2013) proposed the use of a cloud-based DSS service to maintain the data related to diagnose and manage chronic diseases. Castaneda et al. (2015) proposed the use of a DSS in operational and strategic objectives of the healthcare organization. Belard et al. (2017) proposed the use of CDSS in the Precision diagnosis for personalized medicine while Heise et al. (2018) proposed a CDSS for improving Antibiotic Safety, Borab et al. (2017) proposed a CDSS for preventing Venous Thromboembolism in Surgical Patients and Sim et al. (2017) proposed a decision support system for diabetes care. Rawson et al. (2017) carried out a systematic review of CDSS for antimicrobial management, while Jacob et al. (2017) carried out a systematic review of the cost and benefit of CDSS for cardiovascular disease prevention. Oh et al. (2015) proposed an application of Cloud computing in CDSS.

Although there is the widespread adoption of cloud computing globally, it is starting to exist in the healthcare industry, and majority of the stakeholders in the industry believe that its adoption will improve the services rendered by changing the existing trends with the current health information systems. Cloud computing is the use of "providers infrastructure" for storage, processing, and sharing the data "as and when needed." (Arora et al. 2013).

Strukhoff et al. (2009) noted the improvement of patient care and reduction in cost of medical transcriptions during the usage of MedTrak System -a cloud-based software solution with the partnership of IBM Corporation at American Occupational Network (AON). On the creation of a European Union Consortium, there are several studies reported by IBM (2010) about the development of a cloud-based DSS for healthcare industry with the primary objective of securing and protecting patient's data. "Dhatri" an initiative proposed and implemented by Rao et al. (2010) to integrate the functionalities of cloud computing and wireless technology to provide easy access of the patients' health information to physicians at any point in time. Rolim et al. (2010) suggested a cloud-based data automation for real-time data collection for legitimate users to access the data in 24 * 7 fashion and from any location. Koufi et al. (2010) proposed the integration of personal health record system with the emergency medical system. According to Korea IT Times (2010), an eHealth cloud system was developed to provide hosting platform for application development in the healthcare industry under a collaborative agreement amid the Royal Australian College of General Practitioners (RACGP) and Telstra. Guo et al. (2010) discussed cloud computing as a solution for the massive data of EHR including data of images from radiology and genomic. Samuel et al. (2013) discussed the security and regulatory compliance that the cloud computing service providers and vendors need to ensure that the government regulations adhere in handling of healthcare data and records along with the standards and regulations of the industry.

Healthcare Industry is facing significant challenges of data integrity, data confidentiality, availability of data to authorized persons in 24 * 7 fashion, authentication of data user, ease of use, the excessive amount of time and money consumption in healthcare data processing, and data handling.

Thus, there is a necessity for developing a cloud-based clinical decision support system (CCDSS) that could be helpful in overcoming the time, efforts and cost of running its operations; and helpful in keeping and transferring the EMRs to the cloud where patients, an authorized person can easily access to the records irrespective of the geographical location, also, helpful to healthcare professional (Doctors, Nurses etc.) to make the best decision about the patient care. A CCDSS can be referred as an expert system, which can be used in imitating the behavior and knowledge of medical practitioners to generate advice and suggestions for patient health based on respective patient's health record analysis while validating all processes involved in decision making (Berner 2007).

4 Methodology

The methodology used includes a comprehensive and extensive literature review on the Protection and security of data in the health-care organizations and questionnaire-based quantitative surveys.

- A literature review was used to recognize the problems linked with the security of data and the benefits of creating health-care security policies and strategies.

- A questionnaires-based survey was conducted to examine and evaluate the policies for the security of data and compliance regulation in Health Organizations.
- Software as a Service (SaaS) cloud model was implemented as a Web-Based Service to ensure the provision of Software functionality with a Cloud-Based Health Management Software Operated Internally.
- The DSS was developed using a rule-based CDS System and Data Flow.

A rule-based DSS was developed using a Data Flow Modelling tool that facilitated the Cloud deployment of applications and in-House private cloud deployment to ensure the security of data, protection of data, and the regulatory compliance policies in the health organizations. In order to evaluate the proposed system, the User Acceptance Testing was carried out to determine how well the proposed CCDSS meets the requirements. The CCDSS was be built and employed at St. Williams Hospital, Lagos, Nigeria and the user acceptance performance by Timason Technology Nigeria Limited.

4.1 Cloud-Based Clinical Decision Support System Analysis

The analysis of the proposed CDSS deals with functionalities (expected by the healthcare experts and patients) ease of use, the security of data, and cost-effective operational solutions offered by the available CDSSs in the market. EHR is used as a DSS in the healthcare industry, which is capable in providing the solutions that will help the healthcare stakeholder in the decision making, improving patients' safety, strengthening revenue and cost optimization, better communication and understanding of business activities and in resolving operational, tactical and apex level problems through easy to handle ICT based tools.

4.2 CDSS Design and Architecture

The design and development of a CDSS depend upon the requirements and decisions of the healthcare stakeholders; and the layout of the organization (which includes procedures, process, hierarchy, and strategic planning). According to the case study of Penn CARE and the Hendry Ford Health system (Dubbs and Browning 2002), the layout of the organizational design plays a vital role in the development of an effective CDSS.

Further, the three main approaches most popular among the CDSS researchers and developers are heuristics, computational, and mathematical, and clinical algorithms approach (Jain 2015). They designed and built a CDSS using data analytics techniques, which involved data mining and its usage effectively in descriptive, diagnostic and predictive analytics. These data analytics and business intelligence techniques are having their application in cost-effective and qualitative healthcare solutions while avoiding harmful clinical events. The incorporation of data mining based automated and electronic systems helps the healthcare professionals through warning alerts, the planning, and executing their daily tasks of the management and suggestions in medication.

Further, a cloud-based CDSS architecture as an alternative of CDSS is proposed by Oh et al. (2015), which advocates about the cloud-based expert system in the clinical domain, making a computer system process like healthcare professional while interacting with patients. The functionalities of CDSS can be developed or implemented in a cloud-based environment for making decision and providing support to both the health-care professionals and the government. Figure 1 provides the workflow diagram of the proposed CDSS. As seen in Fig. 1, the design of a CDS should be able to provide the necessary information to the right users in the proper format through the appropriate channel at the appropriate time.

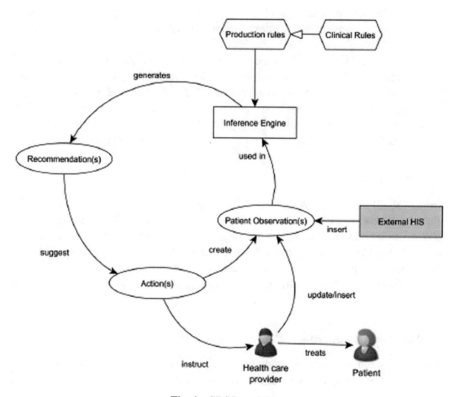

Fig. 1. CDSS workflow

4.3 Quantitative Survey

Adopting the EHR technique aggressively is a vital part of the planning of private owned healthcare organizations, whereas government healthcare organizations are slow. Hence, a survey was carried out to know the proper insight into the effectiveness of healthcare organizations while using IT solutions to deliver healthcare services. Also, 39 participants (healthcare professionals, healthcare stakeholders, IT personnel, Cloud Service Providers and Colleagues) participated in the survey. The questions of the survey were based on description of patient's health and care, privacy, and security of medical records, the communication gap between the healthcare providers and patients, and hospital infrastructures.

In this research, the integration of an existing EHR system with a CDSS tool can be viewed with the implementation of ACAFE CDS (Dinh et al. 2007). Acafe CDS is an open-source web-based tool, used for research of infectious and chronic diseases. The Acafe CDS can handle large volumes of data, which is imperfect, inadequate resources and time devoted to solving complex problems. The Acafe CDS covers the clinical workflow in a healthcare environment to handle the existing daily task. The Acafe will offer the guidelines for treatment of specific diseases and improve the efficiency and flexibility in the documentation of clinical resources. Thus, the implementation of Acafe CDS will deliver efficient and best practices.

The design of a CDSS with Acafe CDS can be accomplished if it is appropriately configured and installed with the minimum system requirements. The Acafe CDS is having portability with Microsoft .Net Framework 2.0. Microsoft .Net Framework also offers better security. In conclusion, the design and development of a cloud-based CDSS from a Doctor's perspective are that the system will be used to keep medical records secure and safe, also will available quickly in the event of an emergency.

4.4   Survey Results Finding

Dawson (2002) and Saunders et al. (2009) described that primary and secondary data collection are two kinds of data are used in research works. In this research work, the primary data is collected through the quantitative survey (online through survey monkey (2015)). The data is collected from 39 participants, who responded until the end of the survey; therefore, the margin of error in the survey results would be $\pm 16\%$. Whereas the secondary was collected from journals, articles, research books, and reliable website sources.

4.5   Data Analysis and Evaluation

The analysis of the data collected from the participants from the online survey is discussed following the survey questions and the screenshots captured from the analysis of the results.

Question 1: How would you describe your own health?

Figure 2(a) shows that only 26% of the participants have an excellent health status, while 53% of the participants have an excellent health status, and 21% of the participants have a good health status.

Hence these results confirm that for the participants to enjoy a robust and excellent health status, there is the need to advance the quality of healthcare service, and proper medical consideration or care is required.

Question 2: How satisfied are you with the communication you have with your doctor or other healthcare staff related to your health?

Figure 2(b) shows the analysis of communication existing between the participants and the healthcare staff. Figure 2(b) depicts that 25% of the participants were delighted with the communication with their doctor, 47% were satisfied, 25% were partly satisfied and 3% partly dissatisfied. Hence, for the participants to enjoy satisfaction of the delivery of

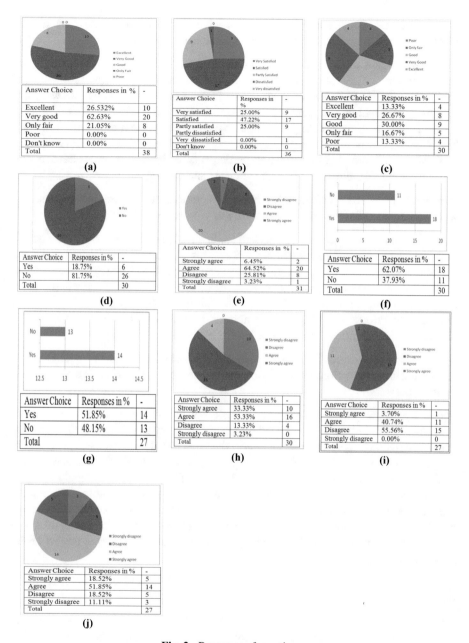

Fig. 2. Response of questions

healthcare services, there is the need for healthcare organizations to create an enabled environment where there is an interaction with the patients outside of the hospital walls by desiring to know about the wellness and health status of patients and a follow up on prescriptions and medications assigned to patients.

Question 3: Do you usually obtain follow-up information and care (test results, medicines, and care instructions) from your healthcare providers?

Figure 2(c) shows the analysis of the information feedback received from healthcare service providers. 13% of the participants had an excellent information follow-up and feedback, 27% had an excellent follow-up and feedback, 30% had good follow-up and feedback, 17% had only a good follow-up and feedback and 13% had a poor follow-up and feedback. Thus, for the participants to receive follow-up information from their healthcare service providers, there is the need for healthcare service providers to create a platform through the CDS system where participants (patients) can always have follow-up information regarding test results, medicines, and care instructions.

Question 4: Have you ever kept information from your healthcare provider because you were concerned about the privacy or security of your medical record?

Figure 2(d) shows the analysis of the lack of confidence in healthcare service providers in ensuring the privacy and security of medical records. 19% of the participants responded that they had kept information regarding the medical records while 81% were genuine in releasing their health information to the healthcare providers. Thus, for the participants to be given a proper healthcare service, it is essential for the patients or participants not to keep their information away from healthcare providers as this may hamper proper diagnosis, assessment, and treatment when the patient visits the healthcare provider.

Question 5: Do your healthcare providers have adequate policies and procedures to protect health information, or are there appropriate steps used to protect the privacy rights of patients?

Figure 2(e) shows the analysis of the provision of adequate policies and procedures to protect health information provided by healthcare providers. 6% of the participants strongly agree that there are adequate procedures provided by healthcare providers to protect health information, 65% of the participants agree to the motion, 26% of the participants disagree with the motion, while 3% of the participants strongly disagree with the motion. Thus, for the participants to strongly agree with the terms of providing adequate policies and procedures to protect health information, there is the need for healthcare service providers to ensure there are laid down or documented rules and regulations guiding the provision of adequate policies and procedures to protect health information.

Question 6: Does the government or organizations collect, use, store, or share health information?

Figure 2(f) shows the analysis of how health information of participants is being shared, collected, and used by government and public organizations. 62% of the participants can testify that their health information is being collected, shared, and used without prior knowledge of the patients while 38% of the participants decline the usage of their health information by the government and public organizations. Thus, in order for the health information of the participants to be used in the public domain, there is the need to notify the patients of the usage of their health information and data to avoid infringement of privacy and breach in data security.

Question 7: Does the government or organizations take appropriate steps to comply with the requirements of HIPAA and other related healthcare regulations?

Figure 2(g) shows the analysis of how the government or organizations take appropriate steps to comply with the requirements of HIPAA and other related healthcare regulations in regards to the protection of data and information. 52% of the participants responded that there is compliance with the requirements of HIPAA and other related healthcare regulations while 48% declined the compliance.

Question 8: I want my healthcare providers to use an EMR to store and manage my health information despite any concerns I might have about privacy and security.

Figure 2(h) shows the analysis of how health information of participants is managed and stored in an electronic medical record system despite the increase in breach in privacy and security of data. 33% of the participants strongly agree for their health information record to be stored on an EMR system; 53% of the participants agree to the opinion, while 13% of the participants disagree with the opinion of having their medical records stored electronically. Thus, in order to transit from the paper-based health record to the EHR, there is the need for the healthcare service providers to enlighten the public and patients of the advantages of transiting to an electronic medical record and how their records will be safeguarded with the best security through encryption measures to enhance full protection and security of data.

Question 9: Healthcare providers have measures in place that provide a reasonable level of protection for EMRs today.

This can be identified and summarized by the different opinions of the participants from the online survey. The measures implemented to provide a reasonable level of protection for electronic medical records of patients can be analyzed from the responses from the participants. Figure 2(i) shows the analysis of how the measures are implemented to ensure a reasonable level of protection for electronic medical records is provided. 4% of the participants strongly agree to the measures provided; 41% agree to the measures provided; 55% of the participants disagree with the measures provided by the healthcare providers. Hence, to provide a reasonable level of protection for electronic medical records, there is the need to set up user policies with authentication methods to checkmate the users having access to the medical records and also providing top-level infrastructural security to secure all the systems and resources to wade off hackers and intruders.

Question 10: I want my healthcare providers to use a computer to share my medical record with other providers treating me despite any concerns I might have about privacy and security.

The collection, sharing, and usage of the health information of the participants can be analyzed from the responses from the participants as also seen in Fig. 2(j), shows the analysis of how the participants want their healthcare providers to use a computer to share their medical record with other providers treating them despite any concerns they might have about privacy and security. 18% of the participants strongly agree with the opinion and want their healthcare providers to use a computer to share their medical record with other providers to treat them, 52% of the participants agree with the opinion, 18% of the participants disagree with the opinion, and 11% strongly disagree with the opinion. Thus, in order to share medical records with other service providers, there is the need to set up user policies with authentication methods to checkmate the users having access to the medical records and also providing top-level infrastructural security to secure all the systems and resources to wade off hackers and intruders and also the need to notify the patients of the usage of their health information and data to avoid infringement of privacy and breach in data security.

Some of the participants identified the privacy, security, and sharing of data as threats to the adoption of an electronic health record system. Thus, the sharing and usage of information by government and healthcare organizations was also identified as an infringement to privacy and protection of data. Hence, there is a need to enhance regulatory policies to ensure adherence to the regulated policies regarding protection and security of patients' data. In conclusion, the protection and security of patients' data will be adequately implemented by healthcare providers if the regulatory compliances are adhered to and the safety of patients is considered.

5   CDSS Implementation

Based on the findings of the quantitative survey, the provision of a framework or guideline that will support the processes involved in decision making during the adoption of a cloud-based healthcare environment is attempted through the development of a CDSS with the installation of Acafe Open source software. As a part of the proposed CDSS design, Acafe CDS is installed on a Windows server with additional utilities, which includes the Microsoft.Net Framework version 2.0, Microsoft SQL Server 2005 Express Edition, Crystal Reports Version

The Acafe CDS after installation offers the option to either sign in as an existing user or sign up as a new user. The legitimate clinical staff must be registered and logged in to the Acafe system before using any features. The Acafe CDS is further started and implemented with a welcome screen showing with the system features layout. Clinical staff can register according to their legitimate role as the Hospital Director, Doctor, Nurse or General staff. The access level for each of these positions and designations can be fully described in Table 1.

In conclusion, the registration of clinical staff as Acafe CDS users allows for patient's details to be created into the CDS system and database. Upon successful login and authentication, the user is directed to the Acafe patient treatment file page for further assessment and treatment. The registration of the patient's data can be implemented by all the registered clinical staff irrespective of the privileges. Once the patient file is created, it is usually stored as a patient record file, and once a patient passes through the Acafe CDS, a new treatment file is created for the patient.

6   System Evaluation

The clear establishment and the successes measured will serve as criteria in the steps in evaluating a system. The criteria for success should be more comprehensive than just the metrics used in IT. Although the IT metrics are necessary and very important, there

Table 1. Acafe CDS User Level Access (Acafe Guidebook 2008)

Designation	Access level
Hospital Director	Can edit all values and fields, including the Suggested Management Plan and Medication databases or any other details which can be changed. Can create and print reports and have full complete access
Doctor	Can create and print reports and have full complete access but cannot change the Suggested Management Plan or Medication database, or Observation Chart details
Nurse	Can only update the observation chart and read-only all other information. They cannot make changes or updates to any information other than the observation chart
General	Can only view a summary of information and reports

are crucial practical performance signals which will support organizational goals and visions to elevate the advantages of the CDS system. Thus, the usage of quantifiable and measurable methods will make the CDS system to be reliable by providing different solutions. The achievable success factors in the adoption of a CDS in a cloud-based healthcare environment needs to be displayed in quantified methods or ways. Some of the quantifiable set of facts for clinical organizations can be derived from the quality of care provided, the processes involved in healthcare improvement and patient satisfaction.

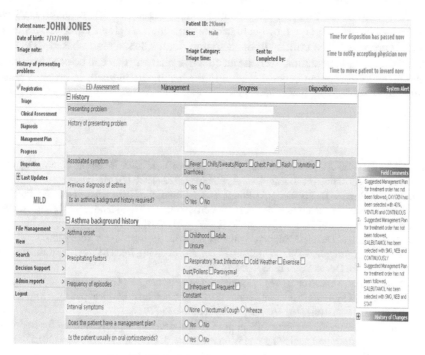

Fig. 3. Clinical Assessment Form (Original)

The system alert feature offered by the Acafe CDS to inform clinical staff of any necessary and immediate guidelines to be taken for specific treatments. An example a patient known as "*Sample Patient Data*" with specific conditions related to Asthma with a Severe Asthma condition.

Furthermore, the clinical forms section of the Acafe CDS comprises the complete integrated treatment and hospital management information such as the Triage, Registration, Management Plans, the Progress Note and the documentation of disposition. All the listed features are made available as forms in the main body of the clinical forms. A screenshot of one of the forms used for clinical assessment for patient "John Jones" is displayed in Fig. 3.

7 Conclusions

The adoption of the technology will help to reduce the cost of operations, providing better interoperability with other providers of healthcare and providing an efficient management system for the health care industry. This research paper advocates the usage of ICT based CDSS in healthcare. The key to successful adoption of cloud computing in a healthcare environment depends upon the healthcare professional/stakeholder and organization layout, which enforce strategy of adopting ICT based healthcare. This research work also highlights that the acceptance of ICT based CDSS are aggressively accepted by private healthcare organization as compare to public healthcare organization of Nigeria. The survey related to IT-based CDSS adoption advocates the protection and security of patients' data must be adequately implemented by healthcare service providers while accomplishing the regulatory compliances. In this research DSS was analyzed with the perspective of healthcare organization. A cloud-based CDSS architecture is proposed. The utilities of open-source Acafe CDS, a cloud-based CDSS was identified and implementation to create an effective platform that can be used in creating an automated information support system in the healthcare industry in order to reduce the cost of operation and maintenance.

References

Arora, R., Parashar, A., Transforming, C.C.I.: Secure user data in cloud computing using encryption algorithms. Int. J. Eng. Res. Appl. 3(4), 1922–1926 (2013)

Belard, A., Buchman, T., Forsberg, J., Potter, B.K., Dente, C.J., Kirk, A., Elster, E.: Precision diagnosis: a view of the clinical decision support systems (CDSS) landscape through the lens of critical care. J. Clin. Monit. Comput. 31(2), 261–271 (2017)

Berner, E.S.: Clinical Decision Support Systems, vol. 233. Springer, New York (2007)

Bonczek, R.H., Holsapple, C.W., Whinston, A.B.: The evolving roles of models in the decision support systems. Decis. Sci. 11, 337–356 (1980)

Borab, Z.M., Lanni, M.A., Tecce, M.G., Pannucci, C.J., Fischer, J.P.: Use of computerized clinical decision support systems to prevent venous thromboembolism in surgical patients: a systematic review and meta-analysis. JAMA Surg. 152(7), 638–645 (2017)

Castaneda, C., Nalley, K., Mannion, C., et al.: Clinical decision support systems for improving diagnostic accuracy and achieving precision medicine. J. Clin. Bioinform. 5, 4 (2015)

Dawson, C.: Practical Research Methods. How to Books Ltd., 20–79, Oxford UK (2002)

Dinh, M., Chu, M., Kwok, R., Taylor, B., Dinh, D.: The asthma clinical assessment form and electronic decision support project (ACAFE): collaborative innovation and development. In: Proceedings of the 12th World Congress on Health (Medical) Informatics; Building Sustainable Health Systems, Medinfo 2007, p. 2506. IOS Press (2007)

Dubbs, N.L., Browning, S.L.: Organizational design consistency: the PennCARE and Henry Ford health system experiences/practitioner application. J. Healthcare Manag. **47**(5), 307 (2002)

Edwards, J.: Cutting through the fog of cloud security. Computerworld, Framingham **43**(8) (2009)

Fernandez-Cardenosa, G., de la Torre-Diez, I., Lopez-Coronado, M., Rodrigues, J.J.: Analysis of cloud-based solutions on EHRs systems in different scenarios. J. Med. Syst. **36**(6), 3777–3782 (2012)

Gibson, M., Arnott, D., Carlsson, S.: Evaluating the intangible benefits of business intelligence: review & research agenda. In: Proceeding of the Decision Support in an Uncertain and Complex World: The IFIP TC8/WG8.3 International Conference, Prato, Italy, pp. 295–305 (2004)

He, C., Jin, X., Zhao, Z., Xiang, T.: A cloud computing solution for hospital information system. In: Proceedings of 2010 IEEE International Conference on Intelligent Computing and Intelligent Systems (ICIS), Xiamen, China, 29–31 October 2010, pp. 517–520 (2010)

Heise, C., Gallo, T., Curry, S., Woosley, R.: The design of an electronic medical record based clinical decision support system in a large healthcare system to improve antibiotic safety using a validated risk score. In: D43. Critical Care: The Rising-Quality Improvement and Implementation of Best Practice, pp. A6826–A6826. American Thoracic Society (2018)

Hussain, M., Khattak, A.M., Khan, W., Fatima, I., Amin, M.B., Pervez, Z.: Cloud-based smart CDSS for chronic diseases. Health Technol. **3**(2), 153–175 (2013)

Hsieh, J.C., Hsu, M.W.: A cloud computing based 12-lead ECG telemedicine service. BMC Med. Inform. Decis. Mak. **12**, 77 (2012)

IBM: European Union Consortium Launches Advanced Cloud Computing Project with Hospital and Smart Power Grid Provider, IBM Press Room, November 2010. https://www.webwire.com/ViewPressRel.asp?aId=127225. Accessed 15 May 2018

Jackson, L.: Leveraging Business Intelligence for Healthcare Providers, Oracle White paper, Oracle Corporation World Headquarters 500 Oracle Parkway Redwood Shores, CA 94065, USA (2009)

Jacob, V., Thota, A.B., Chattopadhyay, S.K., Njie, G.J., Proia, K.K., Hopkins, D.P., Ross, M.N., Pronk, N.P., Clymer, J.M.: Cost and economic benefit of clinical decision support systems for cardiovascular disease prevention: a community guide systematic review. J. Am. Med. Inform. Assoc. **24**(3), 669–676 (2017)

Jain, K.K.: Personalized therapy of cancer. In: Textbook of Personalized Medicine, pp. 199–381. Humana Press, New York (2015)

Kanagaraj, G., Sumathi, A.C.: Proposal of an open-source cloud computing system for exchanging medical images of a hospital information system. In: Proceedings of 2011 3rd International Conference on Trends in Information Sciences and Computing (TISC), Chennai, India, 8–9 December 2011, pp. 144–149 (2011)

Kohn, L.T., Corrigan, J.M., Donaldson, M.S.: Errors in health care: a leading cause of death and injury. In: To Err is Human: Building a Safer Health System. National Academies Press (US) (2000)

Korea IT Times: Telstra Plans Launch of E-Health Cloud Services, Tip of the Iceberg for Opportunity (2010). http://www.koreaittimes.com/news/articleView.html?idxno=9826. Accessed 15 May 2018

Koufi, V., Malamateniou, F., Vassilacopoulos, G.: Ubiquitous access to cloud emergency medical services. In: Proceedings of the 2010 10th IEEE International Conference on Information Technology and Applications in Biomedicine (ITAB), Corfu, Greece, 3–5 November 2010. IEEE, New York (2010)

Guo, L., Chen, F., Chen, L., Tang, X.: The building of cloud computing environment for e-health. In: 2010 International Conference on E-Health Networking, Digital Ecosystems and Technologies (EDT), vol. 1, pp. 89–92. IEEE, April 2010

Liang, H., Xue, T.Y., Berger, B.A.: Web-based intervention support system for health promotion. Decis. Support Syst. **42**, 435–449 (2006)

Makary, M.A., Daniel, M.: Medical error-the third leading cause of death in the US. BMJ: Br. Med. J. (Online) **353** (2016)

Oh, S., Cha, J., Ji, M., Kang, H., Kim, S., Heo, E., Han, J.S., Kang, H., Chae, H., Hwang, H., Yoo, S.: Architecture design of healthcare software-as-a-service platform for cloud-based clinical decision support service. Healthcare Inform. Res. **21**(2), 102–110 (2015)

Osheroff, J.A., Pifer, E.A., Teich, J.M., Sittig, D.F., Jenders, R.A.: Improving Outcomes with Clinical Decision Support: An Implementer's Guide. Health Information and Management Systems Society, Chicago (2005)

Osheroff, J.A., Teich, J.M., Middleton, B., Steen, E.B., Wright, A., Detmer, D.E.: A roadmap for national action on clinical decision support. J. Am. Med. Inform. Assoc. **14**, 141–145 (2007)

Anooj, P.K.: Clinical decision support system: Risk level prediction of heart disease using weighted fuzzy rules. J. King Saud Univ.-Comput. Inf. Sci. **24**(1), 27–40 (2012)

Perrealult, L.E., Metzger, J.B.: A pragmatic frame work for understanding clinical decision support system (1999)

Peter, M., Tim, G.: The NIST definition of Cloud Computing, Version 15. Information Technology Laboratory, 10 July 2009

Rao, G.S.V.R.K., Sundararaman, K., Parthasarathi, J.: Dhatri: a pervasive cloud initiative for primary healthcare services. In: Proceedings of the 2010 14th International Conference on Intelligence in Next Generation Networks (ICIN); The 14th IEEE International Conference on Intelligence in Next Generation Networks (ICIN), Berlin, Germany, 11–14 October 2010. IEEE, New York (2010)

Rawson, T.M., Moore, L.S.P., Hernandez, B., Charani, E., Castro-Sanchez, E., Herrero, P., Hayhoe, B., Hope, W., Georgiou, P., Holmes, A.H.: A systematic review of clinical decision support systems for antimicrobial management: are we failing to investigate these interventions appropriately? Clin. Microbiol. Infect. **23**(8), 524–532 (2017)

Rolim, C.O., Koch, F.L., Westphall, C.B., Werner, J., Fracalossi, A., Salvador, G.S.: A cloud computing solution for patient's data collection in health care institutions. In: Proceedings of the 2nd International Conference on eHealth, Telemedicine, and Social Medicine; 10–16 February 2010. IEEE, New York (2010)

Saunders, M.: Research Methods for Business Students. Prentice Hall, Harlow (2009)

Samuel, O.W., Omisore, M.O., Ojokoh, B.A., Atajeromavwo, E.J.: Enhanced cloud based model for healthcare delivery organizations in developing countries. Int. J. Comput. Appl. (0975–8887) **74**(2) (2013)

Sim, L.L.W., Ban, K.H.K., Tan, T.W., Sethi, S.K., Loh, T.P.: Development of a clinical decision support system for diabetes care: a pilot study. PLoS ONE **12**(2), e0173021 (2017)

Strukhoff, R., O'Gara, M., Moon, N., Romanski, P., White, E.: Cloud Expo: Healthcare Clients Adopt Electronic Health Records with Cloud-Based Services, SYS-CON Media, Inc. (2009)

Survey monkey: Development of a Decision Support System for Implementing Security Strategies for Data Protection and Regulatory Compliance in a Cloud Based Health-Care Environment (2015). https://www.surveymonkey.com/r/3T3YMB6. Accessed 21 Aug 2015

Turban, E., Aronson: Decision Support Systems and Intelligent Systems, 6th edn. Prentice Hall Inc., Upper Saddle River (2001)

Wickramasinghe, N., Geisler, E.: Encyclopedia of Health Care Information Systems. IGI Global (2008)

Zafar Chaudry, M.D., Koehler, M.: Electronic Clinical Decision Support Systems Should be Integral to Any Healthcare System. Gartner Inc. (2013). https://www2.health.vic.gov.au/Api/download media/%7BAB4ADF7B-3A74-4FAE-ACC7-C0CB4FC4B0C6%7D. Accessed 15 May 2018

Performance Measurement of RPL Protocol Using Modified MRHOF in IoT Network

Bhawana Sharma$^{(\boxtimes)}$, Jyoti Gajrani, and Vinesh Jain

Engineering College, Ajmer, Ajmer, India
123.bhawnasharma@gmail.com, {jyotigajrani,
vineshjain}@ecajmer.ac.in

Abstract. Internet of Things (IoT) networks are the type of Low Power and Lossy Networks. Routing protocol for Low Power and Lossy Networks (RPL) is the standard routing protocol for IoT network. RPL uses the Destination Oriented Directed Acyclic Graph (DODAG) to get route from source to destination. DODAG formation is based on different objective functions which optimize various routing metrics to get reliable and more optimized path. Objective function Zero (OF0) which is based on minimum hop count method and Minimum Rank with Hysteresis Objective Function (MRHOF) which optimize one routing metric i.e. Expected Transmission Count (ETX) or energy are standard objective functions defined by Internet Engineering Task Force (IETF). Different routing metrics are combined so far as use of single routing metric by RPL doesn't fulfil overall routing performance requirements. This research proposes a new modified MRHOF objective function which is modification in standard objective function MRHOF. Modified MRHOF objective function uses and optimizes three routing metrics i.e. ETX, Energy and Delay. Implementation of proposed work is done on Cooja Simulator on Contiki OS. Performance evaluation of proposed objective function is done on the basis of these parameters- Packet Delivery Ratio (PDR), Average Power Consumption and Throughput and results show that proposed objective function performs better than other objective functions -0F0 and MRHOF in terms of these parameters.

Keywords: RPL · Objective function · MRHOF · IoT · ETX · Delay

1 Introduction

In constrained Internet of Things (IoT) network, RPL (Routing protocol for Low Power and Lossy Netwoks) [1] is the emerging standardized protocol used for routing. Routing is crucial process as it is used to make connection between devices in IoT network to send or receive data. Performance of constrained IoT network depends on the routing protocol used by it [2]. RPL is constructed such that it can deal with low power and lossy nature of IoT devices. RPL makes use of DODAG (Destination Oriented Directed Acyclic Graph) to find path from source to destination by using its objective function. This objective function is the core of RPL to construct path. Parent is selected by RPL in

M. Tripathi and S. Upadhyaya (Eds.): ICDLAIR 2019, LNNS 175, pp. 235–245, 2021.
https://doi.org/10.1007/978-3-030-67187-7_25

DODAG construction by using objective function on the basis of routing metrics used by it [3]. Objective Function Zero (OF0) [4] and Minimum Rank with Hysteresis Objective Function (MRHOF) [5] are the objective functions standardized by Internet Engineering Task Force (IETF).

In this paper we modify the standard objective function MRHOF which uses single metric and introduce a new objective function which makes use of three routing metrics-Expected Transmission Count (ETX), Energy and Delay instead of single routing metric to obtain path from source to destination. In literature survey we studied a modified version of MRHOF which is proposed earlier in [6] and this objective function uses two routing metrics i.e. ETX and Energy but there is lack of an important parameter- Delay to be used as routing metric along with ETX and Energy. Performance of RPL has great impact of Delay on it. Delay is the time needed to reach the data packet from source to destination. Parameters like throughput, packet loss are majorly affected by delay. To increase throughput of an IoT application delay should be minimized and packet loss also decreases by the minimization of delay. Thus our main goal to use this approach is to get even more optimized and reliable routes by the RPL and to enhance the overall routing performance requirements for IoT applications. Path obtained by using proposed objective function is optimized in terms of three parameters i.e. ETX, Energy and Delay.

Rest of the sections of this paper are arranged as follows. Section 2 discusses the background of RPL protocol. Section 3 describes the literature survey of related work done in past few years. Section 4 discusses the problem statement for the research paper and solution is proposed in Sect. 5. Performance evaluation of proposed objective function and result analysis are discussed in Sect. 6. At last, conclusion of research paper and future scope are described in Sect. 7.

2 Background

RPL Overview

RPL (Routing protocol for Low Power and Lossy Netwoks) [1] is the standard routing protocol for IPv6 low power and lossy networks defined by IETF. RPL builds DODAG (Destination Oriented Directed Acyclic Graph) to make connection between nodes. DODAG root is the border router for DODAG. RPL builds routes using an objective function which optimize various routing metrics to get optimized and reliable path from source to destination. Rank of a node is calculated by using this objective function to form DODAG [1]. IETF has defined various routing metrics which are used by objective function to build optimized routes in low power and lossy networks such as node energy, hop count, latency, ETX etc. [7]. RPL uses control messages to form DODAG which are [8]:

1. DODAG Information Object (DIO)
2. DODAG Information Solicitation (DIS)
3. Destination Advertisement Object (DAO)
4. Destination Advertisement Object Acknowledgement (DAO-ACK) (Fig. 1).

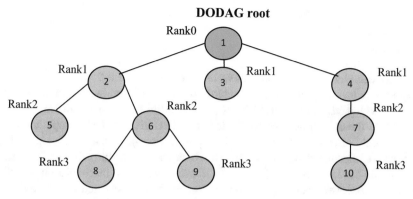

Fig. 1. DODAG graph by RPL

DODAG Construction by RPL

DODAG construction includes two basic operations which are (a) DODAG root broadcasts DIO control message to other nodes to construct route in the downward direction. (b) Client nodes send DAO control message to the DODAG root to construct route in the upward direction [8].

A new DODAG is build by broadcasting a DIO message by the DODAG root. DIO message allows receiving nodes to calculate their rank on the basis of objective function used by the RPL. When a node receives DIO message by the sender node then it performs three operations (1) adds the address of DIO control message sender to its parent list, (2) calculates its rank on the basis of objective function used by the RPL (node selects the DIO message sender node its preferred parent if the rank of DIO message sender is less than the rank of DIO message receiving node) and (3) sends DIO message to other nodes with updated rank information. Node selects that preferred parent node as its most preferred parent whose rank is minimum. When the DODAG is formed each client node has default upward path to the DODAG root [8].

3   Related Work

In [4] authors have defined basic objective function i.e. Objective Function Zero (OF0) which is based on minimum hop count criteria. In DODAG formation preferred parent is selected by a node which is reachable from it and which also has minimum rank among candidate parent nodes. Rank of a node n is calculated as follows in Eq. (1):

$$R(n) = R(p) + RankIncrease \tag{1}$$

Where,
$R(n)$ = new calculated rank of node n
$R(p)$ = rank of preferred parent p
RankIncrease is defined as in Eq. (2):

$$RankIncrease = (Rf * Sp + Sr) * Min\_Hop\_RankIncrease \tag{2}$$

Where,

Rf = rank_factor

Sp = step_of_rank

Sr = stretch_of_rank

Min_Hop_RankIncrease = a constant variable (default value is 256).

In [5] authors has defined another standard objective function i.e. Minimum Rank with Hysteresis Objective Function (MRHOF) which optimizes various additive routing metrics. ETX is the default routing metric used by MRHOF. Path cost computation is done on the basis of routing metric used for optimization in DODAG formation. In preferred parent selection that route is chosen which has minimum path cost in terms of routing metric used for optimization. MRHOF uses 'hysteresis' mechanism in which MRHOF firstly discovers path with minimum cost in terms of selected routing metric and then change current path to new discovered path only if its path cost is lower than that of current path by a given threshold.

In [6] authors proposed a new objective function which is a modification in MRHOF. This new objective function uses two routing metrics for path cost calculation i.e. ETX and energy both in place of one routing metric. In path cost calculation process both energy metric and ETX metric value are received in DIO message and are added for new path cost calculation. Performance of proposed objective function is analysed on the basis of different parameters such as average power consumption, Packet Delivery ratio (PDR), ETX, churn.

In [9] authors proposed a new objective function i.e. OF-EC (objective function based combined metric using fuzzy logic method). In this approach two routing metrics EC (Energy Consumption) and ETX are combined by using fuzzy logic method. Performance of proposed objective function OF-EC is analysed in terms of various parameters like PDR, energy consumption, control traffic overhead by using Cooja simulator.

In [10] a new objective function called as LB-OF (load balanced objective function) is introduced that balances the traffic load of overburdened node to avoid bottleneck problems. This approach is used to improve lifetime of the node. In simulation process performance of proposed objective function is analyzed on the basis of different parameters such as number of children of a node, power consumption, PDR.

In paper [11] authors describe the impact of objective functions on the basic properties of IoT networks. In this paper three objective functions - OF0, MRHOF and Fuzzy Logic Objective Function (OFFL) are compared on the basis of three properties i.e. power consumption, performance and reliability. Simulation is done using Cooja simulator. Results show that power consumption, reliability and performance are affected by objective functions upto 71.2%, 77.5% and 25.15% respectively.

In paper [12] authors proposed a new objective function ERAOF (Energy Efficient and Path Reliability Aware Objective Function). This objective function considers two routing metrics- Energy Consumed (EC) and ETX to get energy efficient and reliable routes by the RPL. In this approach quality of whole route r i.e. Q(r) is determined in terms of EC and ETX and that route is chosen by the RPL among available routes that has the lowest value of Q(r).

In paper [13] performance of RPL is evaluated using two objective functions – OF0 and MRHOF on the basis of parameters like PDR and Power Consumption.

In paper [14] performance of RPL is analysed by using OF0 and MRHOF objective functions in terms of parameters like convergence time, average churn, average power consumption etc. using Cooja simulator.

4 Problem Statement

RPL is designed such that it can use different objective functions and these objective functions can optimize different routing metrics based on routing performance requirements. RPL can cope with different topologies and different type of IoT applications because of its flexibility.

In past few years various research studies proposed different objective functions to deal with different routing performance requirements of IoT applications. Standard MRHOF uses single routing metric i.e. ETX or energy to calculate rank of a node to select preferred parent in DODAG formation. Route obtained by using standard MRHOF is optimized in terms of single routing metric only.

Use of single routing metric is not good enough to satisfy overall routing performance requirements of IoT applications. As use of one routing metric by objective function can optimize the route in terms of single routing metric but another parameters remain unconsidered to meet overall routing performance requirements. So for the purpose of improving overall routing performance requirements different new objective functions are proposed which combine various routing metrics.

A modified version of MRHOF which is proposed earlier in [6] uses two routing metrics i.e. ETX and energy both to form DODAG and it can optimize route in terms of ETX and energy both but another important parameter i.e. Delay is still not taken into account to get more optimized and favourable route. Delay is the time taken to transmit the packet from source to destination.

Our aim is to propose a new objective function which is a modification in standard objective function MRHOF and this new objective function uses three routing metrics i.e. ETX, energy and delay. Delay has significant role to improve routing performance requirement because delay affects many parameters like packet loss, throughput etc. So delay should be considered along with ETX and energy by MRHOF objective function of RPL to obtain even more reliable and optimized path from source to destination during DODAG formation.

5 Proposed Work

RPL can use different objective functions which can optimize more than one metric to deal with different routing performance requirements. This paper proposed a new objective function which is a modified version of standard objective function MRHOF. Standard MRHOF optimize one routing metric but proposed version of MRHOF able to get routes which are optimized in terms of three routing metrics – ETX, energy and delay.

In standard MRHOF only one chosen metric i.e. energy or ETX is advertised through DIO message send by sender node. Path cost calculation is done by adding the value of energy or ETX metric. But in our proposed approach three routing metrics i.e. ETX, energy and delay are used for path cost calculation. Values of these three routing metrics are advertised in DIO message and are added to compute path cost. Path selected by RPL by using proposed objective function is the path which has minimum value of ETX, energy and delay combinely.

Path cost is computed by a node j when it receives DIO control message from node k as follows:

$$Path\text{-}Cost = ETX + Energy + Delay \tag{3}$$

Here in Eq. (3),

ETX is the Expected Transmission Count for the link between node j and node k.

Energy is the Energy Consumption by node j.

Delay is the Delay between node j and node k.

During DODAG construction, preferred parent for node j among candidate parents which send DIO messages to node j is selected which has minimum value of path cost.

Algorithm for preferred parent selection to form DODAG:

Input : sender, DIO, CandidateParentList

begin

Current_node ← DIO;

if CandidateParentList = NULL then

    CandidateParentList ← sender;

     PathCost = ETX + Energy + Delay;

end if

Mulicast the updated DIO to other nodes;

else

NewPathCost = ETX + Energy + Delay;

    if NewPathCost < PathCost then

      set DIO message sender as preferred parent;

    end if

    Mulicast the updated DIO to other nodes;

    else

    CandidateParentList ← sender;

     Mulicast the updated DIO to other nodes;

6 Performance Evaluation

Performance evaluation of proposed objective function is done using Cooja simulator on Contiki OS. In experimental setup on Cooja performance of proposed objective function is evaluated for three different values of Rx which are - 20%, 40% and 80%. In performance evaluation set up three parameters are measured – Packet Delivery Ratio, Average Power Consumption and Throughput for new objective function proposed in this research paper. Proposed objective function is compared with two standard objective functions - OF0 and MRHOF in terms of these parameters. In simulation, experiments are done for 15 and 30 number of nodes in network. Simulation parameters are shown in Table 1.

Table 1. Simulation parameters

Parameter name	Value
Radio medium	Unit Disk Graph Medium (UDGM)
Mote type	T-Mote sky
Mote startup delay	1000 ms
Mote positioning	Random positioning
Transmission range	50 m
Interference range	100 m
Transmission success ratio (Tx)	75%
Reception success ratio (Rx)	20%, 40% and 80%
Network size	15 & 30 no. of nodes

Figure 2 shows the Packet Delivery Ratio (PDR) obtained for network size of 15 nodes for proposed objective function. PDR is calculated also for OF0 and MRHOF to compare it with proposed objective function. PDR is measured for three values of Rx- 20%, 40% and 80%. This figure shows the better value of PDR for proposed objective function than that of OF0 and MRHOF for all three values of Rx. As shown in results PDR raises with the value of Rx. Maximum value of PDR is obtained when proposed objective function is used with 80% value of Rx.

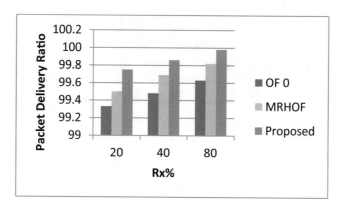

Fig. 2. PDR (Packet Delivery ratio) for 15 nodes

Figure 3 represents the average power consumption measured for 15 nodes in network. This figure depicts that proposed objective function performs better in terms of average power consumption than OF0 and MRHOF when Rx values are 20% and 40%. But we get higher value of average power consumption for proposed objective function than MRHOF when Rx value rises to 80%. This figure also shows that average power consumption drops when Rx is raised.

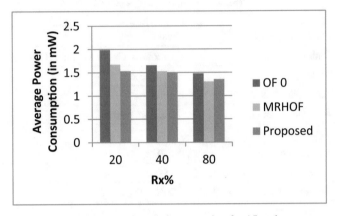

Fig. 3. Average Power Consumption for 15 nodes

Figure 4 exposes the throughput calculated for network size of 15 nodes. As shown in this figure higher throughput is obtained for proposed objective function than two other objective functions i.e. OF0 and MRHOF for 20%, 40% and 80% values of Rx. This figure depicts that throughput increases with the value of Rx. Highest value of throughput for the network is achieved when proposed objective function is used with the 80% value of Rx for 15 number of nodes in the network.

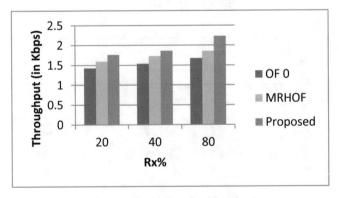

Fig. 4. Throughput for 15 nodes

Figure 5, 6 and 7 exposes the Packet Delivery Ratio (PDR), average power consumption and throughput respectively for the network size of 30 nodes. As shown in Fig. 5, 6 and 7 proposed objective function performs better than OF0 and MRHOF for all three values of Rx.

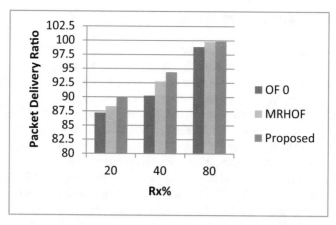

Fig. 5. PDR (Packet Delivery Ratio) for 30 nodes

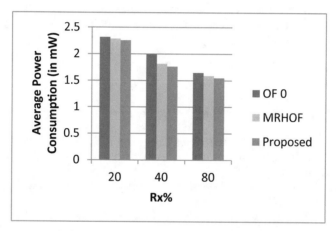

Fig. 6. Average Power Consumption for 30 nodes

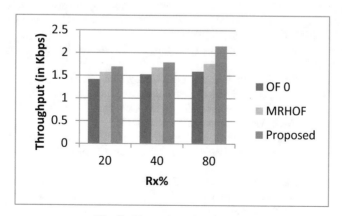

Fig. 7. Throughput for 30 nodes

7 Conclusion and Future Work

In this research paper we introduced a new reliable and energy efficient objective function of RPL which is also used to enhance the throughput in low power and lossy network at the same time. This proposed objective function is based on the alteration in standardized MRHOF as new objective function optimizes three routing metrics in place of one routing metric. The three routing metrics optimizes by this new objective function to find more optimized route are- ETX, Energy and Delay. We use Cooja simulator to evaluate performance of proposed objective functions by using three parameters- Packet Delivery Ratio (PDR), throughput and average power consumption. In results we noticed that proposed objective function provide better PDR, throughput and average power consumption than OF0 and MRHOF in most of the cases experimented in simulation.

For future work, performance of proposed objective function is examined by using various other parameters and to examine the behaviour of this new objective function for different applications of IoT with different network setup.

In addition for future work, proposed objective function can optimized various other routing metrics which is based on the performance requirements that are essential to be fulfilled for an IoT application.

References

1. Winter, T., et al.: RPL: IPv6 Routing Protocol for Low-Power and Lossy Networks. IETF (2012)
2. Umamaheswari, S., Negi, A.: Internet of Things and RPL routing protocol: a study and evaluation. In: 2017 International Conference on Computer Communication and Informatics (ICCCI), Coimbatore, pp. 1–7 (2017)
3. Bhandari, K.S., et al.: CoAR: congestion-aware routing protocol for low power and lossy networks for IoT applications. Sensors (2018)
4. Thubert, P.: Objective Function Zero for the Routing Protocol for Low-Power and Lossy Networks (RPL). IETF (2012)
5. Gnawali, O., Levis, P.: The Minimum Rank with Hysteresis Objective Function. IETF (2012)
6. Abied Hatem, J., Safa, H., El-Hajj, W.: Enhancing routing protocol for low power and lossy networks. In: 13th International Wireless Communications and Mobile Computing Conference (IWCMC), Valencia, pp. 753–758 (2017)
7. Vasseur, J.P., et al.: Routing Metrics Used for Path Calculation in Low-Power and Lossy Networks. IETF (2012)
8. Gaddour, O., Koubaa, A.: RPL in a nutshell: a survey. Comput. Netw. (2012)
9. Lamaazi, H., Benamar, H.: RPL enhancement using a new objective function based on combined metrics. In: 2017 13th International Wireless Communications and Mobile Computing Conference (IWCMC), Valencia, pp. 1459–1464 (2017)
10. Qasem, M., et al.: A new efficient objective function for routing in Internet of Things paradigm. In: 2016 IEEE Conference on Standards for Communications and Networking (CSCN), Berlin, pp. 1–6 (2016)
11. Safaei, B., Monazzah, A.M.H., Shahroodi, T., Ejlali, A.: Objective function: a key contributor in Internet of Things primitive properties. In: 2018 Real-Time and Embedded Systems and Technologies (RTEST), Tehran, pp. 39–46 (2018)

12. Sousa, N., et al.: ERAOF: a new RPL protocol objective function for Internet of Things applications. In: 2017 2nd International Multidisciplinary Conference on Computer and Energy Science (SpliTech), Split, pp. 1–5 (2017)
13. Abuein, Q., et al.: Performance evaluation of routing protocol (RPL) for Internet of Things. Int. J. Adv. Comput. Sci. Appl. **7** (2016). https://doi.org/10.14569/ijacsa.2016.070703
14. Mardini, W., et al.: Comprehensive performance analysis of RPL objective functions in IoT networks. Int. J. Commun. Netw. Inf. Secur. (IJCNIS) **9**(3) (2017)

E-MOC: An Efficient Secret Sharing Model for Multimedia on Cloud

Rama Krishna Koppanati[✉], Krishan Kumar, and Saad Qamar

National institute of Technology, Uttarakhand, Srinagar, India
{krishna.cse16,kkberwal,saad.cse16}@nituk.ac.in

Abstract. In this multimedia era, the security of the multimedia content over Cloud becomes a crucial research area. There are several modes of communication available for transmitting the information over the Cloud from one place to another place around the globe. Such communication networks are open to all, where anyone can access very easily. To abate the crimes, there are so many techniques evolved, such as RSA, DES, Elgamal, etc. However, these techniques seem limited to provide security to the user's data. This work highlights a novel multimedia encryption model over Cloud, where the key is generated dynamically without involving the user by making the use of the original information. It creates difficulty for an attacker to break the cipher in a reasonable time, and there is no chance for an attacker to guess the key because the key is independent of the user's behavior.

1 Introduction

The roots of cryptography are found in Egyptian and Roman civilizations. The word Cryptography is taken from the Greek-Kryptos "Hidden, Secret"; and Graphein "Writing". Cryptography cannot hide the existence of the information, simply by modifying the information into another form using some Mathematical calculations which cannot be understood. It is broadly classified into three types: Symmetric, Asymmetric and Hash functions [1]. There are so many techniques got evolved to offer better protection to the user information which is huge in amount[2–4], however, these techniques seem as too insecure for the user data over Cloud [5,6], the high chance of the data leaking in real time. Therefore, the Cloud users are scared to place their multimedia data on the insecure communication channel.

Even though there are many techniques coming into existence day by day, they are unable to meet all the user requirements. We have discussed the previous techniques for encrypting multimedia content below from the survey we have done.

2 Literature Survey

Iswari et al. [7] have used both algorithms to generate the key in order to overcome the attacks on the key. Individually, RSA and ElGamal algorithms need

M. Tripathi and S. Upadhyaya (Eds.): ICDLAIR 2019, LNNS 175, pp. 246–260, 2021.
https://doi.org/10.1007/978-3-030-67187-7_26

more computation time. There is a 100% chance of getting the more computational time and keys, after combining these two techniques. To get rid of this
problem, it is better to use a single technique for better computational time
and single key usage. Saini et al. [8] have proposed a modified AES algorithm
in order to improve the security of the AES algorithm, where there is a use of
single key and less computational time.

On the other hand, Gowda [9] has proposed model, where the plaintext
is encrypted using Data Encryption Standard (DES) algorithm with the help
of user-defined key which is encrypted using the RSA algorithm. Divide the
encrypted data into n (user-defined)-blocks. Select n+1 random images in order
to hide the encrypted text blocks inside the images, and the positioning of the
blocks and key are stored inside one of the images (the position known to the
receiver and sender alone).

To share n blocks of data, the user has to choose n random frames which
consume more bandwidth and computational time. So, it is better to encrypt
each and every block of information individually. Encrypt the plaintext using
Data Encryption Standard (DES) algorithm with the help of user-defined key
which is encrypted using the RSA algorithm. Finally, applied steganography to
hide the encrypted text inside an image using LSB algorithm. Sender shares the
encrypted image and key with the receiver. After receiving the encrypted image
and key, the receiver needs to decrypt the key using RSA algorithm. Apply
the LSB algorithm on the stego image in order to find the encrypted text in
the stego image. Finally, decrypt the text using the DES algorithm with the
decrypted key [10].

With the support of previous technique, Pillai et al. [11] have proposed a
model based on steganography and cryptography. Initially, encrypt the plaintext using the Data Encryption Standard (DES) algorithm and divide it into
k segments. Subsequently, apply the K-mean algorithm on the cover image to
cluster all the pixel values into groups based on the RGB channels. Hide the
encrypted text segments inside the pixel clusters. Finally, based on the pixel
position rearrange the clusters to regain the cover image.

Keeping the key secure from the unauthenticated users is a big issue in the
previous models. To avoid the use of key, Chen et al. [12] have proposed a
technique using the concept of multi-image secret sharing scheme. Encrypted all
the frames by performing the XOR operation between the original frames. In this
scheme, n secret frames are needed in order to recover all the original frames,
but the loss of at least a frame will make the receiver as well as the attacker
too difficult to recover all the original frames. An attacker can easily recover all
the original frames from (n-1) or fewer secret frames. This made the room for
compromising the threshold security in multi-secret image sharing scheme in the
previous model [12]. To overcome this problem, Yang et al. [13] have proposed
an efficient technique to overcome the drawback and provide a strong threshold,
but the loss of at least a frame will make the receiver as well as the attacker
too difficult to recover all the original frames. In other words, all the frames are
needed in order to recover all the original frames.

There is a possibility of wrapping the secret data in the previous model, In support of this Goel et al.[14] have proposed a model where we divide the plain image into n number of non-overlapping blocks. Then find the median value of each block (i.e. $M_1, M_2, \ldots \ldots M_n$). Finally, encrypt all the pixels which are not close to the median values using the AES algorithm and leave remaining pixel values unchanged. The security of an image in this technique totally depends on the size of the block number [14]. If the size of the block number is less, then the attacker can easily decrypt the data in reasonable time.

Reddy et al. [15] have proposed an efficient model to overcome this problem with the help of Chebyshev and Henon maps. Initially, take an image of size $H \times W$ to perform the XOR operation with the random matrices generated by Chebyshev and Henon maps. In support of the previous model, Rohith et al. [16] has proposed another model where the key for encrypting data is generated by using a sequence of states $K_{1,i}$ of LFSR (Linear Feedback Shift Register) and a sequence of states $K_{2,i}$ of Logistic Map, in order to encrypt and decrypt the image. Initially, using Logistic Map, a sequence of 1-dimension is produced in the range of 0 to 1 and multiplied by 255 and employed a bit by bit XOR operation on 8 bit LFSR to get the final Key K_i. So, with the help of sequence K_i encryption and decryption will be done. To make the previous technique more stronger, Kankonkar et al. [17] have divided the single image into multiple frames of equal size and applied the previous technique on each and every frame.

Even though there are many techniques coming into existence day by day, they are unable to meet all the user's needs because of their limited nature. On the other hand. Attackers started using the malicious softwares to hack into users' personal systems. This paper highlights the Image encryption technique over Cloud. Our model exposes the role of the splitting a single image into different frames and encrypting each and every frame for providing the security to the user data. Algorithm 1 plays a role of the heart in this technique. The salient features of the work are as follows:

– *User independent Key generation:* To prevent attacks on the data; users have to change the key frequently. But in our technique, there is no need for a user to change the key every time. The key is generated by the Data and by the user. Each and every image has its own separate Key sequence.
– *Difficult for an unauthenticated user to break the code without key sequence and mean value in a reasonable time:* There are several types of attacks are possible to break both the key and data. An unauthenticated user can easily get the access to the secure data with the help of a key. In our technique, it is difficult for an attacker to break the data without the key sequence and mean in real time.
– *Provides security to the user data:* The Internet has changed our lives in many countless positive ways, however, it holds a dark side. Data privacy has been lost, leaving the data at risk from shady individuals, companies and security offices. Every day, we can see many types of crimes over the Internet. Our technique provides the security of the user data.

3 Background

Most of our day to day activities depends on computers and the Internet communication, entertainment, transportation, shopping, medicine, and the list go on [18]. Confidentiality, Availability, Integrity, and Non-reputation are the main security goals. Security threats are imminent due to the exposed nature of communication. There are so many techniques got evolved to offer better protection to the user information. We have discussed the most well know techniques below i.e., Linear Feedback Shift Register and Logistic Map.

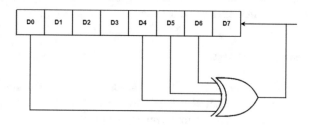

Fig. 1. Linear Feedback Shift Register [19].

8-bit Linear Feedback Shift Register: It is a shift register. Where the seed is given as an input to the 8-bit LFSR. Seed in LFSR is nothing but the initial value given to the LFSR. D0, D1, D2, ..., D7 are given as seed also known as the initial value as shown in the Fig. 1. D0,D4,D5,D6 bits are XORed to get the output bit, and apply Left shift on the Shift register before giving the output bit to the rightmost bit. Linear Feedback shift register generates infinite numbers of the sequence of bits that look random, but has the very long cycle.

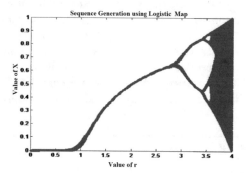

Fig. 2. Logistic Map [19].

Logistic Map: It is a function represented by Eq. 1. It generates a non periodic chaotic sequence of dimension one. Where Xi is random in nature $(0 \leqslant X_i \leqslant 1)$, $X_0 = 0.3$, and bifurcation function 'r' ranges between 0 and 4 given as an input as shown in Fig. 2.

$$X_{n+1} = r.X_n(1 - X_n) \tag{1}$$

In addressing the above concerns, cryptography is playing the vital role in securing the user data. There are two primary problems in Symmetric cryptography: a) the key needs to be stored securely. It is common for symmetric keys to be stored in a safe place and only accessed when required, and b) if another party needs to decrypt the information, a secure channel needs to be used. Asymmetric cryptography is generally slower when compared with Symmetric. In order to get a good mix of performance and security, it is possible to combine both Asymmetric and Symmetric cryptography. In this paper, we have proposed an efficient algorithm where the key is generated by the data, not by the user. It is difficult for an unauthenticated user to break the code in a reasonable time without the key sequence and provides security to the user's data.

4   Proposed Technique

In this paper, we have proposed an efficient algorithm where the key is generated by the data, not by the user. It is difficult for an unauthenticated user to break the code in a reasonable time without the key sequence and provides security to the user's data.

$$A = \begin{bmatrix} A_{11} & A_{12} & A_{13} & \ldots & A_{1n} \\ A_{21} & A_{22} & A_{23} & \ldots & A_{2n} \\ \ldots & \ldots & \ldots & \ldots & \ldots \\ A_{m1} & A_{m2} & A_{m3} & \ldots & A_{mn} \end{bmatrix}$$

4.1   Splitting_Mapping_1

Initially, we perform nibble swapping on the image matrix. Afterwards, we apply Algorithm 1 on the resulting matrix. We proceeded by finding the minimum non-zero element of this matrix. Then we subtract this minimum value of all the non-zero elements of the matrix. Following this, we store the resultant matrix A in a matrix Z. Then we take matrix A and replace all the non-zero elements in it with the minimum value that we found earlier. This A is the first replacement matrix and is written to the cell. After this, we copy the Z matrix back into A and repeat the above process. As we do this process more and a number of

---

Algorithm 1. Proposed Algorithm

1: **procedure** IMAGE
2:     Read the matrix A.
3:     B=A;
4:     Find the $min(A) \neq 0$.
5:     Subtract the min (A) from $\forall$ elements in $B \neq 0$ and store the result in Z.
6:     Replace $\forall$ elements in $A \neq 0$ with the min (A) and store the result in a cell.
7:     A=Z;
8:     Repeat 3 to 8 until $A = nullmatrix$.
9: **end procedure**

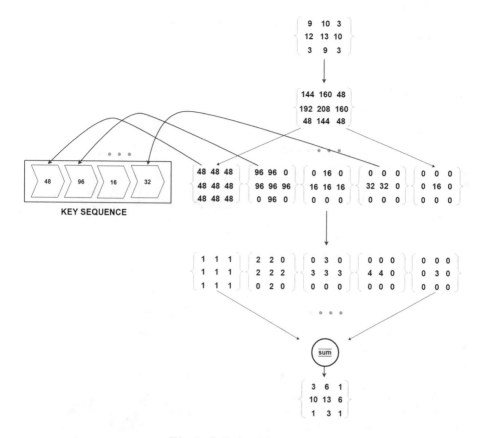

Fig. 3. Splitting_Mapping_1

times we find that the numbers of 0's in the matrix continue to increase. We repeat this process until all the values in the matrix A reduce to 0 as shown in Algorithm 1. After applying Algorithm 1, Matrix got divided into single-valued matrices such that apart from zero, all the rest of the elements in the matrix share a single unique value.

In the first matrix in the cell, we copy the non-zero number found in that matrix and write it into the key array. The index of the key array where it gets written is represented by k and begins at 1. After this, we replace the non-zero elements in the matrix with the key array index k. We also increment the index k so that we can use it to correctly replace the non-zero elements of the later matrices. For all the subsequent matrices, if the non-zero element has not been encountered in any previous matrix we copy that non-zero element in the key array. After this, we increment the key array index k as before. Following this, we replace all the non-zero elements in this matrix with index value k of the key array as shown in Algorithm 2. If the non-zero element has already been encountered before, then we simply replace the non-zero elements in this matrix

with the index of the key array where this current non-zero element was stored
when it was first encountered a process as Splitting and Mapping as shown in
the Fig. 3.

4.2   Generating Key Sequence

For each non-zero element we replace
and store it in a key sequence (array).
We don't store an element which has
been stored in the key sequence or in
other words the non-zero element has
been encountered before.

4.3   Steps Involved
in Encryption Process

Initially, We have taken an image
of size $H \times W \times 3$ and split the
image into three frames based on the
RGB channels (i.e. R, G, B). Split-
ting_Mapping_1 being applied on the
R, G, B channels. The mean value
of all the three channels and the
sum of all the three key sequences
(key sequence generated after applying
Splitting_Mapping_1   operation)   are
being used to generate the seed for the
Logistic Map with the help of Eq. 2.

$$seed = (mean \times sum)\%3.9999 \quad (2)$$

We have performed the XOR operation on R, G and B channel with Logistic Map generated with the seed value
and stored the result in R, G, and B
respectively. Performing the interdependent XOR operation between the
three channels can make the data more
secure (i.e $G = R \oplus G; B = G \oplus B; R =$

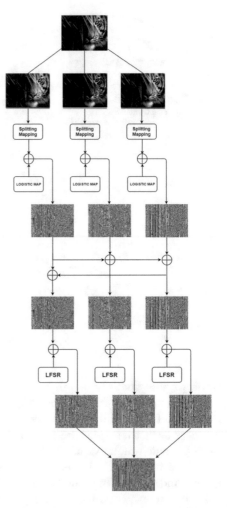

Fig. 4. Encryption process

$R \oplus B$). After performing the interdependent XOR operation on the three channels, we have given the mean value to the LFSR as an input to generate a matrix
of size $H \times W$ to perform XOR operation on the three channels. Finally, we have
combined the three channels to get the final encrypted image as shown in the
Algorithm 2 and Fig. 4.

Algorithm 2. Encryption_Decryption

1: **PHASE - I**
2: **procedure** ENCRYPTION(Image)
3:     Image into 3 RGB channels (i.e. R, G, B)
4:     [R,G,B] = Splitting_Mapping_1(R,G,B)
5:     A=Logistic_Map(seed) (i.e. $seed = (mean \times sum)\%3.9999$)
6:     [R, G, B] = $(R \oplus A, G \oplus A, B \oplus A)$
7:     $G = R \oplus G; B = G \oplus B; R = R \oplus B$
8:     $C = $ LFSR (seed1) (i.e. seed1 = mean)
9:     [R, G, B] = $(R \oplus C, G \oplus C, B \oplus C)$
10:    Encrypted_Image = cat (3, R, G, B)
11: **end procedure**
12: **PHASE - II**
13: **procedure** DECRYPTION(Encrypted_Image, Mean, Key sequence)
14:    Image into 3 RGB channels (i.e. R, G, B)
15:    $C = $ LFSR (seed1) (i.e. seed1 = mean)
16:    [R, G, B] = $(R \oplus C, G \oplus C, B \oplus C)$
17:    $R = R \oplus B; B = G \oplus B; G = R \oplus G$
18:    A=Logistic_Map(seed)(i.e. $seed = (mean \times sum)\%3.9999$)
19:    [R, G, B] = $(R \oplus A, G \oplus A, B \oplus A)$
20:    [R,G,B] = Splitting_Mapping_2(R,G,B)
21:    Decrypted_Image = cat (R, G, B)
22: **end procedure**

4.4   Splitting_Mapping_2

We proceeded by finding the minimum non-zero element of this matrix. Then we subtract this minimum value of all the non-zero elements of the matrix. Following this, we store the resultant matrix A in a matrix Z. Then we take matrix A and replace all the non-zero elements in it with the minimum value that we found earlier. This A is the first replacement matrix and is written to the cell. After this, we copy the Z matrix back into A and repeat the above process. As we do this process more and the number of times we find that that the numbers of 0's in the matrix continue to increase. We repeat this process until the matrix A becomes a null matrix.

We have obtained all the replacement matrices, we take the first replacement matrix and note that the non-zero value in it is actually a reference to the key array index k. So, we copy the non-zero value into k and then replace the non-zero values of this matrix with the value found at that location pointed to by k in the key array. We do this for every replacement matrix to obtain the single-valued matrices as shown in Fig. 5. After we have obtained all the replacement matrices, we sum all of them and perform nibble swapping on all the elements of the matrix to obtain the original matrix. We have obtained all the replacement matrices, we take the first replacement matrix and note that the non-zero value in it is actually a reference to the key array index k. So, we copy the non-zero value into k and then replace the non-zero values of this matrix with the value found at that location pointed to by k in the key array. We do this for every replacement matrix to obtain the single-valued matrices as shown in Fig. 5. After we have obtained all the replacement matrices, we sum all of them and perform nibble swapping on all the elements of the matrix to obtain the original matrix.

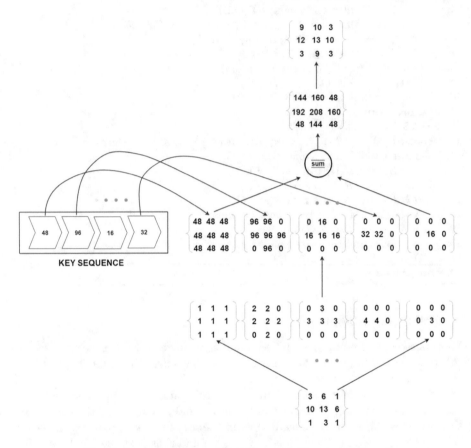

Fig. 5. Implementation of Splitting and Mapping process 2

4.5   Steps Involved in Decryption Process

Split the image into three frames based on the RGB channels (i.e. R, G, B). We have given the mean value to the LFSR as an input to generate a matrix of size $H \times W$ to perform XOR operation on the three channels. After performing the XOR operation with the LFSR, we have performed the opposite interdependent XOR operation on the three channels (i.e. $R = R \oplus B; B = G \oplus B; G = R \oplus G$). The mean and sum of all the three key sequences is being used to generate the seed for the Logistic Map with the help of Eq. 2 to perform the XOR operation with the three channels. Finally, we have combined the three channels to get the final decrypted image as shown in the Algorithm 2.

4.6   Secret Sharing over Cloud

Price reduction, greater flexibility, elasticity and optimal resource utilization are the main uses of cloud computing (use of hardware and software to deliver a service to the user over the internet) [10, 20–22]. We can make use of cloud environment to make the process more effective. For example, first extract the image with size $H \times W \times 3$, where H and W represent the height and width of an image respectively. Then VM-0 is responsible for spitting and assigning the chunks to each and every VMs. After chunks received by each and every VM [23–25]; VMs in the next level (VM_1 to VM_N) performs Splitting and Mapping with coordination and send the frames to VM_{N+1} [26,27]. Finally, VM_{N+1} sums up all the frames to perform LFSR considering key array sum as seed and sends the encrypted image to the receiver as shown in Fig. 6.

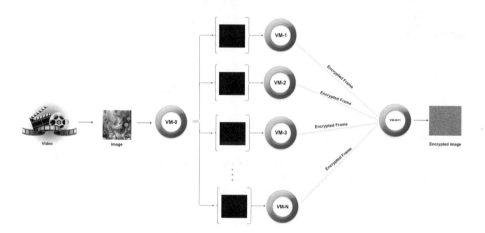

Fig. 6. Secret Sharing secure process in Cloud environment

5   Experiments and Results

This section comprises the following three parts: Experimental results, Limitations, Test Cases. Algorithm 1 plays the role of the heart in this technique. Splitting the image and applying the encryption on each and every individual image will make the data more secure. After applying our technique to images, it has given us efficient results as shown in Table 1 and Fig. 7.

5.1   Qualitative Analysis

After breaking down a video into N frames and then E-MOC concept is employed. Here, we have taken three images to illustrate the plain histogram, encrypted and encrypted histogram results as shown in Fig. 7. The work has been implemented in MATLAB R2016a and analyzed.

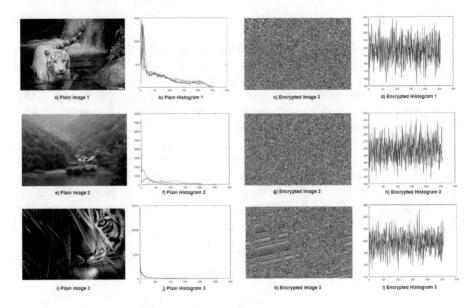

Fig. 7. Plain and encrypted images with their histogram

5.2   Quantitative Analysis

To demonstrate the similarities between original, encrypted and decrypted frame correlation between original, encrypted and decrypted frame are calculated as tabulated in the Table 1. There are the following reasons to attain the correlation between frames, which is used as a parameter:

Table 1. Correlation between Plain, Encrypted and Decrypted image

Frame	Correlation between Plain and Encrypted image	Correlation between Plain and Decrypted image	Key Sequence
a)	−0.00035	1	R: 1 G: 1 B: 1 2 Mean: 121
b)	0.00004	1	R: 1 2 3 4 G: 1 2 B: 1 2 Mean: 114
c)	−0.00023	1	R: 1 G: 1 B: 1 2 Mean: 91

- Correlation denotes the similarity between two digital signals, that is if two signals are more similar than correlation is 1, otherwise 0 for two dissimilar signals.
- If there exist two frames such that they are more similar than the concept of secret sharing becomes obsolete.
- If the correlation is 0 then both frames are important as decryption is not possible if either one is missing as per the concept of secret sharing [28–33].

Mean Square Error: The difference b/w the actual and encrypted image can be computed using the parameter called Mean Square Error using the Eq. 3.

$$MSE = \frac{1}{mn} \sum_{p=0}^{mn-1} [A(p.q) - E(p,q)]^2 \qquad (3)$$

Where A(p,q) is Actual image pixel, E(p,q) is Encrypted image pixel, m and n represent the size of the image. Higher value of MSE is required between actual and encrypted image because it represents how vulnerable to attack.

Peak Signal-to-Noise Ratio: Higher the value, the lower the error between plain and encrypted image. We can calculate the value of PSNR using the Eq. 4.

$$PSNR = 10 \log_{10} \frac{M^2}{MSE} \qquad (4)$$

where MSE is mean square error and M is the maximum variation in the input image data type) reaches infinity, it means that both plain and encrypted images are same there is no error between them (Table 2).

5.3   Limitations and Test Cases

a) Some of the limitations of the proposed model are mentioned below:

Table 2. Comparison between Plain and Encrypted image

Frame		Mode	Median	Mean
a)	MSE	12757	**12775**	12765
	PSNR	7.0733	**7.0671**	7.0705
	Correlation	0.0030	**−0.00017**	−0.00035
b)	MSE	16550	17411	**17662**
	PSNR	5.9427	5.7227	**5.6605**
	Correlation	0.0141	−0.00083	**0.00004**
c)	MSE	8495	15053	**15133**
	PSNR	8.8392	6.3546	**6.3317**
	Correlation	0.2551	−0.0016	**0.0002**

1. *Complex operation:* After applying Algorithm 1 on the image matrix, matrix got divided into single-valued matrices such that apart from zero, all the rest of the elements in the matrix share a single unique value. It is too complex to perform this operation.
2. *Key Sequence Size:* The size of the key sequence depends on the difference between the elements in the image matrix. If the difference between the elements in the image matrix is varying which makes the size of the key sequence to increase.
3. *Managing key sequences:* Managing the key sequence is another limitation in our technique. Each and every time user has to save the Private key generated by the image. For saving key sequence, a separate database needs to be maintained.

b) Test Cases: The size of the key sequence depends on the difference between the elements in the image matrix. If the difference between the elements in the image matrix is varying which makes the size of the key sequence to increase. If the size of the key sequence is known to the attacker, the attacker can break the code in $255 \times 255 P_{(len\_key\_sequence_1)} \times 255 P_{(len\_key\_sequence_2)} \times 255 P_{(len\_key\_sequence_3)}$ attempts as shown in the example with *len_key_sequences* are 4,3 and 1 (i.e. R, G, B) then maximum total no. of attempts $= 255 \times 255 P_4 \times 255 P_3 \times 255 P_1$, But without knowing the key sequence and mean it is difficult for an attacker to break the code in reasonable time.

6   Conclusion

This paper highlighted an efficient algorithm where the key is generated by the data, not by the user. It is difficult for an unauthenticated user to break the code in a reasonable time without the key sequence and mean value. Algorithm 1 plays the main role in encrypting images. Splitting the image and applying the encryption on each and every individual frame will make the data more secure. In the future perspective, the authors would like to implement the technique for long video, and also selective encryption mechanism where only potential/ informative content shall be encrypted. For creating more difficulty to the attacker, stenography may be used for the multimedia content in the Cloud environment as the applications.

References

1. Kunal, K.M., et al.: A mathematical model for secret message passing using stenography. In: IEEE ICCIC, pp. 1–6 (2016)
2. Krishan, K., et al.: Eratosthenes sieve based key-frame extraction technique for event summarization in videos. MTAP **77**, 7383–7404 (2018)
3. Kumar, K., et al.: F-DES: fast and deep event summarization. IEEE TMM **20**, 323–334 (2018)
4. Kumar, K., et al.: ESUMM: event summarization on scale-free network. IETE TITR (2018). https://doi.org/10.1080/02564602.2018.1454347

5. Li, M., et al.: Cryptanalysis and improvement of a chaotic image encryption by first-order time-delay system. IEEE MultiMed. (2018). https://doi.org/10.1109/MMUL.2018.112142439

6. Abdalla, A.A., et al.: On protecting data storage in mobile cloud computing paradigm. IETE TITR **31**, 82–91 (2014)

7. Iswari, N.M.S.: Key generation algorithm design combination of RSA and ElGamal algorithm. In: 2016 8th International Conference on Information Technology and Electrical Engineering (ICITEE), pp. 1–5. IEEE, October 2016

8. Saini, J.K., Verma, H.K.: A hybrid approach for image security by combining encryption and steganography. In: 2013 IEEE Second International Conference on Image Information Processing (ICIIP), pp. 607–611. IEEE, December 2013

9. Gowda, S.N.: Advanced dual layered encryption for block based approach to image steganography. In: International Conference on Computing, Analytics and Security Trends (CAST), pp. 250–254. IEEE, December 2016

10. Gowda, S.N.: Dual layered secure algorithm for image steganography. In: 2016 2nd International Conference on Applied and Theoretical Computing and Communication Technology (iCATccT), pp. 22–24. IEEE, July 2016

11. Pillai, B., Mounika, M., Rao, P.J., Sriram, P.: Image steganography method using k-means clustering and encryption techniques. In: 2016 International Conference on Advances in Computing, Communications and Informatics (ICACCI), pp. 1206–1211. IEEE, September 2016

12. Chen, C.C., Wu, W.J.: A secure Boolean-based multi-secret image sharing scheme. J. Syst. Softw. **92**, 107–114 (2014)

13. Yang, C.N., Chen, C.H., Cai, S.R.: Enhanced Boolean-based multi secret image sharing scheme. J. Syst. Softw. **116**, 22–34 (2016)

14. Goel, A., Chaudhari, K.: FPGA implementation of a novel technique for selective image encryption. In: International Conference on Frontiers of Signal Processing (ICFSP), pp. 15–19. IEEE, October 2016

15. Reddy, V.P.K., Fathima, A.A.: Efficient encryption technique for medical X-ray images using chaotic maps. In: International Conference on Wireless Communications, Signal Processing and Networking (WiSPNET), pp. 783–787. IEEE, March 2016

16. Rohith, S., Bhat, K.H., Sharma, A.N.: Image encryption and decryption using chaotic key sequence generated by sequence of logistic map and sequence of states of Linear Feedback Shift Register. In: 2014 International Conference on Advances in Electronics, Computers and Communications (ICAECC), pp. 1–6. IEEE, October 2014

17. Kankonkar, J.T., Naik, N.: Image security using image encryption and image stitching. In: 2017 International Conference on Computing Methodologies and Communication (ICCMC), pp. 151–154. IEEE, July 2017

18. Krishan, K., et al.: Deep event learning boosT-up approach: DELTA. MTAP **77**, 1–21 (2018)

19. Rohith, S., et al.: Image encryption and decryption using chaotic key sequence generated by sequence of logistic map and sequence of states of Linear Feedback Shift Register. In: IEEE ICAECC, pp. 1–6 (2014)

20. Kumar, K., et al.: Economically efficient virtualization over cloud using docker containers. In: IEEE CCEM, pp. 95–100 (2016)

21. Kumar, K., et al.: Sentimentalizer: docker container utility over cloud. In: IEEE ICAPR, pp. 1–6 (2017)

22. Forouzan, A.B., et al.: Cryptography and Network Security. Tata McGraw-Hill Education (2011)

23. Stallings, W.: Cryptography and Network Security: Principles and Practices. Pearson Education India (2006)
24. Chandrasekaran, K.: Essentials of Cloud Computing. CRC Press (2014)
25. The three ways to cloud compute. https://www.youtube.com/watch?v=SgujaIzkwrEt. Accessed Sept 2017
26. Virtual Machine (VM). http://searchservervirtualization.techtarget.com/definition/virtual-machine. Accessed Oct 2017
27. Smith, J., et al.: Virtual Machines: Versatile Platforms for Systems and Processes. Elsevier (2005)
28. Piyushi, M., et al.: V $\oplus$ SEE: video secret sharing encryption technique. In: IEEE CICT, pp. 1–6 (2017)
29. Shikhar, S., et al.: GUESS: genetic uses in video encryption with secret sharing. In: CVIP 2017, pp. 1–6. Springer (2017)
30. Kumar, K., et al.: Key-lectures: keyframes extraction in video lectures. In: MISP 2017, pp. 1–7. Springer (2017)
31. Kumar, K., Shrimankar, D.D., Singh, N.: V-LESS: a video from linear event summaries. In: CVIP 2017, pp. 1–6 (2017)
32. Kumar, K., et al.: SOMES: an efficient SOM technique for event summarization in multi-view surveillance videos. In: ICACNI 2017, pp. 1–6. Springer (2017)
33. Krishan, K., et al.: Equal partition based clustering approach for event summarization in videos. In: The IEEE SITIS 2016, pp. 119–126 (2016)

Fuzzy Logic Based Clustering for Energy Efficiency in Wireless Sensor Networks

Biswajit Rout, Ikkurithi Bhanu Prasad, Yogita, and Vipin Pal$^{(\boxtimes)}$

National Institute of Technology Meghalaya, Shillong, India
biswajitcs1234@gmail.com, bhanuprasad3214@gmail.com,
thakranyogita@gmail.com, vipinrwr@gmail.com

Abstract. For wireless sensor network (WSN) many routing protocols are proposed to reduce energy consumption. Clustering methodology has been used in WSN for increasing the longevity of the network. Independent clusters are formed by grouping sensor nodes and selecting one node as cluster head (CH) to represent the cluster. Role of CH plays vital part for clustering methodology as CH is more important as compared to member nodes. In the literature fuzzy approaches with clustering methodology have been examined for well load balanced clusters. Our proposed approach in this paper is based on fuzzy logic which considers distance to Base Station (BS), remaining energy of nodes, node density, intra-cluster communication distance, remaining energy of CH and density of CH as input fuzzy sets. In this fuzzy approach, the whole process of clustering is divided into two phases. In the first phase, CH selection is completed by considering the parameters - distance to the base station (BS), remaining energy of node and node density of each node. In the second phase, cluster formation has been taken place by joining of non CH nodes with a particular CH taking parameters distance to CH, the remaining energy of CH and density of the CH. The proposed approach improves the load balancing and node lifetime as compared with the other simulated approaches.

Keywords: Load balancing · Inter and intra cluster communication · Density of nodes · Clustering · Fuzzy · Wireless Sensor Networks

1 Introduction

Wireless Sensor Networks (WSN), a type of infrastructure-less network, have ample challenging aspects in the field of research [1]. A WSNs consist of large number of tiny sensor nodes ranging from hundreds to thousands. These sensor nodes carry finite battery power, limited memory, and less complex processing unit. With the help of radio transceivers component, the sensor nodes communicate information to other nodes in their radio frequency range directly and multi-hop communication can be applied to send information to a node which is not within the communication range [2]. A schematic diagram of wireless sensor network architecture has been demonstrated in Fig. 1. The sensor nodes are

© The Author(s), under exclusive license to Springer Nature Switzerland AG 2021
M. Tripathi and S. Upadhyaya (Eds.): ICDLAIR 2019, LNNS 175, pp. 261–272, 2021.
https://doi.org/10.1007/978-3-030-67187-7_27

used to collect data from the environment by measuring different environmental condition like moisture, pressure and temperature etc. The nodes are working continuous or waiting for an event occurrence to send information. Then nodes send data to sink or base station for further processing. The base station or sink acts as an interface between users and network [3].

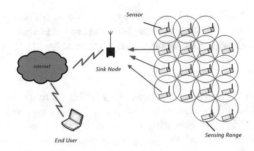

Fig. 1. Architecture of Wireless Sensor Networks

There are few interesting design issues for wireless sensor networks like economical energy consumption, fault tolerance, scalability, network topology, and transmission mode, etc. Because of there infrastructure less architecture and the limited capacity of battery unit (battery unit is also not replaceable or rechargeable in most of the scenarios), efficient energy consumption is a major design issue in WSN. Clustering is one of the techniques to manage energy consumption efficiently. In clustering methodology, nodes are formed in independent groups or clusters with a node acting as head of the cluster. It is the responsibility of the CH to aggregate the data received from the cluster member nodes and send them to the base station [4,5]. Reduced data size and less intra-cluster communication distance as a consequence of hierarchical nature of clustering approach improves the network life time.

Cluster Head plays crucial role to extend the network lifetime. Different methodologies, like Genetic Algorithm, Particle Swarm Optimization, Fuzzy Logic and many more, have been incorporated for selection of CHs which can produce a better set of CHs in each round. Different characteristics of nodes, like remaining energy, node density, intra and inter communication distance and others, make selection of CH very competitive. Fuzzy logic takes full advantage of above mentioned node characteristics and also exploits uncertainty of WSNs to choose CHs. Work of this paper presents a fuzzy logic based clustering approach that considerers distance to Base Station (BS), remaining energy of nodes, node density, intra-cluster communication distance, remaining energy of CH and density of CH as fuzzy inputs for clustering. The proposed fuzzy logic based clustering selects CHs with distance to Base Station (BS), remaining energy of nodes and node density and then choose CH for each node with intra-cluster communication distance, remaining energy of CH and density of CH. The proposed approach maintains a well selected CHs and well formed clusters in each round. The proposed approach is compared with LEACH protocol which selects CHs randomly. Simulation results show that the proposed approach has longer network lifetime as compared to LEACH protocol.

Rest of the paper has been organized as follows: Sect. 2 describes the literature review for various CH selection approaches that have increased the network lifetime, Sect. 3 provides details about the proposed fuzzy system for clustering, Simulation Results have been discussed in Sect. 4, and Conclusion of the presented work has been summarized in Sect. 5.

2 Literature Review

Clustering is used in WSN for increasing the longevity of the network. Efficient energy consumption, load balancing, fault tolerance, scalability are few important challenges for clustering methodology in wireless sensor networks.

Low Energy Adaptive Clustering Hierarchy (LEACH) [6] is one of the well known clustering protocol. LEACH follows the probabilistic approach to choose CH. The probabilistic approach insures that all nodes have been selected as cluster head once in the epoch. Due to the probabilistic CH selection process, CHs selected are random, which results in load unbalancing in the network.

Hybrid Energy-Efficient Distributed (HEED) [7] algorithm is designed for multi-hop network and node equality is the primary assumption. Two-phase parameter check is done to select CHs. In the first phase, the remaining energy of a node is used for the probabilistic selection of the CHs. In the second phase, parameters such as node degree, distance to neighbors, and intra-cluster energy consumption are applied to break the tie in the selection process. However, the HEED algorithm suffers from the hotspots problem and causes unbalanced energy consumption because of its tendency to generate more than the expected number of clusters.

Fuzzy methods have been applied intensively for Cluster Head selection and cluster formation. In CHEF [8] CH election occurs in a distributed manner which does not necessitate the central control of the sink. In this algorithm, two parameters i.e node and local distance are the fuzzy input parameters.

EEUC [9] is a distributed competitive unequal clustering algorithm. In this algorithm, each node has a preassigned competitive radius and CHs are elected by local competition. Competition radius decreases as the nodes approach the sink. In addition to being an unequal clustering algorithm, this method is also a probabilistic approach since for every clustering round, a node probabilistically chooses to attend or not to attend ithe CH election competition. Although EEUC improves the network lifetime over LEACH and HEED, extra global data aggregation can result in much overhead and deteriorate the network performance.

A. K. Singh et al. in [10] a new fuzzy logic CH election algorithm is introduced. For selecting cluster head, node residual energy, number of neighboring alive nodes inside the cluster and node distance from the sink node are taken into consideration. So nodes having high remaining energy, highest number of alive neighboring nodes and minimum distance from sink node are chosen as cluster heads.

N. Kumar et al. in [11] to calculate the chance for a node to become a cluster head and vice-cluster head, the parameters used are residual energy, node density and node distance from base station. The proposed algorithm is used in multi-hop communication. Within a certain transmission range the node having highest chance value is chosen as cluster head the node having second largest value is chosen as vice-cluster head in the same cluster.

B. Baranidharan et al. in [12] the clustering algorithm discussed is distributed in nature. Fuzzy approach is used to elect the cluster heads. Here the concept of unequal clustering is introduced to maintain the energy consumption uniformly among the cluster heads. Node density, residual energy and node distance to sink node are the fuzzy parameter considered for CHs selection. Here two output parameters chance and size are taken. Chance for a node shows the competing power to become a cluster head and size gives the maximum number of member nodes to a cluster have.

P. Kumar et al. in [13] a comparative study of various fuzzy based clustering techniques are reported. Various distinguished properties of WSN are taken for both single and two-level fuzzy based clustering approach. Here centralized algorithm C-means and distributed algorithm like NECHS, F3N, CHEF, EAUCF etc. are discussed. In all the cases fuzzy based clustering algorithm gives better performance than probabilistic based clustering algorithm like LEACH, HEED etc.

Fuzzy C-means [14] clustering technique is centralized in nature. The input parameter node location and node having highest energy are used for cluster head election. Each node have a membership value that is degree of belongingness with respect to each cluster head rather than fully belongs to a particular cluster. In each round a new cluster head is chosen according to the residual energy. It gives better performance than LEACH and K-means.

W. Abidi et al. in [15] the longevity of WSN is analyzed by considering three different fuzzy-based method. The fuzzy parameter node remaining energy, node density and centrality are taken for cluster head election in these approaches. Centrality and energy are considered simultaneously in first approach, node density and energy, centrality and node density used for second and third approach respectively. All these three cases give better performance than LEACH protocol. Also the first fuzzy approach gives best result among all.

Fuzzy approaches for cluster head selection and cluster formation have been applied for energy efficiency in wireless sensor networks. There is always scope for the improvement of energy efficiency with improved fuzzy system.

3   Fuzzy Based Energy Efficent Clustering (FBEEC)

The section presents the working principle of proposed fuzzy based energy efficient clustering (FBEEC). FBEEC operates in the round and each round has been divided into two phases - Set-up Phase and Steady Phase, as happens in the LEACH and other LEACH based protocols. The output of Set-up phase is CH selection, Cluster formation, and TDMA scheduling.

In our proposed FBEEC approach, first by using three fuzzy parameters - distance to BS, the remaining energy of nodes and node density, the CH priority of each node is calculated for selection of cluster heads. After that, the noncluster head nodes are assigned to the cluster head to form a cluster. For this cluster formation, the noncluster head nodes choose the cluster head by taking the fuzzy parameter distance to CH, remaining energy of CH and density of CH.

Distance to BS, remaining energy of nodes and node density are taken for cluster head selection. The minimum and maximum values of Fuzzy input variables for CH priority are given in Table 1. The fuzzy linguist variables used for distance to BS are *close, medium* and *far*. The fuzzy linguist variables for remaining energy are *low, medium* and *high* and for node density, fuzzy linguist variables are *sparse, normal* and *dense*. The input range for the fuzzy variable for selecting cluster head is given in Table 2 and corresponding fuzzy sets are given in Figs. 2, 3 and 4. The trapezoidal membership function is used for *close, far, low, high, sparse* and *dense*. The triangular membership function used for *medium and normal*.

Table 1. Fuzzy input variables and their minimum and maximum values for CH priority

Input variables	Min.value	Max. value
Distance to BS	0	200
Remaining energy	0	0.5
Node density	0	50

The fuzzy output variable is CH priority for being selected as the cluster head. Nine linguist variables are used for output variables. These are *very low, low, rather low, medium low, medium, medium high, high, rather high* and *very high* given in Fig. 5. For *very low* and *very high*, the trapezoidal function is used. The triangular membership function is used for other linguist variables.

Two fuzzy if-then rules for extreme cases are:

1. If the node is *far* from the BS, *high* remaining energy and node density is *sparse*, then the CH priority is *very high*.
2. If the node is *close* to the BS, *low* remaining energy and node density is *dense*, then the CH priority is *very low*.

After the selection of the cluster head, the noncluster head nodes join with the cluster head situated in the radio frequency range to form a cluster. In FBEEC joining of noncluster head nodes with the cluster head node is based on the parameters distance to CH, remaining energy of CH and density of CH rather than considering only one parameter.

Table 2. Range of Fuzzy input variables

Input range	Fuzzy variable
Distance to BS	
0–80	Close
40–120	Medium
80–200	Far
Remaining energy	
0–0.2	Low
0.1–0.3	Medium
0.2–0.5	High
Node density	
0–20	Sparse
10–30	Normal
20–50	Dense

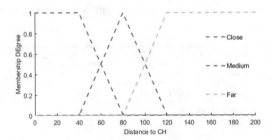

Fig. 2. Fuzzy membership for Distance to BS

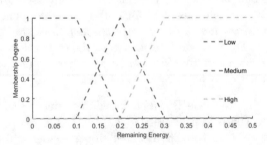

Fig. 3. Fuzzy membership for Remaining Energy

The minimum and maximum values of fuzzy input variables for CH choice are given in Table 3.

Here the linguist fuzzy variables used for distance to CH are *close, medium* and *high*. The linguist variable for remaining energy is *low medium* and *high* and for CH density, the fuzzy linguist variables used are *sparse, medium* and

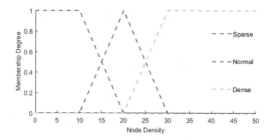

Fig. 4. Fuzzy membership for Node Density

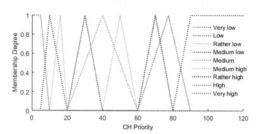

Fig. 5. Fuzzy membership for CH Priority

Table 3. Minimum and maximum of fuzzy variable for CH priority

Input variables	Min. value	Max. value
Distance to CH	0	140
Remaining energy	0	0.5
CH density	0	50

dense. The input range for the fuzzy variable for cluster formation is given in Table 4 and corresponding fuzzy sets are given in Figs. 6, 7 and 8. The trapezoidal membership function is used for *low, high, close, far, sparse* and *dense.* The triangular membership function is used for *medium* and *normal.*

For the output parameter, CH choice, the fuzzy linguist variables used are *very low, low, rather low, medium low, medium, medium high, high, rather high* and *very high* given in Fig. 9. For *very low* and *very high,* the trapezoidal function is used. The triangular function is used for other linguist variables.

Two fuzzy if-then rules for extreme cases are:

1. If the node is *close* to the CH, *high* remaining energy and CH density is *sparse,* then the CH choice is *very high.*
2. If the node is *far* from the CH, *low* remaining energy and CH density is *dense,* then the CH choice is *very low.*

Table 4. Range of Fuzzy input variables

Input range	Fuzzy variable
Distance to CH	
0–60	Close
30–90	Medium
60–140	Far
Remaining energy	
0–0.2	Low
0.1–0.3	Medium
0.2–0.5	High
CH density	
0–20	Sparse
10–30	Normal
20–50	Dense

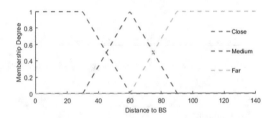

Fig. 6. Fuzzy membership for Distance to CH

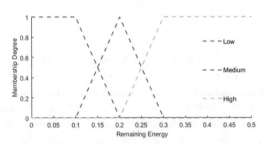

Fig. 7. Fuzzy membership for Remaining Energy

For fuzzy rule evaluation Mamdani Inference Method has been used and the Center of Area(CoA) method has been used for defuzzification. FBEEC is centralized in nature and BS runs the fuzzy system for each round of clustering. BS informs about the complete information of elected CHs, respective CH for each node and TDMA schedule. In steady phase, MN sends data to respective CH and CH sends aggregated data to BS.

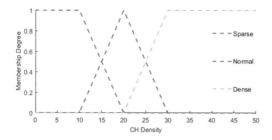

Fig. 8. Fuzzy membership for CH Density

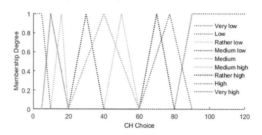

Fig. 9. Fuzzy membership for CH Choice

4 Simulation Results and Analysis

4.1 Network Communication Model

The assumption for the network model are:

1. Initially, all the sensor nodes have the same amount of energy.
2. The sensor network is homogeneous in nature. That means each node has the same amount of memory, processing speed, transmitting and receiving power.
3. Sensor nodes are deployed arbitrarily.
4. Node death is because of energy depletion only.

In WSN different sizes of clusters are formed by sensor nodes. In a cluster the information collected by the cluster member nodes first transmitted to the cluster head. The cluster head aggregates the data and then transmits the compressed data to the base station. The consumption of energy is mainly due to packet transmission and packet reception. The energy consumption model used for transmission and reception is similar to the model used in [6].

4.2 Simulation Analysis

The proposed Fuzzy Based Energy Efficient Clustering (FBEEC) approach has been simulated using MATLAB, 2017(b). In our work 100 sensor nodes are considered in an area of (100×100) m^2. The initial amount of energy is $0.5 Joules$ for each sensor node. The simulation parameters are given in Table 5.

Table 5. Network Parameters

Parameters	Values
Network Size	$(100 * 100)\,\mathrm{m}^2$
Number of Nodes	100
E_0	0.5 J
ETX	$50 * 10^{-9}\,\mathrm{J}$
ERX	$50 * 10^{-9}\,\mathrm{J}$
E_{fs}	$10 * 10^{-12}\,\mathrm{J}$
E_{mp}	$0.0013 * 10^{-12}\,\mathrm{J}$

FBEEC is compared with LEACH and the experimental results show that the proposed FBEEC gives better results compared to LEACH in terms of network lifetime and energy consumption. Here network lifetime is considered for first node death (FND), half of the node death (HND) and last node death (LND). LEACH and proposed FBEEC are simulated for 100 different runs and average is considered for final result analysis.

Figure 10 demonstrates the number of alive nodes over the clustering rounds for LEACH and proposed FBEEC. It can be seen from the figure that stable region of FBEEC is longer than LEACH because the random CH selection in LEACH that leads to unsuitable node as CH that have effect as less load balanced network. It can also be seen that during the most of network lifetime period number of alive nodes are more in case of FBEEC as compared to LEACH.

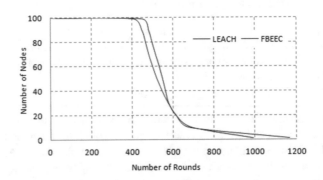

Fig. 10. Nodes alive in Rounds

Figure 11 shows comparison of FND, HND and LND for LEACH and FBEEC. It can be observed that in LEACH, the first node died in 406 round and half of the nodes die in 520 round. But in our proposed FBEEC, node die in 448 round and half of the nodes die in 547 round. This is because of LEACH follows the probabilistic approach while our approach follows the fuzzy parameter for cluster head selection. The proposed FBEEC approach is 10.5% efficient with respect to FND and 5.2% efficient with respect to HND as compared to LEACH. The last node death occurs in 1166 round in LEACH and 989 round in FBEEC.

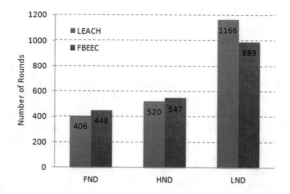

Fig. 11. Network life time

Figure 12 shows a comparative study of the energy consumption of both protocols. It is shown from the figure that FBEEC gives better performance.

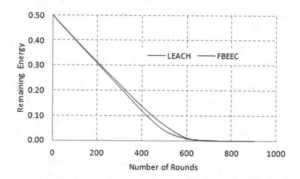

Fig. 12. Energy consumption in Rounds

5 Conclusion

Work of this paper presented a fuzzy based energy efficient clustering approach for cluster head selection and cluster formation. The proposed fuzzy system considers distance to BS, remaining energy of nodes and node density for CH selection priority for each nodes and then with distance to CH, remaining energy of CH and density of CH chooses CH for each node. The proposed FBEEC outperformed competitive LEACH protocol with respect to network lifetime and energy consumption.

Acknowledgment. The work of the paper has been supported by NATIONAL MISSION ON HIMALAYAN STUDIES (NMHS) sanctioned project titled "Cloud-assisted Data Analytics based Real-Time Monitoring and Detection of Water Leakage in Transmission Pipelines using Wireless Sensor Network for Hilly Regions" (Ref. No.: GBPNI/NMHS-2017-18/SG21).

References

1. Pal, V., Singh, G., Yadav, R.P.: Balanced cluster size solution to extend lifetime of wireless sensor networks. IEEE Internet of Things J. **2**(4), 399–401 (2015)
2. Evers, L., Bijl, M.J.J., Perianu, M.M., Perianu, R.M., Havinga, P.J.M.: Wireless sensor network and beyond: a case study on transport and logistic. (CTIT Technical Report Series; No. 05-26). Centre for Telematics and Information Technology (CTIT), Enschede (2005)
3. Anastasi, G., Conti, M., Di Francesco, M., Passarella, A.: Energy conservation in wireless sensor networks: a survey. J. Ad Hoc Netw. 537–568 (2009)
4. Kuila, P., Gupta, S.K., Jana, P.K.: A novel evolutionary approach for load balanced clustering problem for wireless sensor networks. Swarm Evol. Comput. **12**, 48–56 (2013)
5. Kuila, P., Jana, P.K.: Energy efficient clustering and routing algorithms for wireless sensor networks: particle swarm optimization approach. Eng. Appl. Artif. Intell. **33**, 127–140 (2014)
6. Heizelman, W., Chandrakasan, A., Balakrishnan, H.: An application-specific protocol architecture for wireless microsensor networks. IEEE Trans. Wirel. Commun. **1**(4), 660–670 (2002)
7. Huang, H., Wu, J.: A probabilistic clustering algorithm in wireless sensor networks. In: IEEE Vehicular Technology Conference (VTC), VTC-2005-Fall, vol. 62, pp. 1796–1798 (2005)
8. Kim, J., Park, S., Han, Y., Chung, T.: CHEF: cluster head election mechanism using fuzzy logic in wireless sensor networks. In: IEEE 10th International Conference on Advanced Communication Technology (ICACT), pp. 654–659 (2008)
9. Li, C.F., Ye, M., Chen, G.H., Wu, J.: An energy-efficient unequal clustering mechanism for wireless sensor networks. In: 2nd IEEE International Conference on Mobile Ad-hoc and Sensor Systems (MASS), Washington, DC, pp. 604–611 (2005)
10. Singh, A.K., Purohit, N., Verma, S.: Fuzzy logic based clustering in wireless sensor networks: a survey. Int. J. Electron. **100**, 126–141 (2013)
11. Kumar, N., Varalakshmi, P., Murugan, K., Bavadhariny, B.S.: An efficient fuzzy clustering technique in wireless sensor networks. In: 5th International Conference on Advanced Computing, pp. 212–216 (2013)
12. Baranidharan, B., Santi, B.: Distributed load balancing unequal clustering in wireless sensor network using fuzzy approach. Appl. Soft Comput. **40**, 495–506 (2015)
13. Kumar, P., Chaturvedi, A.: Performance measures of fuzzy C-means algorithm in wireless sensor network. Int. J. Comput. Aided Eng. Technol. **9**, 84–101 (2017)
14. Abdolkarimi, M., Adabi, S., Sharifi, A.: A new multi-objective distributed fuzzy clustering algorithm for wireless sensor networks with mobile gateways. AEU - Int. J. Electron. Commun. **89**, 92–104 (2018)
15. Abidi, W., Ezzedine, T.: Fuzzy cluster head election algorithm based on LEACH protocol for wireless sensor network. In: 13th International Wireless Communications and Mobile Computing Conference, pp. 993–997 (2017)

Smart Multipurpose Robotic Car

Daud Ibrahim Dewan[1], Sandeep Chand Kumain[1(✉)], and Kamal Kumar[2]

[1] Department of Computer Science and Engineering, Tula's Institute,
Dehradun, India
ibrahimdaud03@gmail.com, skumain@tulas.edu.in
[2] Department of Computer Science and Engineering,
National Institute of Technology, Uttarakhand, Srinagar, India
kamalkumar@nituk.ac.in

Abstract. In this modern digital world, everyone is moving towards automation. Robotics allows automation where machines perform a well-defined step safely and productively, in autonomous or partial autonomous manners. Many articles have been published in recent times to present automatic/robotic cars which can perform some process such as, smart surveillance, obstacle detection, and collision aversion, etc. The design of the smart robotic car is presented through this paper and demonstrates its operations. The proposed developed model can perform multiple operations such as smart irrigation, obstacle detection, line-following, grass cutting, and vacuum cleaning. All these operations are operated or managed through the smartphone. The proposed smart model work as the expert system which utilized the features of Artificial Intelligence. The smart robotic car can sense the environment and take the proper decision without human guidance.

Keywords: Robot · Sense · Controller · Arduino · Motor · Smart irrigation

1 Introduction

In the past, generally, robotics mainly used for an automated production process in the factory. Presently, robotics finds its application in many fields such as medical science, mining, surveillance, autopilots, etc. Initially, robotics was understood to be a job eater and was seen as a destructive replacement technology. With time, robotics has emerged as a safe and viable technology in complex and unstructured conditions such as automating the number of human activities, automated driving, caring for a sick person, military sector and in the car industry, etc. In robotics design, there is mainly two-points in which the designers are focusing the first one is to build a model that can act autonomously in complex and unstructured environmental conditions. Second, the developed model has the capability of making moral decisions [1].

© The Author(s), under exclusive license to Springer Nature Switzerland AG 2021
M. Tripathi and S. Upadhyaya (Eds.): ICDLAIR 2019, LNNS 175, pp. 273–280, 2021.
https://doi.org/10.1007/978-3-030-67187-7_28

At present, robotics has emerged as a potential technology that can ease human life and enable mankind to tackle several social and ethical issues. Learning, Ambiguous understanding about the problems, Creativity for solving the problems, Reasoning and Deduction, Classification, Ability to build analogies and many more are the common features of intelligent system [2]. In fact, multipurpose systems are the need of the hour and are well accepted in tech-savvy populations.

1.1 Application Area of Robotics

In the 1970s or early day's robotics technology is emerged in the USA, but currently, it spread all over the world. Japan, Europe, Australia, South Korea are the other leading countries that have done a lot of work in this field at present[3]. The brief description of the application areas where the robotics are commonly used as follows.

1. **Robotics in manufacturing:** The involution of robotic technology made a major change in the manufacturing sector. In the harmful scenarios where there always be a life threat possibility for human workers the robots perform efficiently and effectively with great accuracy [1,5]. For example, as compared to the human the robots can able to handle and assemble the tiny part of the manufactured item with good accuracy. The use of robotics increase the efficiency of the manufacturing unit in many way's, another advantage with robotics is it can be operated 24×7 in the light-outs situation. In today's competitive market the manufactures must need to embrace the automation process with robotics [4,5].

2. **Robotics in healthcare:** Another application area where robotics is widely used in the present time is medical and healthcare. These robots are generally named as the medical robot. There are different tasks for which the medical robots are used. Some of them are Surgical robot, Telepresence robots, Companion robot, Disinfection robot.[10,11].

3. **Robotics in military:** In present time robotics play an important role in military operations. Military robots are designed for the task like gun shooting, flying or going underwater, etc. Mostly these robots used the light sensors and infrared touch sensors for performing certain task. The robotic technology has both pros and cons. It is beneficial in terms of keeping and operating continuously because the machine can't tired. The disadvantage is once these robic machine developed the ability of self-learning and become autonomous at that time it may create the threat for human beings.

4. **Robotics in transportation:** Now a day's the robotics used from driving cars to flying an airplane. Even in current times, there is a risk in a fully autonomous system, but at the rate, the technology is growing those days are not so far when almost transportation is done through automation.

In this paper, the author(s) discussed the smart robotic car which is designed for multiple purposes thats include the features of obstacle detection, line-following, grass cutting, vacuum cleaning, and smart irrigation. As the water availability in the earth is decreasing day by day along with the other features, the smart irrigation feature is very useful and helpful for avoiding the water wastage. Further, the paper is described as follows. The related work is discussed in Sect. 2, The proposed designed multipurpose robotic car model is described in Sect. 3, The Working developed prototype results regarding the model is described in Sect. 4. Section 5 is about the conclusion and future scope of the work.

2   Related Work

As the modern world is moving towards automation. Automation in the field of transport is an emerging field of research. As in the field of transportation when the automated vehicle is designed the security of human beings is top in priority. For bringing the car body panel into the accurate position an algorithm is proposed by Markus Herrmann et al. [13], this proposed algorithm was able to detect the critical deviation in the panel as the panel is being grabbed. An autonomous car robot structure is proposed by Denis Varabin et al. the proposed robotic car is capable of moving independently or with the remote control [14]. The reactive replanning problem for autonomous vehicles is addressed by Enrico Bertolazzi et al. [15]. The reactive replanning problem occurs when the unforeseen obstacle is encountered. The solution provided by the author(s) is efficient and can be implemented with the hardware and very helpful for a robotic racing car. For automated racing car the optimal motion planning problem is addressed by Tizar Rizano et al. [16]. For controlling the mecanum wheel robotic car wirelessly by using computer vision approach is proposed by Min Yan Naing et al. [17]. In this proposed method for mecanum wheel robotic car, the location of the robotic car is detected continuously and based on that detected location the robotic car approach towards the target pattern. In automation fault tolerance is important, a mobile robotic car which uses Astrocyte Neuron networks with the self-repairing feature is proposed by Junxiu Liu et al. [18]. A method for simulating the robotic car by time-delay neural network is proposed by Alberto F. De Souza et al. [19]. The car velocity and direction of moving is simulated by these networks.

3 Proposed Multipurpose Smart Robotic Car Model

The proposed multitasking car has the following features like a vacuum cleaner, grass cutting, line-following, obstacle detector and smart irrigation. All these features are smartly operated or controlled through the mobile device (smartphone) for performing the required task. The smart irrigation feature of the robotic car is integrated by keeping in focus to address the real-world problem in India, especially keeping in mind about the water wastage problem.

3.1 Features of Proposed Smart Robotic Car Model

1. **Vacuum cleaner:** Vacum cleaner is useful for removing the dust from the floors, upholstery or the other surfaces by integrating this feature in the car this service can be used whenever required and provide the portability.
2. **Grass cutting:** For keeping the yard looking tidy we generally used the grass cutter. Grasscutter has different sizes and types but all are used for keeping the grass trim.
3. **Line following:** The line follower functionality gives the ability to machine to follow the line drawn on the floor. In this case, generally, the path which will follow is predefined and this path can be visible as a black line on the white surface and the machine which integrates this feature uses Infrared Ray sensors [12]. Some of the items such as radioactive materials which is harmful to carry for the human are easily done by utilizing this feature.
4. **Obstacle detector:** During the development of self-driving cars the obstacle detection is one important feature. The integration of this feature in the car will reduce the accident chance and very useful in foggy conditions (in winters). For integrating this feature the ultrasonic sensors are used and this ultrasonic sensor is attached in front of the car.
5. **Smart irrigation:** The most important feature of the designed smart robotic car is smart irrigation. In our robotic car, the author(s) attached a container for water storage. For achieving the task of smart irrigation the soil moisture sensors are integrated with the developed system. The circuit diagram of the proposed smart multipurpose car is shown in Fig. 1

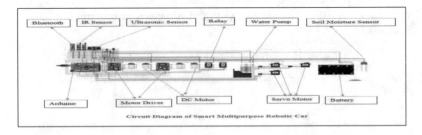

Fig. 1. Illustration of Ciruit diagram of proposed multipurpose robotic car

3.2   Devices Used for the Development

The following major/minor component is shown in Table 1 is used for complete model development.

Table 1. Description of the Used Component

Component name	Quantity
Ardunio Mega 2560	1
BlueTooth Module HC-05	1
Soil Moisture Sensor FC-28	1
DC Water Pump 5V	1
Ultrasonic Sensor	1
Servo Moter	3
Motor Driver	2
IR Sensor	2
Lithium ion Battery 7.4V	1
DC Motor 5V	2
Adapter 7.5 V	1
Bread board	1
Bread board	1
Relay 5V	1
Propaller	1

4   Results

The developed complete robotic car model which integrate the features discussed in previous section is visulaized as per link [20]. The visulaization of different modules are as follows (Fig. 2).

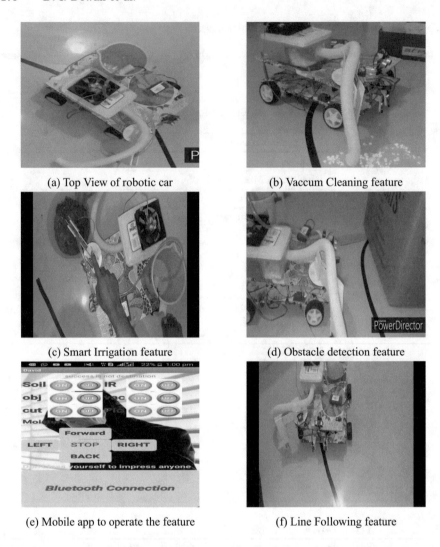

(a) Top View of robotic car

(b) Vaccum Cleaning feature

(c) Smart Irrigation feature

(d) Obstacle detection feature

(e) Mobile app to operate the feature

(f) Line Following feature

Fig. 2. Illustration of different module of robotic car

5   Conclusion and Future Work

The smart robotic car is designed for performing multiple tasks smartly. By adding the feature like the Vacuum cleaner, Grass cutting, Line following, Obstacle detection, and Smart irrigation make the car more productive and all these features are controlled by the smartphone with the help of the Bluetooth. The line following and obstacle detector feature will be very useful when we think about the smart self-driving car and it is also useful for avoiding the accidents in the presence of fog (in winter time). One of the most important features of our proposed smart car model is smart irrigation which is very useful for water in terms of proper water utilization purpose, which is the demand of the current century. The future work will be focused on including more features with the proposed robotic car, especially keeping in mind about road safety.

References

1. Royakkers, L., van Est, R.: A literature review on new robotics: automation from love to war. Int. J. Soc. Robot. **7**(5), 549–570 (2015)
2. Xu, J., Chen, G., Xie, M.: Vision-guided automatic parking for smart car. In: Proceedings of the IEEE Intelligent Vehicles Symposium 2000 (Cat. No. 00TH8511), pp. 725–730. IEEE, October 2000
3. Lin, P., Abney, K., Bekey, G.A.: Robot Ethics: The Ethical and Social Implications of Robotics. The MIT Press (2014)
4. Hedelind, M., Jackson, M.: How to improve the use of industrial robots in lean manufacturing systems. J. Manuf. Technol. Manag. **22**(7), 891–905 (2011)
5. Manufacturing. https://www.acieta.com/why-robotic-automation/robotics-manufacturing/. Accessed 3 Sept 2019
6. Moisture-Sensor. https://www.electronicwings.com/arduino/soil-moisture-sensor-interfacing-with-arduino-uno. Accessed 5 Sept 2019
7. Ultrasonic Sensor. https://www.keyence.com/ss/products/sensor/sensorbasics/ultrasonic/info/index.jsp. Accessed 8 Sept 2019
8. Ultrasonic Sensor. https://randomnerdtutorials.com/complete-guide-for-ultrasonic-sensor-hc-sr04/. Accessed 8 Sept 2019
9. Infrared Sensor. https://www.elprocus.com/infrared-ir-sensor-circuit-and-working/. Accessed 9 Sept 2019
10. Beasley, R.A.: Medical robots: current systems and research directions. J. Robot. **2012** (2012)
11. Mosavi, A., Varkonyi, A.: Learning in robotics. Int. J. Comput. Appl. **157**(1), 8–11 (2017)
12. Pakdaman, M., Sanaatiyan, M.M., Ghahroudi, M.R.: A line follower robot from design to implementation: technical issues and problems. In: 2010 The 2nd International Conference on Computer and Automation Engineering (ICCAE), vol. 1, pp. 5–9. IEEE, February 2010
13. Herrmann, M., Otesteanu, M., Otto, M.A.: A novel approach for automated car body panel fitting. In: 2013 IEEE International Symposium on Robotic and Sensors Environments (ROSE), pp. 190–195. IEEE, October 2013
14. Varabin, D., Bagaev, D.: Autonomous system car robot. In: 2012 IV International Conference Problems of Cybernetics and Informatics (PCI), pp. 1–2. IEEE, September 2012
15. Bertolazzi, E., Bevilacqua, P., Biral, F., Fontanelli, D., Frego, M., Palopoli, L.: Efficient Re-planning for Robotic Cars. In: 2018 European Control Conference (ECC), pp. 1068–1073. IEEE, June 2018
16. Rizano, T., Fontanelli, D., Palopoli, L., Pallottino, L., Salaris, P.: Global path planning for competitive robotic cars. In: 52nd IEEE Conference on Decision and Control, pp. 4510–4516. IEEE, December 2013
17. Naing, M.Y., San Oo, A., Nilkhamhang, I., Than, T.: Development of computer vision-based movement controlling in Mecanum wheel robotic car. In: 2019 First International Symposium on Instrumentation, Control, Artificial Intelligence, and Robotics (ICA-SYMP), pp. 45–48. IEEE, January 2019

18. Liu, J., Harkin, J., McDaid, L., Halliday, D.M., Tyrrell, A.M., Timmis, J.: Self-repairing mobile robotic car using astrocyte-neuron networks. In: 2016 International Joint Conference on Neural Networks (IJCNN), pp. 1379–1386. IEEE, July 2016

19. De Souza, A.F., da Silva, J.R.C., Mutz, F., Badue, C., Oliveira-Santos, T.: Simulating robotic cars using time-delay neural networks. In: 2016 International Joint Conference on Neural Networks (IJCNN), pp. 1261–1268. IEEE, July 2016

20. Developed Multipurpose robotic car model. https://www.youtube.com/watch?v=mLIruZKPmgk

Effect of Arm Posture and Isometric Hand Loading on Shoulders Muscles

Lalit Kumar Sharma[1]($\boxtimes$), Hafizurrehman[2], and M. L. Meena[2]

[1] JECRC, Jaipur, Rajasthan, India
erlksjecrc@gmail.com
[2] MNIT, Jaipur, Rajasthan, India

Abstract. The influence of external factors such as arm posture, hand loading and grip strength on upper arm muscles and shoulder muscles is predicted by this study. Study is done about muscles by using surface electromyography (EMG) in which surface electromyography is collected from three upper extremity muscles on different participants. Objective of this study is to find the effect of arm posture (considering elbow and shoulder angle), hand loading and griping strength on upper arm muscles and also to determine the better arm posture of a worker who works in standing position.

Ergonomics evaluation of posture is done and awkward posture of the workers identified to improve the posture. There are various methods to determine the awkward posture such as RULA, REBA, and EMG Tool. EMG can be applied everywhere in industry for any work to study the better posture of a person in that work. Here EMG recording is done on 16 right hand male who have not any injury of shoulder for determining a good result.

Keywords: ANOVA · Electromyography · Maximum voluntary contraction

1 Introduction

Musculoskeletal disorders (MSDs) are the most important work related health concerns of modern industrialized nations. The Bureau of Labor Statistics USA reported 522,528 total cases of work related MSDs in 2002 [2]. Common body area affected by discomfort is the neck, shoulders, hand and wrists in telecommunication sector [3]. There are eight important muscles of upper arm and shoulder. We considered for study three important muscles Biceps Brachii, Deltoid and Trapezius.

Muscle contraction causes electrical potential difference in muscles in the form of waves that is analyzed by LabChart software. Information about this type of activity in the muscle can be represented graphically or pictorially that is called an electromyogram and its recording is called an electromyographic recording.

2 Related Work

Antony and Keir [1] studied the effect of arm posture, movement and hand load on shoulder muscle activity for isometric and dynamic exertion at neutral elbow flexion

© The Author(s), under exclusive license to Springer Nature Switzerland AG 2021
M. Tripathi and S. Upadhyaya (Eds.): ICDLAIR 2019, LNNS 175, pp. 281–289, 2021.
https://doi.org/10.1007/978-3-030-67187-7_29

and neutral wrist position and at different shoulder angle in different shoulder plane (flexion, abduction, and mid abduction). Farooq and Khan [2] identified that the main effects of shoulder rotation, forearm rotation and elbow flexion angle are significant for the repetitive gripping task. The psychosocial factors were found to be involved in the etiology of almost all types of MSDs and the impact was observed more in the neck and shoulder region [3]. George et al. [4] proposed three models to estimate elbow angle. This paper is very important for designing prosthetics human limb on basis of ergonomics evaluation by using perfect elbow angle relationship with the help of EMG tool. Objective of this research paper is to experiment with common Rehabilitation exercises and to identify those which produce significant integrated electromyography (iEMG) activation of selected shoulder muscles [5]. Young et al. [6] studied to assess postures of the wrists and shoulders and associated muscle activity of these while using touch screen tablet. Reaz et al. [7] gave brief information about electromyography and revealed the different methodologies to analyze the EMG signals. Hossain et al. [8] studied to determine occurrence of WMSDs in various body regions among workers of Readymade Garment industry in Bangladesh. As most of the work in India in most of the industries and at commercial level is carried out manually so that it is very important that to design work condition such that WMSDs is decreased [9]. Electromyography can be a very helpful analytical method if implemented under appropriate circumstances and interpreted in the context of fundamental physiological, biomechanical and recording principles [10]. Increasing the grip arm distance and worktable height significantly reduces the exerted force by the biceps, erector spine and brachioradialis muscle, which helps to decrease the load on the spine [11]. In Indian saw mills most of the work is not carried out automatically so that issues of WMSDs and injury in any part of the body are important [12]. Mogk and Keir [13] conducted a study to compute the response of the forearm musculature with combinations of grip force and forearm and wrist posture.

3   Proposed Work

The purpose of this study is to determine the effect of upper arm posture and isometric hand loading on the only shoulder muscles and how it affects the work related musculoskeletal disorders and also identify that what muscles are affected when adopted any special posture. In India WMSDs is a prominent issue in the worker in any industry, even a computer operator is affected by WMSDs. Muscle contraction is observed in the form of wave. The root mean square (RMS) value is determined by using LabChart. LabChart is software which is used to record the muscle contraction in the form of wave.

3.1   Apparatus Design

As shown below in Fig. 1 apparatus (Gonio Adjuster) is made of iron, similar to human arm and consisting of four links (fixed column, upper arm, lower arm and hand). This apparatus is used to keep the arm posture at a particular angle for a fixed time period that is taken in this study as 10 s and to find out the percentage maximum voluntary contraction (%MVC) by using LabChart.

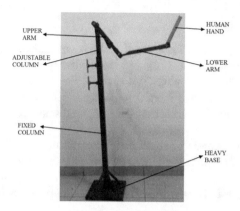

Fig. 1. Gonio adjuster.

3.2 Participants

Sixteen healthy male and right-handed college students (mean age 22.38 years), with no shoulder pain history or injury, participated in this study after providing their informed consent. Body mass, maximum grip strength and height are observed for each participant (Table 1).

Table 1. Mean participant grip strength and anthropometrics (standard deviation).

Grip strength and anthropometrics	Mean	Standard deviation
Body mass	62.57 (kg)	9.19 (kg)
Height	1.57 (m)	.0483 (m)
Grip strength	52.7 (lb)	9.02 (lb)
Age	22.38 (yr)	3.10 (yr)

3.3 Experiment Protocol

Participants performed isometric exertion at twenty specific arm postures by aligning hand along the Gonio Adjuster link. Maximum hand grip is determined by Hand Dynamometer in neutral position of hand. Three Ag-Agcl electrodes are then attached after cleaning the place at which electrode is being hold. After literature review it is found that muscles are affected in both dynamic and isometric exertion but we determined muscle contraction in two shoulder plane (flexion and abduction) and five shoulder angle with different elbow angle for three hand loading conditions no load (NL), hand dynamometer load without grip (W) and with 50% of maximum grip strength (WG). Total readings taken with each participant for each muscle is more than hundred.

3.4 Statistical Analysis

SPSS version 16 is used for various graphs plotting and statistical test that are required during this study. Arm posture and isometric hand loading for a particular work posture can be determined by the two way ANOVA analysis after getting data for those particular postures required for work and identify ergonomic arm posture and isometric hand loading for reducing muscle contraction. The followed steps for this analysis are:

i. A claim is made regarding five different arm posture as taken in this study. This claim is used to determine the following null and alternative hypothesis.

    a. **Null Hypothesis** H0: Arm posture will have no significant effect on muscle contraction.

    b. **Null Hypothesis** H0: Isometric hand loading will have no significant effect on muscle contraction.

    c. **Null Hypothesis** H0: Arm posture and Isometric hand loading will have no significant effect on the muscle contraction.

ii. Select a level of significance generally chosen level of significance 0.05 or 95% confidence level.

iii. Calculate the test statistics square of mean and other values.

iv. Check p-value: If p value is less than 0.05 then null hypothesis is rejected, so alternative hypothesis is accepted.

v. State the conclusion based on the decision made with respect to claim.

4 Result and Analysis

4.1 Statistics of Shoulder Flexion for Trapezius Muscle

As shown in Table 2, p value for each condition is less than 0.05 hence null hypothesis is rejected. It represents that at each angle trapezius muscle is contracted in shoulder flexion.

Table 2. Statistics table of shoulder flexion for trapezius muscle.

Source	Type III sum of squares	df	Mean square	F	Sig.	Partial eta squared
Corrected model	0.051	14	0.004	23.952	<0.001	0.588
Intercept	0.163	1	0.163	993.902	<0.001	0.776
LOAD	0.008	2	0.004	25.157	<0.001	0.119
POST	0.020	4	0.005	30.864	<0.001	0.316
LOAD * POST	0.014	8	0.002	11.425	<0.001	0.213

4.2   Statistics of Shoulder Abduction for Trapezius Muscle

In statistical analysis as shown in Table 3, p value is very low for all conditions hence null hypothesis is rejected. It represents that at each angle trapezius muscle is contracted in shoulder abduction.

Table 3.   Statistics table of shoulder abduction for trapezius muscle.

Source	Type III sum of squares	df	Mean square	F	Sig	Partial eta squared
Corrected model	0.024[a]	14	0.002	14.795	<0.001	0.351
Intercept	0.097	1	0.097	837.640	<0.001	0.689
LOAD	0.003	2	0.001	11.925	<0.001	0.086
POST	0.026	4	0.007	126.040	<0.001	0.423
LOAD * POST	0.009	8	0.001	9.25	<0.001	0.126

4.3   Statistics of Shoulder Flexion for Deltoid Muscle

In Table 4, p value is very low for each combination hence null hypothesis is rejected. It means deltoid muscles are affected by the arm posture and isometric hand loading.

Table 4.   Statistics table of shoulder flexion for deltoid muscle.

Source	Type III sum of squares	df	Mean square	F	Sig	Partial eta squared
Corrected model	0.041[a]	14	0.003	15.186	<0.001	0.486
Intercept	0.124	1	0.124	636.352	<0.001	0.739
LOAD	0.005	2	0.003	13.155	<0.001	0.105
POST	0.032	4	0.008	41.310	<0.001	0.438
LOAD * POST	0.004	8	0.001	2.623	0.009	0.091

4.4   Statistics of Shoulder Abduction for Deltoid Muscle

As it is shown in Table 5, p value is very low so that null hypothesis is rejected. It results that this posture for work is not good as at each angle deltoid muscle is contracted in shoulder abduction.

Table 5. Statistics table of shoulder abduction for deltoid muscle.

Source	Type III sum of squares	df	Mean square	F	Sig	Partial eta squared
Corrected model	0.155[a]	14	0.011	106.968	<0.001	0.762
Intercept	0.361	1	0.361	3.48E + 03	<0.001	0.939
LOAD	0.006	2	0.003	29.253	<0.001	0.108
POST	0.015	4	0.004	352.086	<0.001	0.218
LOAD * POST	0.003	8	<0.001	3.838	<0.001	0.089

4.5   Statistics of Shoulder Flexion for Biceps Brachii Muscle

In Table 6, p value is greater than 0.05 for load with posture so that null hypothesis is accepted. Hence there is minimum effect on biceps brachii muscles in shoulder flexion.

Table 6. Statistics table of shoulder flexion for biceps brachii muscle.

Source	Type III sum of squares	df	Mean square	F	Sig	Partial eta squared
Corrected model	0.018[a]	14	0.001	6.903	<0.001	0.285
Intercept	0.113	1	0.113	564.103	<0.001	0.715
LOAD	0.004	2	0.002	11.184	<0.001	0.090
POST	0.012	4	0.003	14.747	<0.001	0.208
LOAD * POST	0.002	8	<0.001	1.352	0.227	0.040

4.6   Statistics of Shoulder Abduction for Biceps Brachii Muscle

In Table 7, p value is very low for load with posture so that null hypothesis is rejected. It indicates that biceps brachii muscles are contracted at each angle.

Table 7. Statistics table of shoulder abduction for biceps brachii muscle.

Source	Type III sum of squares	df	Mean square	F	Sig	Partial eta squared
Corrected model	0.033[a]	14	0.002	9.387	<0.001	0.451
Intercept	0.184	1	0.184	721.073	<0.001	0.798
LOAD	0.011	2	0.005	21.368	<0.001	0.160
POST	0.016	4	0.004	15.970	<0.001	0.221
LOAD * POST	0.006	8	0.001	3.129	0.002	0.109

From the Table 8, we can conclude that for any arm posture p value is less than 0.05 except shoulder flexion for biceps brachii, for that p value is 0.227. Hence we are failed to reject null hypothesis and accepted this hypothesis. It is result out from this experiment that 95% persons can do work with minimum biceps brachii muscle contraction in flexion.

Table 8. ANOVA.

Posture	Muscles	F values	P values
Shoulder flexion	Biceps brachii	1.352	0.227
Shoulder abduction	Biceps brachii	3.129	0.002
Shoulder flexion	Deltoid	2.623	0.009
Shoulder abduction	Deltoid	3.838	<0.001
Shoulder flexion	Trapezius	11.425	<0.001
Shoulder abduction	Trapezius	9.25	<0.001

Result from the statistical analysis for selected all three shoulder muscles in flexion and abduction is as discussed and summarized below in Table 9.

Table 9. Summary of results.

Chart Title	Chart Image	Summary
Bar Chart for shoulder flexion, hand loading, and RMS value of Trapezius Muscles	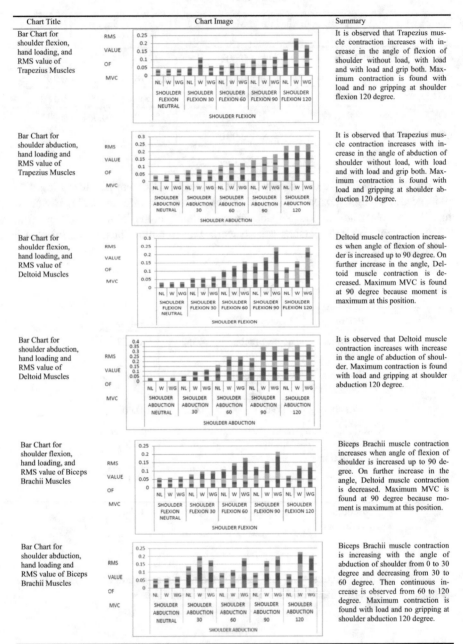	It is observed that Trapezius muscle contraction increases with increase in the angle of flexion of shoulder without load, with load and with load and grip both. Maximum contraction is found with load and no gripping at shoulder flexion 120 degree.
Bar Chart for shoulder abduction, hand loading and RMS value of Trapezius Muscles		It is observed that Trapezius muscle contraction increases with increase in the angle of abduction of shoulder without load, with load and with load and grip both. Maximum contraction is found with load and gripping at shoulder abduction 120 degree.
Bar Chart for shoulder flexion, hand loading, and RMS value of Deltoid Muscles		Deltoid muscle contraction increases when angle of flexion of shoulder is increased up to 90 degree. On further increase in the angle, Deltoid muscle contraction is decreased. Maximum MVC is found at 90 degree because moment is maximum at this position.
Bar Chart for shoulder abduction, hand loading and RMS value of Deltoid Muscles		It is observed that Deltoid muscle contraction increases with increase in the angle of abduction of shoulder. Maximum contraction is found with load and gripping at shoulder abduction 120 degree.
Bar Chart for shoulder flexion, hand loading, and RMS value of Biceps Brachii Muscles		Biceps Brachii muscle contraction increases when angle of flexion of shoulder is increased up to 90 degree. On further increase in the angle, Deltoid muscle contraction is decreased. Maximum MVC is found at 90 degree because moment is maximum at this position.
Bar Chart for shoulder abduction, hand loading and RMS value of Biceps Brachii Muscles		Biceps Brachii muscle contraction is increasing with the angle of abduction of shoulder from 0 to 30 degree and decreasing from 30 to 60 degree. Then continuous increase is observed from 60 to 120 degree. Maximum contraction is found with load and no gripping at shoulder abduction 120 degree.

5   Conclusion

It is observed that when shoulder exertion is performed simultaneously with hand grip, it leads to posture specific differential distribution of shoulder muscle activity. From statistical data, charts and ANOVA table it is clear that all three muscles Biceps Brachii, Trapezius and Deltoid are affected by the arm posture, isometric hand loading in each posture except in shoulder flexion, Biceps Brachii muscle was contracted minimum because p value is very high for this muscle.

This study can be used as reference for further study to determine a better work posture using EMG. For further improvements EMG recordings can be taken from workers in their working positions and then best fit posture may be identified for any particular work.

References

1. Antony, N.T., Keir, P.J.: Effects of posture, movement and hand load on shoulder muscle activity. J. Electromyogr. Kinesiol. **20**(2), 191–198 (2010)
2. Farooq, M., Khan, A.A.: Effect of shoulder rotation, upper arm rotation and elbow flexion in a repetitive gripping task. Work **43**(3), 263–278 (2012)
3. Crawford, J.O., Laiou, E., Spurgeon, A., McMillan, G.: Musculoskeletal disorders within the telecommunications sector—a systematic review. Int. J. Ind. Ergon. **38**(1), 56–72 (2008)
4. George, K.S., Sivanandan, K.S., Mohandas, K.P.: Estimation of elbow angle using surface electromyographic signals. J. Intell. Fuzzy Syst. **34**(6), 4191–4201 (2018)
5. Kumar, M., Srivastava, S., Das, V.S.: Electromyographic analysis of selected shoulder muscles during rehabilitation exercises. J. Back Musculoskelet. Rehabil. **31**, 947–954 (2018)
6. Young, J.G., Trudeau, M.B., Odell, D., Marinelli, K., Dennerlein, J.T.: Wrist and shoulder posture and muscle activity during touch-screen tablet use: effects of usage configuration, tablet type, and interacting hand. Work **45**(1), 59–71 (2013)
7. Reaz, M.B.I., Hussain, M.S., Mohd-Yasin, F.: Techniques of EMG signal analysis: detection, processing, classification and applications. Biol. Proced. Online **8**(1), 11 (2006)
8. Hossain, M.D., Aftab, A., Al Imam, M.H., Mahmud, I., Chowdhury, I.A., Kabir, R.I., Sarker, M.: Prevalence of work related musculoskeletal disorders (WMSDs) and ergonomic risk assessment among readymade garment workers of Bangladesh: a cross sectional study. PLoS ONE **13**(7), e0200122 (2018)
9. Ansari, N.A., Sheikh, M.J.: Evaluation of work posture by RULA and REBA: a case study. IOSR J. Mech. Civil Eng. **11**(4), 18–23 (2014)
10. Marras, W.S.: Overview of electromyography in ergonomics. In: Proceedings of the Human Factors and Ergonomics Society Annual Meeting, vol. 44, no. 30, pp. 534–5–536. SAGE Publications, Los Angeles (2000)
11. Balasubramanian, V., Swami Prasad, G.: An EMG-based ergonomic evaluation of manual bar bending. Int. J. Ind. Syst. Eng. **2**(3), 299–310 (2007)
12. Ali, A., Qutubuddin, S.M., Hebbal, S.S., Kumar, A.C.S.: An ergonomic study of work related musculoskeletal disorders among the workers working in typical Indian saw mills. Int. J. Eng. Res. Dev. **3**(9), 38–45 (2012)
13. Mogk, J., Keir, P.: The effects of posture on forearm muscle loading during gripping. Ergonomics **46**(9), 956–975 (2003)

Cancer Detection Using Convolutional Neural Network

Ishani Dabral[✉], Maheep Singh, and Krishan Kumar

Department of Computer Science and Engineering,
National Institute of Technology, Srinagar, Uttarakhand, India
{ishani.cse16,maheepsingh,kkberwal}@nituk.ac.in

Abstract. Cancer is a dreadful heterogeneous disease that refers to abnormal growth of cell tissue. Millions of people die every year due to unnoticed recognition or late detection of the disease. Hence, detecting presence of cancer cells correctly becomes important. Putting into attention the number of people affected by cancer, we need to find better ways of diagnosis and treatment. Neural networks are a burning research area in medical science, especially in the areas of radiology, cardiology, oncology, urology and etc.. A variety of these techniques as Decision Trees (DTs), Support Vector Machines (SVM) and Artificial Neural Network (ANN) have been widely applied in cancer detection. This paper presents a method of classification of cancer cells into Benign and Malignant using deep learning Convolutional Neural Network.

1 Introduction

Cancer is one of the major reason to cause death in the world [1]. American Cancer Society performed a survey that showed that around 600,920 people would die from cancers in USA in 2017 [2]. Thus, fighting against cancers is one of the important and a big challenge faced by clinic doctors as well as research scientists. A study on Breast Cancer Care shows that 42% of National Health Service (NHS) trusts say that they hardly have any staffs to assign individual with limited cancer specialist nurse [3]. It is a major reason that causes low survival rate of cancer all over the world. Due to lack of specialist doctor or nurse, the cancer remains undetected for a longer duration thus reducing chances of optimal treatment and recovery. Early identification and detection plays a key role in cancer diagnosis and can lead to long survival rate. Manual interpretation and assemblance of enormous medical images is a time consuming and tedious task. Therefore, cancer detection techniques using neural networks are being introduced that would result in early diagnosis and treatment. Various methods of feature extraction have been investigated for different cancer types. In past few years, advanced analysis of medical imaging using deep learning, machine learning, including Convolutional Neural Networks (CNNs), has been explored [4]. These approaches offer great promises for future applications for both diagnostic and predictive purposes.

M. Tripathi and S. Upadhyaya (Eds.): ICDLAIR 2019, LNNS 175, pp. 290–298, 2021.
https://doi.org/10.1007/978-3-030-67187-7_30

In this paper, we have used deep learning technology as the main algorithm so as to recognize cancer cells. A classification occurs mainly when an object needs to be assigned to a predefined group or class based on a number of observed attributes related to that object. Mainly in this deep learning technology, Convolutional Neural Network has been used for classifying the images. CNN is basically a kind of multilayer perceptron that uses small sub-regions also called as receptive fields. These fields are then tied to cover the entire image and then produce the feature maps to share same weights and biases. Hence this architecture reduces the number of parameters which are needed for a neural network learning process and thus increases the learning efficiency. Recognition of cancer cells is usually solved in two steps- preprocessing the dataset images and subsequent classification. For classification purpose, we have used ConvNet model that was trained and tested on the Invasive Ductal Carcinoma (IDC) dataset [5]. For example-Fig. 1 shows a cancer cell image.

The rest of the paper is organized as follows: Sect. 2 that provides information about the related work done in this specific area. Section 3 gives description about the architecture used to implement classification along with the proposed implementation, while the Experimental Results carried out on Invasive Ductal Carcinoma (IDC) Dataset are presented in Sect. 4. Finally the conclusions and future work are drawn in Sect. 5.

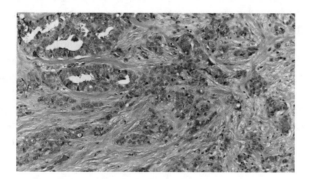

Fig. 1. Cancer cell

2 Literature Survey

There are many researches in the literature dealing with Cancer detection and recognition problem. Albayrak et al. developed a deep learning based algorithm that used CNN model to extract features which were then used to train a Support Vector Machine (SVM), to detect mitosis in breast histopathological images [6]. Spanhol et al. used AlexNet to classify benign and malignant tumors from the histopathological images [7]. A skin classification algorithm was proposed by Pomponiu et al. that is by applying a pre-trained CNN, AlexNet, to generate high-level feature representation of the skin samples which where then used

to train a k nearest neighbour (KNN) classifier [8]. A deep cascade network
was proposed by Chen et al. for detection of mitosis in breast histology slides
[9]. They initially trained a FCN model to extract mitosis candidates and then
tuned a CaffeNet model. Xu et al. proposed a Stacked Sparse Auto-Encoder
(SSAE) algorithm to classify nuclei in the cancer cell images [10]. Wichakam et
al. proposed a system combining SVM and deep CNN on digital mammograms
for mass detection [11]. Along with using pre-trained models, some CNN models
have been developed for classification. Demyanov et al. proposed deep CNNs
method to detect two types of patterns regular globules and typical network) [12].
The CNN was trained by stochastic gradient descent algorithm. A CNNs based
method was presented by Sabouri et al. for skin lesions recognition using border
detection system. Swiderski et al. presented a way to prevent the overfitting of
CNN models when the data is limited [13]. They used statistical self-similarity
and non-negative matrix factorization (NMF) to enrich the training data.

Apart from Convolutional Neural Network, various other deep learning techniques have been used to implement the same on mammographic images. A mass
detection algorithm was proposed by Dhungel et al. for mammograms. In the
proposed method, a cascade of random forest classifiers and deep learning were
used. Kim et al. proposed a 3-D multiview deep ConvoNet model to learn latent
bilateral feature representation of digital breast tomosynthesis (DBT) volume.
Higher level features were extracted from two VOIs seperately using two CNNs.
Masood et al. proposed learning model for melanoma detection that was semi-
supervised. In the proposed system [14], two support vector machines (SVM) and
a deep network with polynomial kernel and radial basis function (RBF) kernel
were trained on three different datasets generated from unlabeled and labeled
data inclusively. Majtner et al. proposed a classification system combining deep
features and hand-crafted features for melanoma recognition. A deep neural
network was proposed by Sabbaghi et al. that learned and thereafter mapped
high-level image representation into bag-of-features (BoF) space to enhance the
classification accuracy. Yu et al. proposed deep residual networks for recognizing
melanoma in dermascopy images using a Fully Convolutional Residual Network
(FCRN).

3   Architecture

The architecture used in the present work departs from traditional ConvoNets
by the type and number of convolve and pooling layers used.

The architecture uses the ConvoNet architecture layers that are convolution
layer, dropout, fully connected layer for definite epochs to train the model.

3.1   Convolutional Neural Network

A neural network is a mathematical model, similar to biological neural network,
based on neural units-artificial neurons. Typically, neurons are organized in layers and the connection are set up between neurons only from adjacent layers.

The input feature vector is given to the first layer and, moving from layer to layer, is transformed to a high level features vector. The output layer neurons finally lead to classification and there number is equal to number of classifying classes. Each convolutional layer consists of a set of trainable filters and an activation map is obtained by computing dot product between these filters and layer input. Filters are also called kernels. They allow detection of same features in different locations. The activation function used is RELU function and Sigmoid function [15].

Table 1. Layered Structure of proposed implementation

Layer	Description
Input	$50 \times 50 \times 3$
Convolution 32	3×3, Valid padding, 1×1 stride
Relu	
MaxPooling	2×2
Convolution 64	3×3, Valid padding, 1×1 stride
Relu	
MaxPooling	2×2 , stride $= 2$
Convolution 128	3×3, 1×1 stride
Relu	
Convolution 256	3×3, 1×1 stride
Flatten	
Dropout	
Relu	
Fully Connected	128
Relu	
Dropout	
Fully Connected	128
Relu	
Fully Connected	2
Sigmoid function	

3.2 Proposed Implementation

- Initially, all the dataset images are resized. This is done as few of the images in dataset are not of size $50 \times 50 \times 3$.
- Images are normalized by dividing it by 255 to ensure that all the values are between 0 and 1. This is done to increase the model's training efficiency and to train the model faster. It also prevents us from falling into the exploding gradients problem.

- Data imbalancing is handled by randomly undersampling the majority class. The samples from the majority class are removed to make them equal to minority class.
- Data Augmentation is done to generalize the model by reducing network's capacity to overfit the training data. Rotation, horizontal and vertical flipping techniques are performed to ensure the same. These steps are necessary to ensure that all classes have sufficient number of images so that it is easy for the network to predict all kinds of classes and not just a few of them on which the network is trained.

Training Process: We have used deep learning library Keras. Training and testing was done using the Invasive Ductal Carcinoma (IDC) dataset. The developed method can classify the cells into benign and malignant. Table 1 describes the network architecture. The architecture includes several convolutional layers, Pooling, Fully connected, Dropout layers as shown in Table 1.

Convolution Layer: It comprises of a set of independent filters such that each filter is independently convolved with the image.

$$h_{i,j} = \sum_{k=1}^{m} \sum_{l=1}^{m} w_{k,l} . x_{i+k-1,j+l-1} \tag{1}$$

where, m-Kernel width and height
h-Convolution Output
x-Input
w-Convolution Kernel

Pooling Layer: It progressively reduces the size of representation to reduce the amount of computation and parameters in the network. The approach used here is MaxPooling [1].

$$h_{i,j} = max\{x_{i+k-1,j+l-1} \forall 1 \le k \le m \ and 1 \le l \le m\} \tag{2}$$

Relu and Sigmoid: These are the non linearity added to the network.

$$RELU = \begin{cases} x, & \text{if } x > 0 \\ 0, & \text{if } x \le 0 \end{cases} \tag{3}$$

$$Sigmoid = \frac{1}{(1 + e^{-(ax)})} \tag{4}$$

Dropout: To reduce the chances of overfitting, some of the neurons are dropped [16].

Fully Connected Layer: This layer have full connections to all the activations in the previous layer and their activation is computed with a matrix multiplication followed by a bias offset. This layer takes an input volume from the preceding layers and outputs a n dimension vector where n is the number of classes.

Flatten: This layer is used to convert 3D feature maps to 1d feature vectors.

The model was trained in 37 epochs, using a batchsize of 256. The optimizer used is Adam Optimizer to update weights and minimize loss, with initial learning rate as 0.00001.

4   Experimental Evaluation and Results

This section present the experimental results carried out using ConvNets for detecting cancer cell. We first discuss the dataset used for the experiment and then the results. We have used Python language for the implementation of this architecture.

4.1   Invasive Ductal Carcinoma Dataset Visualization

The Invasive Ductal Carcinoma Dataset is comprised of two different cancer cell types specified as Benign and Malignant. Each image can be represented as a 50 × 50 × 3 array of RGB pixels in the range of 0 to 255. The details of dataset are shown in Table 2 and Fig. 2 shows the classes of the dataset.

Table 2. Dataset summary

Class 0	44478
Class 1	15522
Shape of images	50 × 50 × 3
Number of unique classes	2

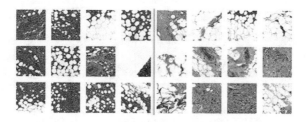

Fig. 2. Classes of Invasive Ductal Carcinoma dataset

4.2   Testing Results

Data Resizing and Data Augmentation was done on the dataset images to ensure proper training efficiency. The training of the proposed model took 7 h on an intel CORE i3 using google colab platform. The model implementation gave the following quantitative measures as shown in Table 3 and Fig. 3 gives idea about training and testing set loss:

The confusion matrix is a binary classification matrix having four quadrants, True Positive (Predicted malignant and actually malignant), False Positive (Predicted malignant but actually benign), False Negative (Predicted benign but actually malignant), True Negative (Predicted benign and actually benign) as shown in Table 4. Model's performance is evaluated using confusion matrix as shown in Table 5.

Table 3. Experimental result

Batch size	Epochs	Test accuracy
256	37	79.10%

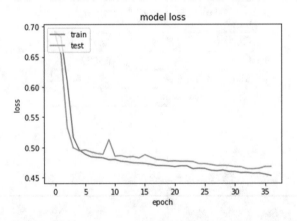

Fig. 3. Training and Testing set loss

4.3   Comparison with the State-of-Art

CNNs were used by Araujo et al. to perform classification with an overall accuracy of 66.7% with an increased model complexity (5 convolutional layer model, 50 epochs, back-propagation algorithm for weight update) for four classes and 77.6% for two classes [16].

In our work, the overall accuracy gets increased to 79.10% for two classes classification using 37 epochs. Enhancing the hyperparameters further can give better results and improved accuracy.

Table 4. Resultant confusion matrix

1761	551
416	1896

Table 5. Confusion Matrix representation in four quadrants

Predicted malignant and actually malignant	Predicted malignant and actually benign
Predicted benign but actually malignant	Predicted benign and actually benign

5 Conclusion

This paper presented an architecture that represents an effective way of detecting cancer cells whether they are benign or malignant. The experimental results achieved after testing on Invasive Ductal Carcinoma Dataset, conclude that the model is invariant to scale and viewing angle. The quantitative results validated that the model can be used to detect the cancerous cells on an average scale.

Apart from the promising results presented in the paper, there are different aspects of future research. As a future work, the model can be enhanced by using other non linear activation functions (such as ELU function) as well as parallel layers that would result in an increase in accuracy.

References

1. Chen, W., Zheng, R., Baade, P.D., Zhang, S., Zeng, H., Bray, F., Jemal, A., Yu, X.Q., He, J.: Cancer statistics in China, 2015. CA: Cancer J. Clin. **66**(2), 115–132 (2016)
2. Cancer facts & figures 2017. American Cancer Society (2017)
3. Stevens, C., Vrinten, C., Smith, S.G., Waller, J., Beeken, R.J.: Acceptability of receiving lifestyle advice at cervical, breast and bowel cancer screening. Prev. Med. **120**, 19–25 (2019)
4. Lambin, P., Rios-Velazquez, E., Leijenaar, R., Carvalho, S., Van Stiphout, R.G., Granton, P., Zegers, C.M., Gillies, R., Boellard, R., Dekker, A., Aerts, H.J.: Radiomics: extracting more information from medical images using advanced feature analysis. Eur. J. Cancer **48**(4), 441–446 (2012)
5. Cancer Dataset. https://www.kaggle.com/paultimothymooney/breast-histopathology-images
6. Albayrak, A., Bilgin, G.: Mitosis detection using convolutional neural network based features. In: Proceedings of IEEE Seventeenth International Symposium on Computational Intelligence and Informatics (CINTI), pp. 000335–000340 (2016)
7. Krizhevsky, A., Sutskever, I., Hinton, G.E.: ImageNet classification with deep convolutional neural networks. In: Advances in Neural Information Processing Systems, pp. 1097–1105 (2012)

8. Shiraishi, J., Katsuragawa, S., Ikezoe, J., Matsumoto, T., Kobayashi, T., Komatsu, K.I., Matsui, M., Fujita, H., Kodera, Y., Doi, K.: Development of a digital image database for chest radiographs with and without a lung nodule: receiver operating characteristic analysis of radiologists' detection of pulmonary nodules. Am. J. Roentgenol. **174**(1), 71–74 (2000)

9. Jia, Y., Shelhamer, E., Donahue, J., Karayev, S., Long, J., Girshick, R., Guadarrama, S., Darrell, T.: Caffe: convolutional architecture for fast feature embedding. In: Proceedings of the 22nd ACM International Conference on Multimedia, pp. 675–678. ACM, November 2014

10. Xu, J., Xiang, L., Hang, R., Wu, J.: Autoencoder, Stacked Sparse, (SSAE) based framework for nuclei patch classification on breast cancer histopathology. In: 2014 IEEE 11th International Symposium on Biomedical Imaging (ISBI), Beijing, pp. 999–1002 (2014)

11. Wichakam, I., Vateekul, P.: Combining deep convolutional networks and SVMs for mass detection on digital mammograms. In: 2016 8th International Conference on Knowledge and Smart Technology (KST), Chiangmai, pp. 239–244 (2016)

12. Demyanov, S., Chakravorty, R., Abedini, M., Halpern, A., Garnavi, R.: Classification of dermoscopy patterns using deep convolutional neural networks. In: 2016 IEEE 13th International Symposium on Biomedical Imaging (ISBI), Prague, pp. 364–368 (2016)

13. Kim, D.H., Kim, S.T., Ro, Y.M.: Latent feature representation with 3-D multi-view deep convolutional neural network for bilateral analysis in digital breast tomosynthesis. In: 2016 IEEE International Conference on Acoustics, Speech and Signal Processing (ICASSP), Shanghai, pp. 927–931 (2016)

14. Mahbod, A., Schaefer, G., Wang, C., Ecker, R., Ellinge, I.: Skin lesion classification using hybrid deep neural networks. In: ICASSP 2019-2019 IEEE International Conference on Acoustics, Speech and Signal Processing (ICASSP), pp. 1229–1233. IEEE, April 2019

15. Hu, Z., Tang, J., Wang, Z., Zhang, K., Zhang, L., Sun, Q.: Deep learning for image-based cancer detection and diagnosis a survey. Pattern Recogn. **83**, 134–149 (2018)

16. Araújo, T., Aresta, G., Castro, E., Rouco, J., Aguiar, P., Eloy, C., Polónia, A., Campilho, A.: Classification of breast cancer histology images using convolutional neural networks. PloS One **12**(6), e0177544 (2017)

POCONET: A Pathway to Safety

Pankaj Pundir$^{(\boxtimes)}$, Shrey Gupta, Ravi Singh Patel, Rahul Goswami,
Deepak Singh, and Krishan Kumar

National Institute of Technology, Uttarakhand, Srinagar, India
{pankaj369.cse16,shreyg29.civ16,ravisingh.cse16,rahulg.cse16,
deepaksingh16.cse,kkberwal}@nituk.ac.in

Abstract. Potholes are an ever-increasing problem in India leading to several accidents and wear & tear in the vehicles. As the size of potholes increases with time, therefore, detection and redressal of these potholes at early stages is essential which can avert accidents and provide a comfortable ride to the road users. In this paper, a robust, fully automated, versatile pothole detection system POCONet- POthole COnvolution Network, is proposed for improving the existing pothole detection methods and accurately detecting a pothole. An object detection model, trained by using a multi-layered convolution network YOLOv2, is used for detecting the potholes in the video feed collected from smartphones and is then marked on the maps using their respective GPS location. The trial runs, using the model depicts an accuracy level above 88% at different speeds of observing vehicle. Thus, the proposed system provides a cost-efficient, fully automated road assessment technique, which includes the detection of potholes and their marking on the map services.

Keywords: Automatic road assessment · Deep learning · YOLOv2 · Pothole detection · Computer vision · GPS

1 Introduction

India is a populous developing country where roads serve as the main mode of transportation, and with an increasing population, there has been significant vehicular growth on roads. This increase not only generates problems related to traffic congestion but also aggravates the formation of potholes. Their formation initiates with the development of cracks that are formed due to fatigue from traffic, thermal stresses, and warping stresses [1]. The freezing and thawing action of water that seeps through these cracks leads to their widening and ultimately, results in the formation of potholes, which can turn into a major purpose behind high-chance of mishaps and loss of human lives. According to a report "Road Accidents in India, 2017", by the Ministry of Road Transport and Highways (MORTH), potholes across the nation have claimed 3,597 lives in the year 2017 and are responsible for around 10 deaths each day across the country. Due to the action of water and high traffic density on the potholes, the cracking worsens and the hole gets bigger. Therefore, detection and redressal of these potholes

© The Author(s), under exclusive license to Springer Nature Switzerland AG 2021
M. Tripathi and S. Upadhyaya (Eds.): ICDLAIR 2019, LNNS 175, pp. 299–306, 2021.
https://doi.org/10.1007/978-3-030-67187-7_31

at an early stage is essential which can avert accidents and provide a comfortable ride to the road users. In the current practice, potholes detection is mostly performed manually, which is labor-intensive and time-consuming. Furthermore, minor potholes are generally ignored which may grow bigger with time. To overcome these limitations, several automated detection techniques are proposed in the past which include 3D pavement reconstruction methods [2], 3D laser scanning and stereo-vision [3], and vibration-based approaches [4]. However, these techniques come with their own set of limitations. 3D laser scanning systems are very expensive that makes them unfeasible to be installed at the individual vehicle level. Techniques based on stereo-vision require accurate alignment of two digital cameras which may be affected due to vibration and vehicular-motion. Vibration-based approaches may produce vague outcomes by assuming road joints as potholes. Also, this technique would allow only the detection of potholes that will have an encounter with the wheels. Thus, a robust, fully automated, versatile pothole detection system POCONet- **PO**thole **CO**nvolution **Net**work, is proposed for improving the existing pothole detection methods and accurately detecting a pothole. In this system, we have considered the pothole detection problem as object detection and used YOLOv2 deep learning architecture [5]. The proposed system estimates the intensity of a road segment by taking into account the count of potholes in the area and marks it on google maps using GPS (Global Positioning System) location of potholes.

The rest of the paper is organized as follows: In Sect. 2, a brief overview of past research has been provided followed by the proposed work in Sect. 3. Section 4 and Sect. 5 provide results of the trial runs and conclusion respectively.

2   Related Work

Various techniques have been developed in the past to identify and track the potholes. These techniques are based on using classical machine learning, Image Processing techniques and sensors based. Some of the major work related to this is discussed below.

2.1   Using Image Processing and Classical Machine Learning

In the paper [6], the authors developed a model in which LED light and cameras are used to identify the 3D cross-section of potholes in the pavement. The limitation of this model is that the outcomes get affected by LED light intensity and natural elements. [7] proposed a model based on Support Vector Machine to distinguish potholes from another type of distortion on roads, like split and cracks. Another model proposed in [8] uses 2D vision-based system for the detection but fails to spot the defects if images are not correctly enlightened.

One of the detection systems proposed by the authors of the [9] uses image processing and spectral clustering for detection and giving rough estimation of potholes. It specifically uses image segmentation and histogram based data collected from segmentation of gray-scaled images to detect potholes.

2.2   Using Sensors and Smartphone

Most of the smartphones nowadays come with inbuilt accelerometer and gyro-scope sensors which are used to find the speed and alignment of the device. These sensors can be used to track the motion of the vehicles and identify the pothole using the continuous sensors data. [4] suggested a pothole detection model making use of Android phones with accelerometers giving results in real-time. In another study by [10] an android platform was created for detecting road hazards based on three components; Sensing, Analysing and Sharing. The data gathered from sensors is analyzed and then disseminated to the central application.

One another approach by [11] utilizes the GPS framework and 3-axis accelerometer to recognize the correct area of potholes, and the output of the system is passed to a data cleaning algorithm and roughness of the pothole and its severity level is decided.

3   Proposed Work

In this paper, we have proposed a complete system to autonomously detect the pothole and update the information in the database. For detection of pothole a deep learning based object detection model is used, to make this module robust and more accurate. Web service is maintained to store the data and make it publicly available to the citizens.

This section is divided into two parts. The first section explains dataset collection along with training of our object detection model, while the second section demonstrates the working of our system.

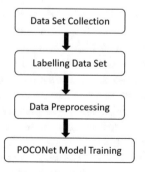

Fig. 1. Model training pipeline

3.1   Model Training

In this section complete pipeline for training the POCONet model is explained. The major set of processes involved in the training is shown in Fig. 1.

Dataset Collection. The pothole images were gathered using web scraping. To decrease false detection and increase robustness, varying conditioned pothole images were preferred. The dataset comprises of 342 images with 817 potholes.

Labelling Data Set. The images were labeled and annotated with the bounding box representing the size and location of the pothole. A graphical image annotation open-source library labelImg [12] is used, to mark the potholes in road images and determine the coordinates of the potholes for making the bounding box around the potholes. This automates the process of storing the coordinates and creating the XML [13] file for feeding coordinate information for training the model.

Data Preprocessing. Gathered images are preprocessed to maintain a similar dimension and enhance the features. Images are resized to width 1280 px and height 720 px, this proportion is estimated by using the camera in landscape mode and capture quality data within the frame. To enhance the image and adjust the contrast, Contrast limited adaptive histogram equalization (CLAHE) [14] enhancement method is used. As the normal histogram equalization technique tends to amplify the noise in the homogenous region, CLAHE [14] (adaptive equalization technique) computes the histogram for several regions within the image and redistributes the lightness value within the section.

Training Model. For training the object detection model YOLOv2 (You Only Look Once) [5] architecture is used. It is a multi-layered convolution neural network used for general purpose object detection. It can detect objects and provide bounding boxes for the detected object in real-time with high accuracy. This darknet architecture-based model can detect a wide variety of over 9000 different object classes and can process the images at a rate of about 40 frames per second. Therefore, it is suitable for use in processing and providing bounding boxes around objects in live video streams in real-time. In our system, we have used YOLOv2 pretrained weights on imagenet dataset [15] and fine-tuned the weights using these 342 images, with the train test split of 80–20. The learning rate was set to 0.0001 with a limit of 10 maximum bounding boxes per image. The performance measure used is mAP (mean average precision) and the testing score is 67.

3.2   POCONet Implementation

To use the power of crowd-sourcing and make this system robust we proposed a system to track the active potholes within the cities and make this data available

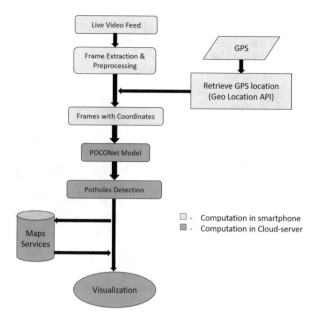

Fig. 2. Proposed architecture

to the citizens and the concerned authorities. The overview of our system pipeline is given in the Fig. 2.

Smartphone Based Computation. The proposed system uses a collection of the input data in real-time in the form of the road video feed. The smartphone is mounted on the windshield and the angle of device is maintained such that the major road portion is covered. An android app is developed to capture the video feed and record the location coordinates at each instant using GPS service, which is available in the smartphone. A general camera in a smartphone captures 30 frames per second (FPS), which results in a lot of redundant information to process. By conducting several trial experiments, using a 16 Mega Pixel camera of the smartphone-"Realme 3 Pro", at different vehicle speeds (20–80 kmph), it was observed that extracting three frames per second (FPS) provides optimal video data, which in addition to 90% data savings, is able to capture potholes images even at high speed. The frames are processed as explained in Sect. 3.1. The image frame along with its corresponding geolocation is uploaded to the server as shown in Fig. 2.

Server Based Computation. Each frame is fed into the YOLOv2 model and the output bounding box vector is stored and mapped with the image geolocation. This data frame is sent to the visualization module to display the image and maps the location with google maps API [16]. To convey the road condition more efficiently, color coding is used to mark the location on the maps. Pothole count is used to classifying the road condition, this red color is used to show high pothole count within the region as shown in Fig. 3. Users can also view the actual pothole image using the service.

(a)

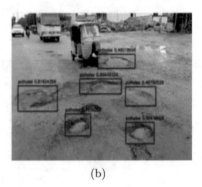
(b)

Fig. 3. a. Green and red marker are used to plot the intensity of pothole on the maps. b. the image of the pothole on the marker location.

4   Results

We evaluated the system by feeding a video with its corresponding GPS location, of a road, captured using a smartphone. Three test runs with different speed ranges were carried out in an urban road in India which has a sufficient number of potholes. The speed range was chosen on the observation of the usual speed of vehicles when passing over the potholes. As shown in Table 1, the system is able to spot 89.8%, 89.2%, 88.6% of the potholes in the video in the test runs with speed of 20–30 kmph, 30–40 kmph, 40–50 kmph respectively. This shows that the accuracy in detection is almost independent of the speed of the observing vehicle. Other pothole detection system using mobile sensors [17] resulted in precision of 88% and [18] image processing based method showing precision 81.8%, shows the object detection technique provides better performance for pothole detection. Different stages of our system includes CLAHE equalization and predicted bounding boxes are visualized in the Fig. 4.

Table 1. Accuracy in detection of potholes at different speed range

Speed range (kmph)	20–30	30–40	40–50
Accuracy (%)	89.8	89.2	88.6

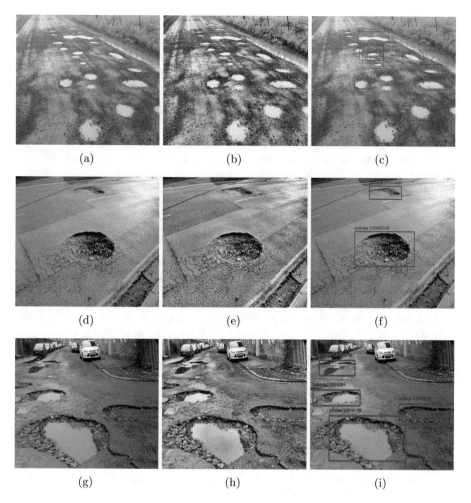

Fig. 4. (a, d, g) Input image, (b, e, h) CLAHE equalization, (c, f, i) predicted potholes

5   Conclusion

Our proposed system provides a fully automated road assessment technique, which includes the detection of potholes and their marking on the map services. This is a cost-efficient method as the only hardware device is the user's smartphone. When used together with Google Maps routes, it will help in producing more accurate and desirable paths for the smooth travel experience. Future works include extending our model for crowd-sourcing, since our model has a bare minimum requirement of a camera smartphone with GPS services, we can use video feeds from various people, traveling at different locations. Thus in return uploading local road condition to server. The system can be improved further, but pothole's information needs to be updated periodically, to check if any pothole has been repaired by the concerned authorities. To extend our systems capability, the model can be further trained to identify and mark the speed-breakers location to warn the drivers of the upcoming breakers, as the lack of their visibility is also one of the reasons behind major road accidents.

References

1. Choubane, B., Tia, M.A.N.G.: Nonlinear temperature gradient effect on maximum warping stresses in rigid pavements. Transp. Res. Rec. **1370**(1), 11 (1992)
2. Jog, G.M., et al.: Pothole properties measurement through visual 2D recognition and 3D reconstruction. Comput. Civ. Eng. **2012**, 553–560 (2012)
3. Zhang, Z., et al.: An efficient algorithm for pothole detection using stereo vision. In: 2014 IEEE International Conference on Acoustics, Speech and Signal Processing (ICASSP). IEEE (2014)
4. Mednis, A., et al.: Real time pothole detection using android smartphones with accelerometers. In: 2011 International Conference on Distributed Computing in Sensor Systems and Workshops (DCOSS). IEEE (2011)
5. Redmon, J., Farhadi, A.: YOLO9000: better, faster, stronger. In: Proceedings of the IEEE Conference on Computer Vision and Pattern Recognition (2017)
6. Youquan, H., et al.: A research of pavement potholes detection based on three-dimensional projection transformation. In: 2011 4th International Congress on Image and Signal Processing, vol. 4. IEEE (2011)
7. Madli, R., et al.: Automatic detection and notification of potholes and humps on roads to aid drivers. IEEE Sens. J. **15**(8), 4313–4318 (2015)
8. Murthy, S., Bharadwaj, S., Varaprasad, G.: Detection of potholes in autonomous vehicle. IET Intell. Transp. Syst. **8**(6), 543–549 (2014)
9. Buza, E., Omanovic, S., Huseinovic, A.: Pothole detection with image processing and spectral clustering. In: Proceedings of the 2nd International Conference on Information Technology and Computer Networks, vol. 810 (2013)
10. Orhan, F., Eren, P.E.: Road hazard detection and sharing with multimodal sensor analysis on smartphones. In: 2013 Seventh International Conference on Next Generation Mobile Apps, Services and Technologies. IEEE (2013)
11. Chen, K., et al.: Road condition monitoring using on-board three-axis accelerometer and GPS sensor. In: 2011 6th International ICST conference on communications and networking in China (CHINACOM). IEEE (2011)
12. Tzutalin. LabelImg. Git code (2015). https://github.com/tzutalin/labelImg
13. Bray, T., et al.: Extensible markup language (XML) 1.0 (2000) 2000-10
14. Yadav, G., Maheshwari, S., Agarwal, A.: Contrast limited adaptive histogram equalization based enhancement for real time video system. In: 2014 International Conference on Advances in Computing, Communications and Informatics (ICACCI). IEEE (2014)
15. Krizhevsky, A., Sutskever, I., Hinton, G.E.: ImageNet classification with deep convolutional neural networks. In: Advances in Neural Information Processing Systems (2012)
16. Svennerberg, G.: Beginning Google Maps API 3. Apress, New York (2010)
17. Jo, Y., Ryu, S.: Pothole detection system using a black-box camera. Sensors **15**(11), 29316–29331 (2015)
18. Nienaber, S., Booysen, M.J., Kroon, R.S.: Detecting potholes using simple image processing techniques and real-world footage (2015)

Comparative Study of Object Recognition Algorithms for Effective Electronic Travel Aids

Rashika Joshi[1], Meenakshi Tripathi[1], Amit Kumar[2(✉)], and Manoj Singh Gaur[3]

[1] Department of Computer Science and Engineering, MNIT Jaipur, Jaipur, India
er.rashika@gmail.com, mtripathi.cse@mnit.ac.in
[2] Department of Computer Science and Engineering, IIIT Kota, Kota, India
amit@iiitkota.ac.in
[3] Department of Computer Science and Engineering, IIT Jammu, Jammu, India
gaurms@gmail.com

Abstract. Object recognition involves detecting the probability of presence and class of one or more objects in a given image. In this paper we carried out a comparison between different state of art approaches and concluded which approach is more user-friendly, cost-effective and produces more accurate results for object recognition. We deployed various algorithms like You Only Look Once (YOLO), Single Shot Multibox Detector (SSD), Faster RCNN (Region Convolution Neural Network) on diverse Convolution neural network models (CNNs) namely Mobile Net, Inception, Res Net that act as base networks when used with detection algorithms. Subsequently, using varying development boards like Jetson Nano and Raspberry Pi 3 B+ the speed, latency and accuracy values were evaluated for different combinations of base networks and detection algorithms.

Keywords: Jetson Nano · Convolution neural networks · YOLO · SSD · Epoch

1 Introduction

According to information released by the World Health Organization in [1], 285 million individuals globally who make up 4% of the total population are projected to be visually impaired, 246 million of whom are blind and 39 million others have low sight. This implies that every five seconds someone in our globe goes blind. The prevalence of blindness is higher among those in lower socioeconomic status. As, per census 2011, 20% of the disabled persons in India, are having disability in movement, 19% are with disability in seeing and another 19% are with disability in hearing, 8% have multiple disabilities.

Over the past few decades, research has been regulated for new devices to design a reliable and self-sufficient system for visually impaired persons to detect objects in path and warn them at danger places. In 1940s, with the introduction of the long cane [2] and techniques for using the same, it became apparent that mobility of blind pedestrians could be improved significantly with the help of training. To provide the same, specialists of different type called mobility specialists came into view. This long cane had its own

© The Author(s), under exclusive license to Springer Nature Switzerland AG 2021
M. Tripathi and S. Upadhyaya (Eds.): ICDLAIR 2019, LNNS 175, pp. 307–316, 2021.
https://doi.org/10.1007/978-3-030-67187-7_32

disadvantages as it needed rigorous training and concentration to master the skill for one to move independently. Many upgraded variants of long cane also appeared during that era such as foldable cane, kiddie etc. which can be used by children also.

The mobility of a blind pedestrian using an electronic travel aid should approach the mobility of a sighted pedestrian. The electronic travel aid should make a difference that is obvious and indisputably significant. A successful travel aid should enable independent, efficient, effective, and safe travel in unfamiliar surroundings.

The main contribution of this paper is to provide an efficient object recognition algorithm for researchers based on various performance parameters. The rest of the paper is arranged as follows: Section 2 discusses the literature survey of the work done followed by the methodology in Sect. 3; Results are mentioned in Sect. 4; finally, Sect. 5 wraps up the paper with a brief Conclusion.

2   Literature Review

Hardware specific approach for obstacle detection has been widely used over a long time in electronic travel aids (ETAs). Many ETAs have already captured the market. In [3], water sensors and ultrasonic sensors have been used as a means to detect the presence of water and obstacles. Thereby, notify the blind person via buzzer and vibrator. RF transmitter and receiver in it help to search the stick easily. The discussion in [4] shows that with the use of ultrasonic sensors distance from obstacle can easily be calculated and also about the presence of bumps and potholes. And the feedback is provided with the help of vibration output. User can also capture picture using Smartphone camera to know about surroundings. This picture is processed using various image processing techniques to recognize objects in image.

IoT based smart walking cane [5] took advantage of 5 ultrasonic sensors placed at different heights to recognize obstacles of each level. It also has 3 motion sensors each with a view angle of 95 that is placed left, right and front that covers a total range of 285. Based on motion detected by sensors or not controller informs whether it's safe to cross the road or not via vibratory feedback as well as voice assistance. In [6], the system is attached to a belt tied around the waist with 2 modes record and playback. In recording mode, blind person walks path of interest and when reaches decision point i.e. when need to take a turn or cross rode, etc. then user presses key on aid to record decision word within speech synthesizer. So, distance traveled is stored in microcontroller and instruction is stored in speech synthesizer. In playback mode, aid measures distance traveled by the user if it is similar to one stored, a decision word will be generated at decision point by speech synthesizer. There are a lot of other techniques that cater to the needs of blind people. Below is the comparative study of various techniques as seen in Table 1.

Table 1. Comparison of various travel aids

S. No.	ETA	Year	Object recognition	Output type
1	Smart stick for blind and visually impaired people [3]	2018	No	Vibration sequences and buzzer
2	Smart phone-based obstacle detection for visually impaired people [4]	2017	Yes	Audio output
3	IoT based smart walking cane for typhlotic with voice assistance [5]	2016	No	Buzzer and vibrator, also through headphones
4	The development of a pedestrian navigation aid for the blind [6]	2006	No	Output via speech
5	An assistive system of walking for visually impaired [7]	2018	No	Buzzer
6	Electronic travel aid system for visually impaired people [8]	2017	Yes	Output via speech
7	Smart glove for visually impaired [9]	2017	No	Output via vibrating motor
8	A wearable portable electronic travel aid for blind [10]	2016	No	Bluetooth headphone through speech
9	Indriya - A smart guidance system for the visually impaired [11]	2017	Yes	Voice feedback
10	The 2SEES smart stick [12]	2018	No	Provides output via vibrating motor

3 Methodology

3.1 Component Description

This section comprises of the detailed description of components involved along with the figures as seen in Fig. 1.

Jetson Nano - This is NVidia developer kit low powered, small computer that is easy to use. It is an AI computer which makes it possible to bring the power of Artificial Intelligence to resource-constrained devices and can run multiple neural networks in parallel [13]. Being a GPU enabled board, we speculated this to be the ideal choice for running deep learning models with high processing speed and it is also possible to train models on this.

Raspberry pi3 B+ - The Raspberry Pi 3 Model B+ has Cortex-A53 (ARMv8) 64-bit SoC and 1GB LPDDR2 SDRAM. In addition to that, Bluetooth 4.2, BLE extended 40-pin GPIO header, 2.4GHz and 5GHz IEEE 802.11.b/g/n/ac wireless LAN, and full-size HDMI. Rpi has a CSI camera port to connect Raspberry Pi camera, 4 USB 2.0 ports and DSI display port to interface it to touch screen display also a micro SD port for loading your operating system and storing data [14]. This faster CPU, Ethernet, and smoother power management capabilities encouraged us to choose it to run deep learning models for testing.

Fig. 1. Jetson Nano and Raspberry Pi 3 B+ model

3.2 Collection of Dataset

Dataset collects data instances and is used for distinct purposes such as training the model to perform various actions and testing the performance in order to validate its accuracy. Machine learning is highly dependent on data, in fact dataset is the most crucial aspect to make any kind of training possible. We have collected dataset from various sources, like:

- Web browser, initially we searched images of a class using Google search. Further, we downloaded those images by installing bulk image downloader software and tapping on the icon and opening it in a separate tab. This helps to download multiple images of the same class and later on filter them as per our criteria.
- And for some classes for which we had issues collecting images online, we captured images manually via camera. Also, we kept image quality slow to reduce the image processing time and speed up the training and dataset uploading on the cloud.

We kept 1000 images per class for training purposes with a total of 12 classes including-Background, bicycle, motorbike, bus, car, chair, dog, person, bottle, horse, train, TV monitor. The overall work flow is shown in Fig. 2 below.

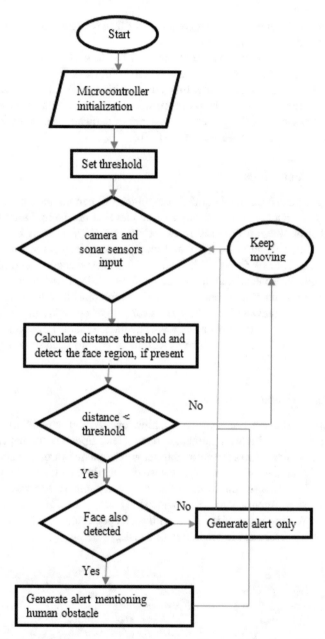

Fig. 2. Flow chart of the proposed work on Jetson Nano and Raspberry Pi 3 B+ model

3.3 Convolution Neural Networks

Convolution Neural Networks are a category of deep neural networks that assign weights and biases to various aspects/objects of an image so as to differentiate them from one another. CNN's have emerged to be a great tool for face recognition, object recognition and thus is a great tool for machine learning.

Mobile Net is a CNN that acts as a base network for the purpose of feature extraction. This is most widely used with resource-constrained devices where there is a lack of computation power [15]. Inception is Image recognition model that is widely used. It uses complex tricks to improve the performance in terms of speed and accuracy [16]. Res Net-Also known as Residual Network is a sort of dedicated neural network used to manage more advanced deep learning tasks [17].

3.4   Detection Algorithms

Detection Algorithms are capable of detecting objects of certain class in digital images and videos. Each object is identified with a certain level of precision. Faster RCNN [18] use the concept of 'Region Proposal Network' (RPN). RPN takes as input the image feature map being generated by CNN and outputs the objectness score. YOLO [19] stands for 'You Only Look Once'. This is an extremely fast detection algorithm capable of detecting multiple objects in an image. It applies neural network to input image that divides it into a grid and outputs bounding boxes around objects of interest. Single Shot multibox Detector abbreviated as SSD [20] is capable of detecting multiple objects in a single shot. In order to improve accuracy, it includes default boxes and multi-scale features. MAVI (mobility assistant for visually impaired) has been reported recently in 2019 in [21].

3.5   Training the Dataset

Training is provided using Keras which is a high level API of neural networks written in python. This process involves numerous parameters: Epoch is the hyper parameter that specifies number of times the entire dataset will be passed through neural network. This controls the number of complete passes we can get through the entire dataset. Loss function is used to enhance the machine learning algorithm performance and ideally keeps on decreasing. Accuracy metric provides the measure of algorithms performance based on the increasing value of accuracy with each epoch step. In other words, it gives the percentage of correctness in prediction of a class.

```
Epoch 8/20
 - 3s - loss: 0.0717 - acc: 0.9800 - binary_crossentropy: 0.0717 - val_loss: 0.4454 - val_acc: 0.8633 -
Epoch 9/20
 - 3s - loss: 0.0589 - acc: 0.9855 - binary_crossentropy: 0.0589 - val_loss: 0.4854 - val_acc: 0.8623 -
Epoch 10/20
 - 3s - loss: 0.0508 - acc: 0.9881 - binary_crossentropy: 0.0508 - val_loss: 0.5248 - val_acc: 0.8590 -
Epoch 11/20
 - 3s - loss: 0.0408 - acc: 0.9921 - binary_crossentropy: 0.0408 - val_loss: 0.5652 - val_acc: 0.8562 -
Epoch 12/20
 - 3s - loss: 0.0329 - acc: 0.9942 - binary_crossentropy: 0.0329 - val_loss: 0.6110 - val_acc: 0.8548 -
Epoch 13/20
 - 3s - loss: 0.0270 - acc: 0.9960 - binary_crossentropy: 0.0270 - val_loss: 0.6458 - val_acc: 0.8528 -
Epoch 14/20
 - 3s - loss: 0.0209 - acc: 0.9974 - binary_crossentropy: 0.0209 - val_loss: 0.6838 - val_acc: 0.8509 -
```

Fig. 3.   Training screenshot 1

Binary cross-entropy initially averages class-wise errors in order to obtain the final loss. This value gives us an idea of how far is the prediction from true value.

```
Epoch 15/20
  - 3s - loss: 0.0162 - acc: 0.9985 - binary_crossentropy: 0.0162 - val_loss: 0.7180 - val_acc: 0.8524
Epoch 16/20
  - 3s - loss: 0.0124 - acc: 0.9991 - binary_crossentropy: 0.0124 - val_loss: 0.7509 - val_acc: 0.8513
Epoch 17/20
  - 3s - loss: 0.0097 - acc: 0.9994 - binary_crossentropy: 0.0097 - val_loss: 0.7819 - val_acc: 0.8510
Epoch 18/20
  - 3s - loss: 0.0076 - acc: 0.9996 - binary_crossentropy: 0.0076 - val_loss: 0.8067 - val_acc: 0.8510
Epoch 19/20
  - 3s - loss: 0.0060 - acc: 0.9998 - binary_crossentropy: 0.0060 - val_loss: 0.8325 - val_acc: 0.8508
Epoch 20/20
  - 3s - loss: 0.0048 - acc: 0.9999 - binary_crossentropy: 0.0048 - val_loss: 0.8561 - val_acc: 0.8502
```

Fig. 4. Training screenshot 2

While loss and accuracy are the parameters for training set, val loss and val acc apprise the same for validation dataset. Images are taken while training the datasets at varying time intervals to see the loss value and accuracy as we proceed further with training seen in Fig. 3 and 4.

3.6   Testing the Dataset

Testing the dataset is used to corroborate the trained model accuracy by evaluating it on testing dataset as seen in Fig. 5.

```
Train on 60000 samples, validate on 10000 samples
Epoch 1/5
60000/60000 [==============================] - 9s 150us/step - loss:
1.0790 - acc: 0.7676 - val_loss: 0.5100 - val_acc: 0.8773
Epoch 2/5
60000/60000 [==============================] - 9s 143us/step - loss:
0.4401 - acc: 0.8866 - val_loss: 0.3650 - val_acc: 0.9011
Epoch 3/5
60000/60000 [==============================] - 12s 194us/step -
loss: 0.3530 - acc: 0.9032 - val_loss: 0.3136 - val_acc: 0.9127
Epoch 4/5
60000/60000 [==============================] - 16s 272us/step -
loss: 0.3129 - acc: 0.9124 - val_loss: 0.2868 - val_acc: 0.9188
Epoch 5/5
60000/60000 [==============================] - 12s 203us/step -
loss: 0.2875 - acc: 0.9194 - val_loss: 0.2659 - val_acc: 0.9246
Test Loss: 0.2659078140795231
Test accuracy: 0.9246
```

Fig. 5. Testing screenshot

4   Results

We used distinct combination of CNNs and detection algorithms and trained custom datasets for the same. Training and testing were executed on both platforms Jetson Nano as well as Raspberry pi to observe the possible outcomes. It can be spotted in Fig. 6, that by increasing model size the accuracy of Mobile Net keeps on increasing. Also in Table 2, we noticed that Mobile Net SSD on Jetson Nano hardware offers the highest level of accuracy compared to Raspberry Pi with faster speed and reduced loss while other versions fall behind in terms of accuracy, which is the most important aspect of object detection. Based on these criteria, we opted for Mobile Net SSD on Jetson Nano platform for object detection.

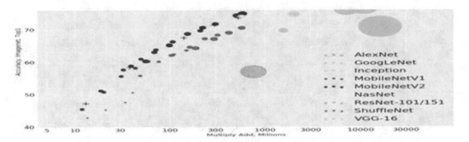

Fig. 6. Comparison of CNNs based on accuracy vs model size

Table 2. Results of running various object detection models on Jetson Nano and RPi3 B+ development boards. JN: Jetson Nano RP: Rasberry Pi 3 B+ MU1A: Micro USB 5v-1A, MU2A: Micro USB 5v-2A

H/W	Power	OS	CNN	Detection Algo.	Classes	Accuracy	Time Comp	Speed (fps)	Loss
JN	MU2A	Linux	Mobile Net	SSD	12	94%	10 epoch 2.5 h	48	0.1278
JN	MU2A	Linux	Mobile Net	SSD	20	90.7%	10 epoch 3 h	45	0.1524
RP	MU1A	Raspbian Stretch	ResNet	SSD	20	72%	10 epoch 5.5 h	20	0.2895
RP	MU1A	Raspbian Stretch	Mobile Net	SSD	20	92.4%	20 epoch 8 h	25	0.3200
RP	MU1A	Raspbian Stretch	Mobile Net	YOLO	90	45.5%	5 epoch 2 h	20–30	0.0048
RP	MU1A	Raspbian Stretch	Inception	Faster RCNN	20	92%	5 epoch 1.5 h	25	1.777

We run the trained MobileNet SSD model after powering on the Jetson Nano and interfacing it to the monitor via HDMI cable. As the camera is turned on trained model starts recognizing the class of objects it has been trained on. The output provides the class and accuracy value of each object detected, accuracy value ranges between 0 to 1 as seen in Fig. 7. This model is capable of simultaneously detecting multiple objects belonging to distinct classes.

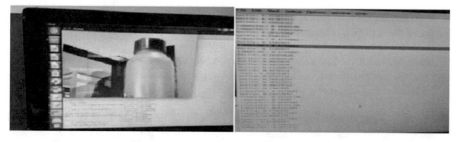

Fig. 7. Obstacle recognition and accuracy

5 Conclusion and Future Scope

This research was initiated to compare different object recognition algorithms on various parameters, keeping in mind the independent and free navigation of the visually impaired. In accordance with the outcomes of the study, it can be deduced that the Mobile Net Convolution neural network, when used with SSD detection algorithm, gives the most accurate results with a high speed. Also, Jetson Nano being a GPU enabled device is capable of processing multiple sensors in parallel. It provides a support for varying popular AI frameworks that makes it easy to integrate preferred frameworks and models into the board. Since experimental results were in initial phase, it would appropriate to compare prospects and future feasible development options.

Future work includes extensive field testing accompanying indoor as well as outdoor navigation for the blind people. We further aim to modify the classes like street sign board, tree branch, lamp post etc. for outdoors while table, door, almirah, bed etc. for the indoors based on the navigation mode and needs of the user.

References

1. World Health Organization: World health statistics 2014. World Health Organization (2014). https://apps.who.int/iris/handle/10665/112738
2. Electronic Travel Aids: New Directions for Research, National Research Council (US) working group on mobility aids for visually impaired and blind, Washington (DC), National Academies Press (US) (1986)
3. Agrawal, M.P., Gupta, A.R.: Smart stick for the blind and visually impaired people. In: 2018 Second International Conference on Inventive Communication and Computational Technologies (ICICCT), Coimbatore, pp. 542–545 (2018). https://doi.org/10.1109/ICICCT.2018.847 3344
4. Patel, S., Kumar, A., Yadav, P., Desai, J., Patil, D.: Smartphone-based obstacle detection for visually impaired people. In: 2017 International Conference on Innovations in Information, Embedded and Communication Systems (ICIIECS), Coimbatore, pp. 1–3 (2017). https://doi.org/10.1109/ICIIECS.2017.8275916
5. Narayanan, S., Deepan, G., Nithin, B.P., Vidhyasagar, P.: IoT based smart walking cane for Typhlotic with voice assistance. In: 2016 Online International Conference on Green Engineering and Technologies (IC-GET), Coimbatore, pp. 1–6 (2016). https://doi.org/10.1109/GET.2016.7916687

6. Bousbia-Salah, M., Fezari, M.: The development of a pedestrian navigation aid for the blind. In: 2006 IEEE GCC Conference (GCC), Manama, pp. 1–5 (2006). https://doi.org/10.1109/IEEEGCC.2006.5686241

7. Islam, M.N., Khan, N.S., Kundu, S., Ahsan, S.A.: An assistive system of walking for visually impaired. In: 4th International Conference on Computer, Communication, Chemical, Material and Electronic Engineering (IC4ME2), Bangladesh (2018). https://doi.org/10.1109/IC4ME2.2018.8465669

8. Ranaweera, P.S., Madhuranga, S.H.R., Fonseka, H.F., Karunathilaka, D.M.: Electronic travel aid system for visually impaired people. In: 5th International Conference on Information and Communication Technology (ICoIC7), Malacca City, pp. 1–6 (2017). https://doi.org/10.1109/ICoICT.2017.8074700

9. Linn, T., Jwaid, A., Clark, S.: Smart glove for visually impaired. In: 2017 Computing Conference, London, pp. 1323–1329 (2017). https://doi.org/10.1109/SAI.2017.8252262

10. Laubhan, K., Trent, M., Root, B., Abdelgawad, A., Yelamarthi, K.: A wearable portable electronic travel aid for blind. In: International Conference on Electrical, Electronics, and Optimization Techniques (ICEEOT), Chennai, pp. 1999–2003 (2016). https://doi.org/10.1109/ICEEOT.2016.7755039

11. Kallara, S.B., Raj, M., Raju, R., Mathew, N.J., Padmaprabha, V.R., Divya, D.S.: Indriya: a smart guidance system for the visually impaired. In: International Conference on Inventive Computing and Informatics (ICICI), Coimbatore, pp. 26–29 (2017). https://doi.org/10.1109/ICICI.2017.8365359

12. Connier, J., et al.: The 2SEES smart stick: concept and experiments. In: 11th International Conference on Human System Interaction (HSI), Gdansk, pp. 226–232 (2018). https://doi.org/10.1109/HSI.2018.8431361

13. Jetson Nano Developer Kit. https://developer.nvidia.com/embedded/jetson-nano-developer-kit

14. Raspberry Pi 3 Model B+. https://www.raspberrypi.org/products/raspberry-pi-3-model-b-plus/

15. Howard, A.G., et al.: MobileNets: efficient convolutional neural networks for mobile vision applications. eprint arXiv:1704.04861 (2017)

16. Szegedy, C., Vanhoucke, V., Ioffe, S., Shlens, J., Wojna, Z.: Rethinking the inception architecture for computer vision. In: IEEE International Conference on Computer Vision and Pattern Recognition (CVPR), pp. 2818–2826 (2016)

17. Antonio, G., Sujit, P.: Deep Learning with Keras. Packt Publishing Ltd., Birmingham (2017)

18. Ren, S., et al.: Faster R-CNN: Towards Real-Time Object Detection with Region Proposal Networks. Packt Publishing Ltd. (2015). eprint=1506.01497

19. Joseph, R., Divvala, S., Ross, G., Ali, F.: You only look once: unified, real-time object detection. In: IEEE International Conference on Computer Vision and Pattern Recognition (CVPR), pp. 779–788 (2016)

20. Liu, W., et al.: SSD: single shot multibox detector. In: Computer Vision – ECCV. Springer International Publishing (2016)

21. Kedia, R., et al.: MAVI: mobility assistant for visually impaired with optional use of local and cloud resources. In: 32nd International Conference on VLSI Design and 18th International Conference on Embedded Systems (VLSID), pp. 227–232 (2019). https://doi.org/10.1109/VLSID.2019.00058

Unmanned Vehicle Model Through Markov Decision Process for Pipeline Inspection

Chika O. Yinka-Banjo[1], Mary I. Akinyemi[1], Charity O. Nwadike[1], Sanjay Misra[2(✉)], Jonathan Oluranti[2], and Robertas Damasevicius[3]

[1] University of Lagos, Lagos, Nigeria
cyinkabanjo@unilag.edu.ng, maryi.akinyemi@gmail.com,
charitableindeed@gmail.com
[2] Covenant University, Ota, Nigeria
{Sanjay.misra,Jonathan.oluranti}@covenantuniversity.edu.ng
[3] Kaunas University of Technology, Kaunas, Lithuania
robertas.damasevicius@ktu.lt

Abstract. Frequent inspection and proactive monitoring are crucial in monitoring the health of a pipeline else, leakages because of inner corrosion, pipeline wear out or vandalism of pipeline may lead to loss of lives and properties. This research addresses the challenges or limitations of pipeline inspection methods. We demonstrated how a simulation of pipeline inspection can be managed by Markov decision process (MDP). The proposed policy selection was controlled by an algorithm that manages how the mobile agent (unmanned ground vehicle) responds to observed conditions of the pipes in its immediate vicinity. Based on various simulated experiments the ground vehicle correctly detects defects in pipes without false alarm and stores details for the maintenance team to carry out necessary actions. The size of pipeline corrosion was measured by two different robots. Statistical tests were hence conducted to compare the performance of the 2 robots. The result show that variation in the size of corrosion for both robots is not statistically difference.

Keywords: Associated probability · Associated reward · Associated cost · Unmanned ground vehicles · Virtual robot experimentation platform · Markov decision process

1 Introduction

Pipeline installations are massive economic infrastructures and capitally intensive, therefore it is of necessity on the part of business owners to provide adequate security measures to avoid major environmental consequence that could result from the release of usually high volatile contents conveyed through them perhaps due to bunkering, partial or total material failure. Environmental consequence comes with a very large economic cost and from experience, business owners prefer prevention than a partial or total environmental breakdown; usually an environmental reversal process might take several months and even years. There is a significant economic loss both to environment dependent

© The Author(s), under exclusive license to Springer Nature Switzerland AG 2021
M. Tripathi and S. Upadhyaya (Eds.): ICDLAIR 2019, LNNS 175, pp. 317–329, 2021.
https://doi.org/10.1007/978-3-030-67187-7_33

professions (e.g. fishing, farming) and pipeline business owners. A notable environmental consequence can be found in the Niger Delta Region of Nigeria where there has been incessant sabotaging of pipeline installations for selfish interest; oil spillage and bunkering caused fishermen and farmers several billions of Dollars a year due to perpetual environmental degradation or release of pipeline contents. The recent resurgence of restiveness in Niger Delta Region of Nigeria in 2016, have brought damning economic consequences to pipeline business owners; where they have to spend several billions of dollars to repair spoilt pipe installations. Militant groups in the region have always claimed that the region has been neglected and their environment dependent jobs have been taken away by perpetual oil spillage. In an attempt to cushion the perceived grievances of people of the region; President Muhammad Buhari of the federal republic of Nigeria flagged-off environmental cleaning process [6] that will necessarily take several years to be completed. Pipeline business owners have adopted various security and inspection approaches in time past to secure these assets especially through physical security and human security agents.

Pipeline leakage is defined as the unwanted outflow of fluid such as oil, gas or water [11]. Leakage is the major cause for indeterminate losses in every pipe network around the world. Over the years, the percentage of gas lost as a result of leaks in the gas pipe system is usually higher than people's expectation [6]. This leakage may be owned to Pipeline vandalization: is the deliberate destruction of pipelines for selfish gain [1], Pipeline corrosion; natural deterioration of some pipelines. Pipelines are engineering plant used for the delivering of energy [15]. They are widely used to transfer natural gas and oil to destinations all through the world. Rigid, Composite, and Flexible Pipelines cannot be completely relied on because like other engineering plant, they can fail. Recently, several pipelines have failed leading to some tragic effect. Companies maintaining and supplying oil faces lot of challenges and spend huge money to carry out inspection or check the oil pipeline. The process involved in maintenance is tedious because employees are exposed to the risk of choking while carrying out maintenance but may end up not discovering any fault leading to loss of money [2]. Hence, the ideal to design a robot for pipe line inspection to inspect or detect problems or any malfunctions in pipeline. This would help companies solve the problems they face during inspection [14]. The frequent maintenance is important which help to prevent abnormalities such as cracks or leakages. Several methods exist for detecting of leakages, these include PIGS, Acoustic sensor, wheel, caterpillar, wall-press, walking, inchworm and screw type [9].

AI and IT based support systems are applied in several domains [18–20]. In this work, the objective is to simulate pipeline inspection using MDP in VREP, Leak (corrosion) detection in the pipeline system using a simulated environment, Combination of sensor technologies and autonomous robotic agent with pipeline monitoring system for appropriate event localization. The paper is organized as follows. In the next Sect. 2- background of the work is provided. The proposed model, experimental results, discussion and conclusion and future work is provided in Sect. 3, 4, 5 and 6 respectively.

2   Background and Related Work

Pipeline Inspection Gauge (PIGS) are large, and they can inspect and clean pipes but difficult or almost impossible to PIG sharp bends with dirty inner surfaces, smaller diameter or irregular turns [10]. There are some non-destructive testing methods used by PIGS for carrying out inspections, such as magnetic flux leakage, ultrasound, visual and optical testing, etc. [3]. Acoustic sensors operate in active and passive mode for pipeline transmission, monitoring and inspection distributed Pipeline Systems by Wireless Sensor Networks [7]. A system called wireless sensor networks for pipeline monitoring (PIPENET) which localize, quantify and detect burst and leaks and other anomalies in the transmission of water through pipeline [13]. Pipeprobe A mobile sensor system called for mapping water pipelines under floor coverings or inside cement walls [4].

2.1 Formal Description of MDP

Saranga et al. described the five stages of MDP [16].

- Decision epochs: are the point of times at which decisions are made in a discrete time MDP, the total decision-making period (decision horizon) is divided into intervals which are called decision intervals, and at the beginning of each decision interval, a decision epoch occurs. The set of decision epochs is given by $D = \{1, 2, 3, \ldots, N\}$. In a finite horizon MDP, N is finite and according to the convention, decisions are not made at the Nth decision epoch.
- States: are the sets of possible positions the agent can assume after an action has been taken.
- Actions: are set of possible actions or decisions that can cause a change in state of the process or learning agent.
- Transition probabilities: It is the probability associated with choosing an action that resulted to a change in state of a process or learning agent.
- Reward: At each decision epoch $< N$, the agent receives a reward, as a result of choosing an action. The reward received upon choosing action in state i at the tth decision epoch is denoted by r. The reward received at the Nth decision epoch is assigned based on the state that the equipment is being found at the Nth decision epoch.

2.2 Markov Process and It's Discrete Time Finite State

Kallen, described a discrete time finite-state Markov process as a stochastic process which encapsulates the movement between a finite set of states and that the Markov property holds. The Markov property in Eq. 1 states that, given the current state, the future state of the process is independent of the past states. [17].

Markov property is given as

$$Pr\{X_{t+1} = x_{t+1} | X_t = x_t, X_{t-1} = x_{t-1}, \ldots, \ldots X_1 = x_1, X_0 = x_0\} \tag{1}$$

$$Pr\{X_{t+1} = x_{t+1} | X_t = x_t\}$$

Such that the set of possible states is finite and represented by a sequence of positive integers: $X_k \in S$ for $S = \{0, 1, 2, .., n\} \bigvee k$.

A discrete time finite state Markov process relies on the time and probabilities between two states, such that the possible probabilities among states can be captured on a probability matrix.

$$P = \begin{bmatrix} p00 & p01 & p02 \\ p10 & p11 & p12 \\ p20 & p21 & p22 \end{bmatrix} \tag{2}$$

The matrix **P** in Eq. 2 is stochastic, which means that $0 < P_{ij} < 1$ for i, j $= 0, 1, 2,$ ... n and $\sum_{j=0}^{n} P_{ij} = 1$ for all i. That is the probability values exist between 0 and 1 and the sum of probabilities is less or equal to unity.

The transition probabilities can also be defined as $P_{ij} = \Pr\{P_i = j\}$, where P_i is the random variable describing the probability of the destination state if currently in state i. The transition probability matrix **P** not only defines the randomness of the process in time, but it also defines the structure of the model.

3 Proposed Methodology (Mdp)

The proposed solution is modeled as a control theory problem using Markov decision process (MDP). MDP is one of the many combinatorial design models. It is a feedback compliant system that ensures every action is rewarded and the reward helps the learning agent to take smarter decisions as the learning activity progresses.

Our MDP approach to this inspection problem combines properties and features of both the agent and pipe network. The conditions of the pipes are taken to be the MDP states and the reactions of the robotic agent as MDP actions. By this we rely on the change in state of the robotic agent sufficient enough to model the possible conditions of the pipes.

3.1 MDP State

The MDP states are the possible conditions the Pipeline installations can be. The conditions of the pipeline installations at any particular time are Normal and Corroded. A pipe with a normal condition is one that is not corroded or punctured, while a Corroded condition is the exact opposite of the normal condition. The MDP states as presented in Table 1 are actually some form of input for the MDP managed robotic agent; they trigger different actions from the robotic agent.

Table 1. Possible conditions of the pipe (MDP state)

S/N	State	Notation
1	Normal	NORM
2	Corroded	CORD

3.2 MDP Actions

MDP actions are the possible reactions of the robotic agent to the observed conditions of the inspected pipes. The MDP actions also represent the states the robotic agent (UGV) can be at any point in time during inspection. These actions are Start, Scan and Mark. Mainly Scan and Mark are to be triggered based on observed conditions of the pipeline. All these MDP actions constitute the state space S for the robotic agent. Table 2 outlines the possible action & their corresponding rewards at any given state:

Table 2. MDP action

S/N	STATE	POSSIBLE ACTION
1	NORMAL	START, MOVE, SCAN,
2	CORROSION	MARK

Table 3. State – action transition matric

PIPE CONDITION	ACTION	NORMAL	CORROSION
NORMAL		START, MOVE, SCAN	MARK
CORROSION		SCAN	MARK

3.3 MDP State Transition Matrix

There are two transition matrixes in this section. Table 3 relates the pipe detectable conditions and corresponding possible actions that can be taken as the robotic agent carries out the inspection task while having the MDP states in mind.

More so, the possible actions of the robotic agent constitute set of states in itself; thus we are interested in showing the allowable action or state transition matrix for the MDP actions (agent-states). Table 4 represents the possible actions or successions of actions that can be taken by the robotic agent or the possible agent-state transition while carrying out the task. The intersections of columns represent a connection between any two actions. An "F" in a column means that the robotic agent cannot take such actions in succession, while a "T" means such actions can be taken in succession. The state transition probabilities of the UGV are displayed in Table 5.

Table 4. UGV state transition matric

UGVSTATE|TRANSITION	START	SCAN	MARK
START	F	T (NORMAL)	F
SCAN	F	F	T (If Corrosion is detected)
MARK	F	T (If agent is done marking or recording)	F

Table 5. UGV state transition probabilities

STATE	START	SCAN	MARK
START	0.1	0.7	0.2
SCAN	0.1	0.2	0.7
MARK	0.1	0.7	0.2

3.4   A Model for Markov Decision Processes

- After each transition ($i = 0, 1, ..., M$), a state I of a discrete time Markov Chain is observed
- A decision (action) k is chosen from a set of K possible decisions ($k = 1, 2, ... K$) after each observation. (Some of the K decisions may not be relevant for some of the states.)
- If a decision k is made in a state i, an immediate cost is allocated with an expected value Cik.
- The decision k in state I determines what the transition probabilities1 will be for the next transition from state i. Denote these transition probabilities by P_{ij} (k), for $j = 0$, 1, ..., M.
- A specification of the decisions for the respective states ($k_0, k_1, ..., km$.) prescribes a policy for the Markov decision process.
- The objective is to find an optimal policy according to some cost criterion which considers both immediate costs and subsequent costs that result from the future evolution of the process. One common criterion is to minimize the (long-run) expected average cost per unit time.

3.5   MDP Associated Action-Reward

The MDP is a reward based control theory model. It is expected that for every successive action in a policy there are associated cost or reward; this reward scheme tends to project and manage how a robotic agent will behave if those successive actions were selected (Tables 6 and 7).

Table 6. Associated cost

S/N	STATE	REWARD
1	START	10
2	SCAN	20
3	MARK	30

Table 7. Pseudocode of the control algorithm

a)   Start Robotic Agent
b)   Compute the associated cost
c)   Select policy option
d)   If Defect is observed repeat b, Else
e)   Repeat (a-c) until inspection task is completed, then Shutdown

Thus, the associated cost for any action taken by the MDP agent is:

$$ AC_s = AP_s * AR_s \tag{3} $$

Where:

AC_s: Associated cost at state, s. AP_s: Associated probability at state, s

AR_s: Associated reward at state, s

3.6 MDP Actions

The recurrence formula below describes the learning policy for the robotic agent. The policy prescribes the suitable conditions that will lead to an optimal value by selecting next states using the appropriate schemes.

$$\pi(s) = \begin{cases} Max\{AC_{s1},\ AC_{s2} \dots \dots \dots \dots\},\ AC_s > 0 \\ Shutdown\ (Goal\ reached), \qquad Max\{AC_s\} = 0 \end{cases} \qquad (4)$$

4 Experiment

Virtual robot experimentation platform (VREP) was adopted as the simulator to model the proposed solution. The presentation is in 3D and all objects appear as it were in real live. The platform is based on multi-robot and distributed control architecture which makes it very suitable for the proposed approach.

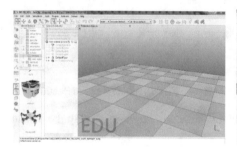
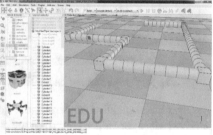

Fig. 1. Landing page in VREP   **Fig. 2.** Pipeline showing corroded parts

Once the VREP application has been launched a page as depicted in Fig. 1 will be loaded; a blank page with hard-terrain canvass as the floor. This whole view is usually referred to as the World or Simulation World or Environment.

4.1 Construction of Pipelines

The experimental pipeline layout is represented in Fig. 2. It is constructed by right clicking on the scene hierarchy (VREP) and selecting primitive cylindrical object from a list of drop down, then change the orientation to a suitable direction for the pipe outline. Double clicking on the primitive object in the scene hierarchy to edit its dynamic properties and enable static option. Click on the common button to assign solid properties to the primitive shape and enable the following (renderable, measurable, collidable and detectable) then close the form. This object is copied and pasted in its buffer in about sixty (60) times. Select the object and click selection control to align the cylindrical Objects like the presentation in Fig. 2.

4.2   Design of Corroded Parts

The corroded part is given in Fig. 2. Which is added to the pipeline by right clicking on the scene, add primitive object and select a cuboid. From the scene hierarchy on the left pane of the simulator right click, select Edit modes and select Enter shape edit mode. From the opened dialog box, deselected every other plane and reduced the size by double clicking on the edited shape and click on common button, then set the scaling value to 0.5. Click on the shape button and select show dynamic properties button to enable static option then we clos the form.

From Fig. 3, VREP has a built in unmanned ground vehicle model known as epuck, which requires little modification to meet our expected robotic agent. We added more features to enable the model to suit the inspection purpose. The floating view on the scene displays what the epuck robot sees at any point in time. The floating view is a canvass that can be associated with any sensing or readable object. To add the a floating view right click on the scene and then go to Add & select floating view (Fig. 4).

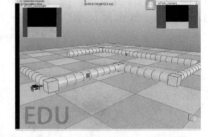

Fig. 3. E-puck support on the scene   **Fig. 4.** Single robot inspection

Once the simulation is started, the floating view will begin to display the sensing details or views or values read from the environment.

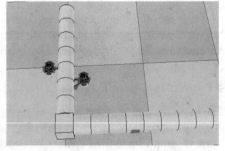

Fig. 5. Simulation scene showing corrosion   **Fig. 6.** Multi-robot inspection
detected with a single robot

The Fig. 5 shows the windows that opens up when a simulation instance starts. The two main windows are the simulation World and the output window that reports the inspection exercise. The World consists of the pipe network, UGV (robot), corroded parts

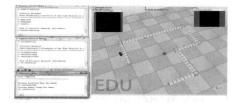

Fig. 7. Inspection Scene with two robots and output windows.

and floating view. The floating view only displays a red object when a corroded part is detected by the vision sensor. The debug window shows details of the inspection exercise by marking and recording locations of the pipes relative to the simulation world. The principal language for scripting the behaviors of objects in the world (VREP platform) is LUA. LUA is a multi-paradigm programming language; its building concepts cut across structural, object oriented and functional programming (Figs. 6 and 7).

5   Discussion

Tables 8 and 9 show the Full Output of the Inspection Exercise by the Single Robot. The Output Window Indicates that a Corroded Part is Detected and Takes the Records of the Affected Pipe Relative to the Simulation World. It also Records the Size of the Corrosion, Average Time and Memory for the Robot.

Table 8. Inspection output by a single UGV

Robot Marking (Location of the Pipe Relative to the World)			Corrosion Detected
x	y	z	Size of Corrosion (mm)
0.9453125	0.80078125	0.1015625	0.00604248046875
0.9921875	0.8046875	0.0078125	0.00604248046875
0.421875	0.791015625	0.06640625	0.0016632080078125
0.46875	0.837890625	0.08203125	0.0019683837890625

Table 9. Performance output by a single UGV

Average Execution Time Per Robot	58.801288604736
Average Memory Usage Per Robot	70.5966796875

Tables 10 and 11 show the output of the inspection exercise by the double robot. The output window indicates that a corroded part is detected and takes the records of the affected pipe relative to the simulation world. It also records the size of the corrosion, Average time and memory for the robot.

Table 10. Inspection output by double UGV's

Robot Marking (Location of the Pipe Relative to the World)			Corrosion Detected	
x	y	z	Size of Corrosion (mm)	UGV's
0.46484375	0.73046875	0.05078125	0.0010986328125	UGV 1
0.3671875	0.755859375	0.05859375	0.0014190673828125	UGV 1
0.421875	0.791015625	0.06640625	0.0016632080078125	UGV 2
0.46875	0.837890625	0.08203125	0.0019683837890625	UGV 2

Table 11. Performance output by double UGV's

Average Execution Time Per Robot	28.749704360962
Average Memory Usage Per Robot	92.53076171875

Table 12. Performance report for inspection model

Number of robot	Execution time	Memory usage	Number of observed defects
1	58.80	70.60	4
2	28.74	92.53	4

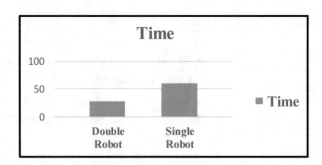

Fig. 8. Performance evaluation based on execution time

Table 12, Figs. 8 and 9 show the outcomes of the simulation exercise and compare the performances of the robots that use our model, based on execution time and memory usage. Our approach does not degrade with increased number of robots but performs better with higher number of agents.

The multi robot scenario, scale gracefully by not increasing the memory overhead as the number of agents increases. It also reduces the execution time by almost half because the agents are well distributed, and the task is collectively and efficiently allocated. The

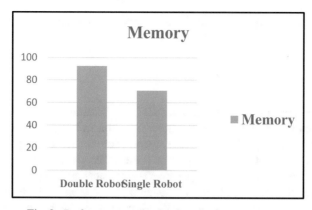

Fig. 9. Performance evaluation based on memory usage

slight increase in memory is as a result of the additional resources used by the vision sensor.

The model and V-rep support distributed multi-robot architecture which makes it possible to fully demonstrate the full capability of our model.

The inspection simulation did not give a false alarm by indicating detection where there is none; it only triggers an appropriate action as prescribed by the MDP model. The simulation can be stopped by pressing the stop button (Table 13).

Table 13. T-test and Levene's test to check Homogeneity of Mean and Variance of the 2 Robots.

Test	Test statistic	p-value
T-test	−1.6536	0.1014
Levene's test	2.7832	0.09684

From the plots of size of corrosion as measured by each robot, we observe some outliers in the robot 2. Statistical tests were hence conducted to compare the performance of the 2 robots. A t-test was conducted to compare the average performance of the 2 robots while, the Levene's test of homogeneity of variances was carried out to compare the variation in the size of corrosion as measured by each of the robots. The following hypothesis were tested;

T-test-H0: average measurement of size of corrosion is the same for both robots.

Levene's test-H0: variation of in the measurement of size of corrosion for both robots are homogeneous.

The results are presented in Table 13. The tests were carried out at 5% level of significance (i.e. $\alpha=0.05$), the p-values of each of the tests were compared to the level of significance. It was observed that for both the t-test (p-value = 0.1014) and the Levene's test (p-value = 0.09684), the p-values are both greater that 0.05 so that in both instances we decide in favour of the null hypothesis hence the difference in average size of corrosion measured by both robots is not statistically different, furthermore, variation in size of corrosion for both robots is not statistically difference.

6 Conclusion

In this paper, we have demonstrated how MDP can be used to manage the inspection task. More importantly, the high-end results from the simulation experiments justify the need to adopt automated inspection at pipeline dependent industries. Notably, there are many more reasons to corroborate these results. The UGV have higher chances of detecting leakages or corrosion in pipes than the physical eyes because it relies on the sniffing abilities of cameras and sensors. Sometimes, corrosion might be very small that human eyes might not see it, but an UGV will usually pick it up; thereby making the maintenance team more proactive and not wait until a major damage had resulted due to neglected corrosion. UGV's is naturally reusable and cost effective when compared to human inspectors. The conducted study shows that the statistics of vandalization has the highest percentage cause of pipeline leakage over corroded pipeline.

Acknowledgment. The authors gratefully acknowledge the financial support of Covenant University, African Institute for Mathematical sciences (AIMS) Alumni small research grant (AASRG), the Organisation for Women in Science for the Developing World (OWSD), and L'oreal-Unesco for Women in Science.

References

1. Agwu, O.A.: The op eleven challenges facing the exploration and production industry in Nigeria. Int. J. Res. Eng. Technol. **v**(6), 203–204 (2016)
2. Al-Hajry, H.: Design and testing of pipeline inspection robot. Int. J. Eng. Innov. Res. **2**(4), 319 (2013)
3. Center, N.R.: Pipeline Inspection (2014). https://www.nde-ed.org/AboutNDT/SelectedAppl ications/PipelineInspection/PipelineInspection.htm
4. Chang, Y.L.: Pipeprobe: mapping spatial layout of indoor water pipelines. In: Mobile Data Management: Systems, Services and Middleware, MDM 2009, pp. 391–392. IEEE (2009)
5. Chijioke, A.-N.: NetNaiga. https://www.netnaija.com/forum/general/news/25353-buhari-set-visit-niger-delta-june-2-flag-clean-ogoni-land. Accessed 12 Sept 2016
6. Chraim, F.E.: Wireless gas leak detection and localization. IEEE Trans. Ind. Inform. **xii**(2), 768–779 (2016)
7. Jin, Y., Eydgahi, A.: Monitoring of distributed pipeline systems by wireless sensor networks, Maryland (2008). ISBN
8. Kim, J.-H.: Sensor-based autonomous pipeline monitoring robotic system (2011)
9. Beller, M.: Inspection of challenging pipelines. In: Pipeline Technology Conference, Berlin (2017)
10. Mulder, J., Wang, X., Ferwerda, F., Cao, M.: Mobile sensor networks for inspection tasks in harsh. Sensors **X**(3), 1600–1611 (2010)
11. Nawaz, F.: Remote pipeline monitoring using wireless sensor networks. J. Netw. Technol., **v11**(4), 113 (2016)

12. Safety, E.: Ultrasonic devices improve gas leak detection in challenging environments. In: World Oil, pp. 133–135 (2014)
13. Stoianov, I.N.: PIPENET: a wireless sensor network for pipeline monitoring. In: Information Processing in Sensor Networks, pp. 264–273. IEEE (2007)
14. Tsang, K.: Guest editorial industrial wireless networks: applications, challenges, and future directions. **xii**(2), 755–757 (2016)
15. Zhang, L.D.: The design of natural gas pipeline inspection robot system. In: Information and Automation, pp. 843–846 (2015)
16. Saranga, K.A., Xu, H.: Adaptive maintenance policies for aging devices using a Markov decision process. IEEE Trans. Power Syst. **28**(3), 3194–3203 (2013)
17. Kallen, M.J.: Markov processes for maintenance optimization of civil infrastructure in the Netherlands. Ph.D. thesis, Delft University of Technology, Delft (2007)
18. Iheme, P., Omoregbe, N., Misra, S., Ayeni, F., Adeloye, D.: A decision support system for pediatric diagnosis. In: Innovation and Interdisciplinary Solutions for Underserved Areas, pp. 177–185. Springer, Cham (2017)
19. Adewumi, A., Taiwo, A., Misra, S., Maskeliunas, R., Damasevicius, R., Ahuja, R., Ayeni, F.: A unified framework for outfit design and advice. In: Data Management, Analytics and Innovation, pp. 31–41. Springer, Singapore (2020)
20. Jonathan, O., Misra, S., Ibanga, E., Maskeliunas, R., Damasevicius, R., Ahuja, R.: Design and implementation of a mobile webcast application with Google analytics and cloud messaging functionality. J. Phys.: Conf. Ser. **1235**(1), 012023 (2019)

Face Spoofing Detection Using Dimensionality Reduced Local Directional Pattern and Deep Belief Networks

R. Srinivasa Perumal[1], G. G. Lakshmi Priya[1],
and P. V. S. S. R. Chandra Mouli[2(✉)]

[1] School of Information Technology and Engineering, Vellore Institute of Technology,
Vellore, Tamilnadu, India
r.srinivasaperumal@vit.ac.in, lakshmipriya.gg@vit.ac.in
[2] Department of Computer Science, Central University of Tamil Nadu,
Thiruvarur, Tamil Nadu, India
mouli.chand@gmail.com

Abstract. Face spoofing detection is considered a mandatory requirement for a robust face recognition system. This paper presents a framework to detect whether a given face is spoofed face or not using dimensionality reduced local directional pattern descriptor (DR-LDP) combined with color textured features. Deep Belief Networks (DBN) classifier has been used to classify the real faces and spoofed faces. The efficiency of our system elucidates performance by comparison of the results with the state-of-the-art method.

Keywords: Face spoofing · Deep learning · Local descriptor · Deep belief networks · Image classification

1 Introduction

Biometric systems help to identify persons based on their physiological and behavioral characteristics. It is mostly useful for various security applications. Despite their advantages, biometric systems fail to provide better security in some situations for the reason that a person can fool with spoofed data. A spoofing attack takes place when people try to cheat face recognition system by presenting a photo of a person or a video or a 3D mask in front of the acquisition device [8]. These attacks have stimulated consent on liveness detection and hence anti-spoofing techniques have become a mandatory requirement in improving the security systems by classifying a genuine and imposter trait [6].

The features extracted can be categorized into either dynamic features, global features, or local features for face liveness detection. Dynamic features extract the information of the facial appearance through eyes movement, lips, and motion analysis. Global features analyze the complete image to extract discriminative information. Local features extract the information by dividing the

M. Tripathi and S. Upadhyaya (Eds.): ICDLAIR 2019, LNNS 175, pp. 330–338, 2021.
https://doi.org/10.1007/978-3-030-67187-7_34

image to fixed-size blocks and extract features from each block and thereby form a feature vector. If the gray level face image quality is good, the texture analysis of the face image is sufficient to expose the recapturing artifacts of fake faces. Textural analysis of the face image cannot discriminate the fake user if image quality is not good. In this case, the color texture analysis of the face image is useful to distinguish whether the given face is genuine or fake [5].

This work exploits color texture analysis and DR-LDP for face spoofing detection. Perumal et al. introduced DR-LDP to extract the feature vector of a face image [14]. DR-LDP has limitations in facial expression recognition. It is further extended to modified dimensionality reduced local directional pattern (MDR-LDP) [15] and dimensionality reduced local directional number pattern [16]. The deep belief network (DBN) has been employed to classify the genuine and imposter user from the face image.

The significant contribution in this work is analyzing the face spoofing databases by combining color texture analysis and micro features obtained from DR-LDP. Similarity metrics are used for real and fake face classification. Based on the results, it is observed that DBN followed by a backpropagation network serves as a better anti-spoofing feature. The performance of the proposed system is compared extensively with state-of-the-art methods.

2   Related Works

Face Spoofing approaches are classified into two categories namely texture based and motion based approaches. Texture based approaches exploit the shape, color, etc., to differentiate the real face with fake face. Motion based approaches analyze the movement clues, such as eye blink for every two to four seconds and lip movement. Motion based approaches are time consuming and it requires user cooperation. Even if the user cooperates, the authentication failure rate may be high due to motion blur induced, if any, while acquisition. Texture analysis gives encouraging results in face spoofing than motion based approaches.

Li et al. [13] used 2-D Fourier spectra for detecting the real faces. It is easy to realize the fake face, due to its small size. While a face image is shown in front of the camera, since the poses and expressions of the face are invariant, the frequency components in the sequence are very small. It is not robust to changes in illumination conditions. Tan et al. [20] introduced a non-intrusive method, which classifies the real face and fake face by using binary classification method. They extracted the information about surface properties of the photograph on a real face and applied sparse regression model for accurate spoof detection with user involvement or extra device.

Benlamoudi et al. [4] used overlapping block LBP to extract the texture information. The histogram of each block was computed to form a single feature vector. Then they applied Fisher score to reduce the feature vector. Finally, they applied non-linear support vector machine (SVM) for classification. The same authors used multi-level local phase Quantization (ML-LPQ) to extract features of face from the region of interest [3]. Arashloo et al. [2] used histograms of multi-scale dynamic binarized statistical image features (MBSIF-TOP) and multi-scale dynamic local phase quantization (MLPQ-TOP) on three orthogonal planes to extract the micro texture of a face image. The histograms of dynamic texture were computed, then the kernel discriminant analysis (KDA) was used to classify the real faces. The author used efficient KDA based on spectral regression to improve the computational time and performance. Shikhar Sharma et al. [18] proposed a Deep Facial recognition system (D-FES). In this paper, the authors used recurrent neural networks to detect facial emotions based on changes occur in the lips structure periodically. Harman Singh et al. [19] proposed a habit detection model using machine learning concepts. Krishan Kumar [11] proposed an event summarization technique using deep learning framework. He has exploited graph theoretic concepts like highly connected subgraphs and clusters for detecting the keyframes in an event video. Krishan Kumar et al. [12] proposed a fast and deep event summarization. Deep learning framework is used to extract multi-view features from the given video sequence and finally objects are tracked from frames.

Boulkenafet et al. [5] recently used color texture information, to detect whether the face image is a real or fake from the chrominance and luminance channels by extracting the low-level features from various color spaces. The histogram was computed for each color channel separately and it was compared with each color channel to classify the real and fake face.

Some of the existing approaches were not able to detect all types of attacks (photograph, video and 3D Mask).

To overcome these difficulties, a color texture combined with micro pattern local features is proposed such that it withstands to different types of attacks.

3   Proposed Method

The proposed method for face spoofing detection uses dimensionality reduced local descriptor (DR-LDP) [14] to extract the micro texture patterns of the face image. The general architecture of the proposed system is shown in Fig. 1.

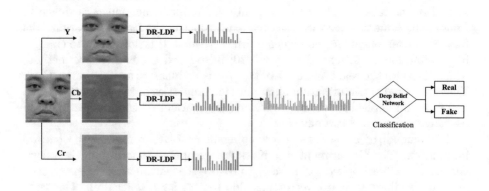

Fig. 1. Architecture of the proposed approach

1. The input images are normalized and cropped into the size of 112×112 by using face detector tool.
2. The cropped face is converted into YCbCr color space from RGB color space because the chrominance and luminance in the color space are used to detect the fake face from the real face image.
3. The color channels are separated based on the color space.
4. The DR-LDP descriptor is applied on each channel to extract the micro texture information.
5. The image is divided into 8×8 regions to compute the histograms of each region.
6. The histogram of each region is concatenated to form a single feature vector.
7. The face is checked to find whether it is a real or fake from the extracted feature vector using deep belief network, followed by back propagation network.

The proposed method improves the performance and invariant to pose, illumination conditions and orientation of face are compared to the other descriptors like variations of LBP and LDP. The deep belief network followed by back propagation network is more efficient method for classification. DR-LDP [14] is a similar descriptor like LDP [9] and it extracts the micro texture information of the face image. In DR-LDP, the face image is divided into 3×3 overlapped region for encoding the face image. It extracted the local texture information of the face image using Kirsch masks. It extracted the information about edges, spots. DR-LDP assigns an eight bit binary code for each sub region to represent the texture information of the sub region. The resultant image of the DR-LDP is further divided into 8×8 regions to compute histogram of each region. Finally, the histogram feature vector of each region is concatenated to create a single feature vector for the face image. The DR-LDP is different from LDP in encoding the eight bit code for each sub region. DR-LDP reduces the dimensionality of the image by $9 : 1$ pixel ratio.

Recently [7], fast parallel neural net code for graphics cards has overcome the above issues. It is necessary to design GPU code carefully for image classification so that it can be up to n^{th} orders of magnitude faster than its CPU counterpart. Hence the large scale data can be trained in hours or days. Deep learning using a set of algorithms in machine learning attempts to create high level abstraction in data by using deep architecture [17]. Deep belief networks (DBNs) are deep architectures that use stack of restricted Boltzmann machine (RBM) to generative model that can learn a probability distribution over training data. DBN is an alternative model of deep neural network. It can extract features of the image as well as classify the image effectively [10].

The reasons for choosing DBNs are, it has many hidden layers; it can be fine tuned and it can be used for feature extraction and dimensionality reduction on unsupervised learning data set [1]. The performance of DBN motivated to use DBN for classification in the paper. In this paper, the RBM layer is pre trained step by step and the entire network is fine tuned to achieve best classification rate. Then back propagation network is applied, followed by deep belief network to reduce the error rate.

4   Experimental Results

In face spoofing, the performance of the system is measured with equal error rate (EER), which is equal to false rejection rate and false acceptance rate. Both FAR and FRR are measured based on threshold value. The experiments are carried out on standard benchmark face spoofing databases such as CASIA Face Anti- Spoofing Database, NUAA Database and MSU Mobile Face Spoof Database. Then, the complementary facial color analysis method is compared with State-of-the-art methods.

4.1   CASIA Face Anti- Spoofing Database

The CASIA Face Anti-Spoofing database [22] contains 50 genuine subjects and fake faces were made from the high quality records of the genuine faces. Different types of image qualities are considered. They are low, normal and high qualities. Three types of spoofing attacks are designed, which include cut photo attack, warped photo attack and video attack. Each subject contains 12 videos (9 fake and 3 genuine), totally 600 videos in the database. The test protocol consists of 7 scenarios for a thorough evaluation for all possible phases. The database is divided into training set and testing test. The training set consists of 20 subjects and rest of 30 subjects are used for the testing set.

The face is localized and normalized by face detector tool, the resultant image is normalized to 112×112 size. The performance of the descriptors on different color spaces is presented in Table 1. In color space, the HSI comparatively yields better performance than YCbCr and RGB. The results show that the proposed method outperformed than other existing methods.

Table 1. Equal Error Rate of DR-LDP descriptor on CASIA face anti-spoof database

Descriptor	Gray	RGB	HSI	YCbCr
LBP [5]	22.60	21.00	13.60	12.40
LPQ	23.20	14.40	7.40	16.20
BSIF [5]	26.21	21.00	6.70	17.00
SID [5]	19.90	15.80	11.20	11.60
CoALBP [5]	14.80	11.00	5.50	10.00
CoALBP+LPQ	14.80	8.60	4.00	8.70
DR-LDP	13.60	8.00	3.00	7.50

4.2   MSU Mobile Face Spoof Database

The MSU mobile face spoof database [21] contains 280 video recordings of real and fake faces. These recordings are captured from 35 subjects using two different types of cameras (MacBooK Air 13-inch laptop and Google Nexus 5 Android phone). The resolution of the videos is 640 × 480 and are captured from laptop camera and the resolution of video captured through Android camera is 720×480. The duration of the each video is approximately 9 s.

The training set contains 15 subjects and remaining 20 subjects are used for the testing set. Table 2 shows the performance in terms of equal error rate on DR-LDP descriptor in mobile database. From the results, it is observed that the proposed method attained better results than state-of-the-art methods.

Table 2. Equal Error Rate of DR-LDP descriptor on MSU MFS Database

Method	Gray	RGB	HSI	YCbCr
LBP [5]	35.00	12.30	13.90	13.00
LPQ	23.90	23.20	12.20	7.40
BSIF [5]	24.20	23.50	12.10	7.50
SID [5]	24.40	22.30	13.50	8.50
CoALBP [5]	19.90	17.70	9.80	8.10
CoALBP+LPQ	19.90	12.50	11.40	4.90
DR-LDP	17.35	11.80	7.81	4.45

4.3 Cross Database Validation

To check the capability of the method, the method is evaluated by conducting cross-database validation. Here, the images in one database is considered as trained images and remaining databases images are considered as test images for evaluation. The performance of cross validation on different database results are shown in Table 3.

In this paper, every database is used as train dataset and test dataset to estimate the performance of the method by average half total error rate (HTER) to prove the ability of the proposed method with other methods. CASIA is considered as train set then NUAA and MSU-MFD is considered as test set. The average HTER is 27.12% for NUAA and 32.53% for MSU-MFSD. Similarly, NUAA is trained then the average HTER is 35.12% for CASIA and 46.67% for MSU-MFSD. When MSU-MFSD is trained, the average HTER is 44.32% for CASIA and 34.20% for NUAA. From the results CASIA database is generally good than other database, because the database consists more variations of images compared with other two database. The enrollment process of face may affect the performance of the cross validation scheme in face anti-spoofing methods.

Table 3. Performance of cross validation on different databases

Method	Train Set	Test Set	Average HTER
LBP	CASIA	NUAA	29.12
		MSU-MFSD	36.60
	NUAA	CASIA	43.26
		MSU-MFSD	54.24
	MSU-MFSD	CASIA	49.6
		NUAA	43.71
LDP	CASIA	NUAA	25.41
		MSU-MFSD	33.63
	NUAA	CASIA	38.56
		MSU-MFSD	51.27
	MSU-MFSD	CASIA	47.68
		NUAA	35.20
DR-LDP	CASIA	NUAA	27.12
		MSU-MFSD	32.53
	NUAA	CASIA	35.12
		MSU-MFSD	46.67
	MSU-MFSD	CASIA	44.32
		NUAA	34.20

5   Conclusion

This paper has used dimensionality reduced local directional pattern to extract the micro texture information of the face image on color spaces. The different color spaces was examined in DR-LDP for distinguishing the genuine user and fake user. The color texture analysis is used to represent the low level descriptors of the images. From each channel, the features are extracted and it is fused to form single feature vector for classification. Deep belief networks have classified the face timely and accurately. DBN is fast in classification stage, so the operation time of the system will be reduced than existing methods. The experiments are carried out on the standard benchmark databases namely CASIA face anti-spoofing database, NUAA database and MSU mobile face spoof database. This system has attained better performance rate compared with the state-of-the-art methods. It is robust to illumination conditions, facial expressions, aging conditions and pose variations.

References

1. Abdel-Zaher, A.M., Eldeib, A.M.: Breast cancer classification using deep belief networks. Expert Syst. Appl. **46**, 139–144 (2016)
2. Arashloo, S.R., Kittler, J., Christmas, W.: Face spoofing detection based on multiple descriptor fusion using multiscale dynamic binarized statistical image features. IEEE Trans. Inf. Forensics Secur. **10**(11), 2396–2407 (2015)
3. Benlamoudi, A., Samai, D., Ouafi, A., Bekhouche, S., Taleb-Ahmed, A., Hadid, A.: Face spoofing detection using multi-level local phase quantization (ML-LPQ) (2015)
4. Benlamoudi, A., Samai, D., Ouafi, A., Bekhouche, S.E., Taleb-Ahmed, A., Hadid, A.: Face spoofing detection using local binary patterns and fisher score. In: 2015 3rd International Conference on Control, Engineering & Information Technology (CEIT), pp. 1–5. IEEE (2015)
5. Boulkenafet, Z., Komulainen, J., Hadid, A.: Face spoofing detection using colour texture analysis. IEEE Trans. Inf. Forensics Secur. **11**(8), 1818–1830 (2016)
6. Chen, F.M., Wen, C., Xie, K., Wen, F.Q., Sheng, G.Q., Tang, X.G.: Face liveness detection: fusing colour texture feature and deep feature. IET Biometrics (2019)
7. Ciresan, D.C., Meier, U., Masci, J., Maria Gambardella, L., Schmidhuber, J.: Flexible, high performance convolutional neural networks for image classification. In: IJCAI Proceedings-International Joint Conference on Artificial Intelligence, vol. 22, p. 1237 (2011)
8. Gragnaniello, D., Poggi, G., Sansone, C., Verdoliva, L.: An investigation of local descriptors for biometric spoofing detection. IEEE Trans. Inf. Forensics Secur. **10**(4), 849–863 (2015)
9. Jabid, T., Kabir, M.H., Chae, O.: Local directional pattern (LDP) for face recognition. In: Proceedings of the IEEE International Conference on Consumer Electronics, pp. 329–330 (2010)
10. Keyvanrad, M.A., Homayounpour, M.M.: A brief survey on deep belief networks and introducing a new object oriented toolbox (deebnet v3. 0) (2014)
11. Kumar, K.: EVS-DK: event video skimming using deep keyframe. J. Visual Commun. Image Representation **58**, 345–352 (2019)
12. Kumar, K., Shrimankar, D.D.: F-des: fast and deep event summarization. IEEE Trans. Multimedia **20**(2), 323–334 (2017)
13. Li, J., Wang, Y., Tan, T., Jain, A.K.: Live face detection based on the analysis of Fourier spectra. In: Defense and Security, pp. 296–303. International Society for Optics and Photonics (2004)
14. Perumal, R., Chandra Mouli, P.: Dimensionality reduced local directional pattern (DR-LDP) for face recognition. Expert Syst. Appl. **63**, 66–73 (2016)
15. Ramalingam, S.P., Mouli, P.C.: Modified dimensionality reduced local directional pattern for facial analysis. J. Ambient Intell. Hum. Comput. **9**(3), 725–737 (2018)
16. Ramalingam, S.P., Sita, C.M.P.V.S., et al.: Dimensionality reduced local directional number pattern for face recognition. J. Ambient Intell. Hum. Comput. **9**(1), 95–103 (2018)
17. Schmidhuber, J.: Deep learning in neural networks: an overview. Neural Netw. **61**, 85–117 (2015)
18. Sharma, S., Kumar, K., Singh, N.: D-FES: deep facial expression recognition system. In: 2017 Conference on Information and Communication Technology (CICT), pp. 1–6. IEEE (2017)

19. Singh, H., Dhanak, N., Ansari, H., Kumar, K.: HDML: habit detection with machine learning. In: Proceedings of the 7th International Conference on Computer and Communication Technology, pp. 29–33. ACM (2017)
20. Tan, X., Li, Y., Liu, J., Jiang, L.: Face liveness detection from a single image with sparse low rank bilinear discriminative model. In: Daniilidis, K., Maragos, P., Paragios, N. (eds.) European Conference on Computer Vision, pp. 504–517. Springer, Heidelberg (2010). https://doi.org/10.1007/978-3-642-15567-3_37
21. Wen, D., Han, H., Jain, A.K.: Face spoof detection with image distortion analysis. IEEE Trans. Inf. Forensics Secur. **10**(4), 746–761 (2015)
22. Zhang, Z., Yan, J., Liu, S., Lei, Z., Yi, D., Li, S.Z.: A face antispoofing database with diverse attacks. In: 2012 5th IAPR International Conference on Biometrics (ICB), pp. 26–31. IEEE (2012)

Prediction of Liver Disease Using Grouping of Machine Learning Classifiers

Shreya Kumari[✉], Maheep Singh, and Krishan Kumar

Department of Computer Science and Engineering, National Institute of Technology,
Uttarakhand, Srinagar, India
{shreya.cse16,maheepsingh,kkberwal}@nituk.ac.in

Abstract. Machine Learning today in data analysis field robotize inter-
perative model framework. Classification algorithms in Machine learning
have grown to be among the leading research topics and its utilization
in therapeutic datasets are in discussion all over. Acknowledging the
fact that combining multiple predictions leads to more accurate results
than merely depending on a single prediction, a single dataset has been
trained on various algorithms and the highest voted class is predicted as
the result. Liver disease is the only major instigation of death still peren-
nial, hence early detection and treatment of not very symptomatic liver
disease is must which can significantly reduce the chances of death. In
the proposed work, the dataset of Indian Liver Patient has been utilized
and it clearly states that grouping classification algorithms efficiently
improves the rate of prediction of illnesses.

1 Introduction

The notion behind Machine Learning is to apprentice a machine through knowledge, experience and precedents so to make it proficient in making decisions as
good as humans or even surpass the human experts. With the growing modern-
ization, the rate at which the diseases are increasing is a cause of concern and
the most critical aspect of human lives, their health is at a great risk. Rigorous
cataloging or labeling of a disease in its early stage is a prerequisite for proper
treatment. In many cases, the methodology of carrying out several diagnostic
tests is complicated followed by skeptical results. In my work, I have particularly focused on the precise prediction of Liver Disease, the tenth most common
cause of death in India as per the World Health Organization. The dilemma is
that it is an organ that competent to function even when partially damaged.
Around 10 lakh patients diseased with Liver are newly diagnosed every year in
India. Liver disease may affect every one in 5 Indians. The lack of specialized
doctors is another agony. Hence, the need of mechanized and accurate systems
for classification of healthy versus non-healthy is crucial for human survival.
However, as the body type of Indians and western people is different, the proposed methodology might prove to be unsuitable to them. Many recent disease
diagnostic methods use clinical decision support system (CDSS), better than

M. Tripathi and S. Upadhyaya (Eds.): ICDLAIR 2019, LNNS 175, pp. 339–349, 2021.
https://doi.org/10.1007/978-3-030-67187-7_35

clinical diagnosis and therefore increases the disease prediction rate and human well being.

For the present study, the dataset is taken from UCI vault. It is often difficult to choose the best classifier. The study aims to evaluate seven well known Supervised Machine Learning classification algorithms many a times with varying feature selection techniques on each. The classifiers that have been addressed are Logistic Regression, Naive Bayes, KNN, Decision Trees, Random Forest Classifier, SVM and Artificial Neural Network (ANN). Selecting the best feature extraction technique for each algorithm is the most critical part. The highest predicted class by combining the results of all classifiers is the final outcome. In the light of the proposed method I am trying to forsee if the patient is having Liver disease or is healthy with highest precision.

The paper is organized as follows. Section 4 gives a brief introduction of the techniques used. The detailed explanation of the suggested methodology based on feature extraction and grouping of basic machine learning approach in Sect. 5. All the achieved results are presented and discussed in Sect. 7. Finally, the paper is concluded in Sect. 6.

2   Literature Survey

Han et al. [1] utilized patient recordings for prediction of four specified classes of CVDs using deep learning. The technique addressed for feature extraction is Autoencoder and Softmax has been used for classification. Trends with respects to various features in the dataset have also been observed. Dinesh Kumar G et al. [2] proposed various parameters such as sensitivity, specificity, etc. in order to evaluate the model performance. Mere accuracy is not a sufficient measure for model performance when the data is non-uniform. Classifiers used are SVM, Gradient Boosting, Random Forest, Naive Bayes and Logistic Regression.

Auxilia [3] used Pearson Correlation along with Decision Trees, Naive Bayes, Random Forest and ANN as machine learning classifiers with decision tree outperforming the rest of the algorithms. Sana Bharti and Shailendra Naraana Singh [5] applied and analyzed Genetic algorithm, Particle Swam optimization and Artificial Neural Network for predicting cardio diseases.

Dhomse et al. [6] applied Principal Component Analysis for Feature Selection. Finding the best features in the dataset is necessary because it helps prevent over-fitting, reduces training time and many a times provides better accuracy. Jagdeep Singh et al [8] presented associative classification techniques and various data mining algorithms for foretelling heart diseases. Jabbar [9] utilized Principal Component Analysis to determine the important features followed by Alternate Decision Trees in order to find heart diseases. Chieh-Chen Wu [10] used random forest, Naïve Bayes, artificial neural networks and logistic regression to predict Fatty Liver Disease. Mustafa Akin [12] utilized the magnetic resonance images to classify the most common benign lesions, cysts and hemangiomas in liver. They were further normalized and thresholded using histogram equalization. Anisha [13] verified the liver cancer by using image processing and data mining. The methods utilized are MRI, CT and USG scan imagery and K-means algorithm for clustering. In the next stage by using the Haar Wavelet the threshold values have been calculated and the accuracy comes out to be 82.

In [14] different algorithms in the IBM SPSS Modeler and Rapid Miner software have been compared. Eight algorithms were implemented on liver disease data and the results proved that the C5.0, had the best performance with 87.91 accuracy. Weng, Huang in [15] examined four datasets, namely Wisconsin Diagnostic Breast Cancer dataset, the Indian Liver Patient dataset, Vertebral Column dataset and Heart Disease dataset, using ANN algorithm. Their study revealed that the best accuracy in the Individual Classifier for the ILDP was 79.38.

3   Dataset Description

The term "Liver Disease" deals with Cirrhosis, hepatitis, Tumors or metabolic disorders. It has been found that a small amount of Bilirubin in human blood is normal but in excess is a sign of Liver disease. Aspartate Amino transfer ase is an enzyme our liver produces. Aspartate Aminotransferase test is a blood test that analysis for Liver contamination. Lower level of AST signifies a healthy Liver. Alkaline Phosphate is an enzyme raised in all body tissue, chiefly in bile ducts and Liver. Its high level indicates Liver disease. Conventional values are frequently higher for pregnant women or those with gallstone conditions. Total protein reveals number of diverse proteins in the blood. It is further branched into Albumin and Globulin fractions. Minor levels of total protein in the blood developes impaired utility of the liver. The A/G ratio normal range 0.8–2.0.

All the records were obtained from UCI Machine Learning repository for this study. It represents a sample of the unified Indian population collected from Andhra Pradesh region and has the data of 583 patients amongst which 416 are the liver patient records and 167 non liver patient records. Machine Learning achieves best result when we have plenitude of data and hence as the dataset for non liver patient is comparatively less, it has been oversampled and made equal to the liver patient record class. The dataset is named as The Indian Liver Patient Dataset (ILPD) and has 11 attributes. The last attribute is the target class, i.e. 1 represents diseased patients and 2 represents non diseased. It isolates the disease patterns. The Table 1 shows the list of all the attributes on which the methodology is working.

4   Various Techniques Used

Various feature Selection techniques available to find the most important attributes of the dataset so as to reduce the computation time to train and test the model. Also, various Supervised Machine Learning Classifiers are now applied in healthcare department for the fast diagnosis of the disease. Below mentioned are the Feature Extraction techniques and Classifiers that are grouped for testing.

S. No	ATTRIBUTES	ATTRIBUTE TYPE
1	Age	Numeric
2	Gender	Nominal
3	Total Bilirubin	Numeric
4	Direct Bilirubin	Numeric
5	Alkaline Phosphatase	Numeric
6	Alamine Aminotransferase	Numeric
7	Aspartate Aminotransferase	Numeric
8	Total Proteins	Numeric
9	Albumin	Numeric
10	Albumin and Globulin Ratio	Numeric
11	Result	Numeric1,2

4.1 Feature Selection Techniques

Univariate Selection. Each feature is reviewed distinctively and the vitality of each with the target is determined. The class used is SelectKBest and the statistical test is chi-squared(chi2) as the features have non negative value.

$$chi^2 = \frac{N - E}{E} \tag{1}$$

N: Number of Observations of Class
E: Number of Expected Observations of Class if there exist no relationship among feature and the target Higher the value of chi2, more strong is the feature.

Feature Importance. The Extra Trees Classifier of sklearn library with Gini Index is used for the determination of importance of features. Score for each attribute is returned. Higher the score,more important is the feature. *Recursive Feature*

Elimination. Iteratively each feature is removed to frame a model inculcating remaining features and accuracy of the model is calculated. The features in the model with the highest accuracy are considered true and others are interpreted as false.

4.2 Machine Learning Algorithms

Logistic Regression. Logistic Regression is one of the elementary Classification Technique. The target variables can be Binary(2 classes), Multinational (multiple classes) or Ordinal(Category with a score). It is a parametric model and can be normally characterized by a set of vectors. Let, the dataset has 'n' features and 'm' observations.

$$X = \begin{pmatrix} 1 & x11 & ... & x1n \\ 1 & x21 & ... & x2n \\ . & . & ... & .. \\ . & . & ... & .. \\ 1 & xm1 & ... & xmn \end{pmatrix}$$

The hypothesis function for a particular observation, here i is: $h(x_i) = g(B^T x_i) = \frac{1}{1+e^{(-BTx_i)}}$, $g(z) = \frac{1}{1+e^{(-z)}}$ g: logistic function. B: regression coefficients in a compact form. The output given by function g is compared with the actual output to calculate the error and accordingly the regression coefficients are changed to again compute the output. The procedure is repeated till the error is not small enough.

Naive Bayes. Naive Bayes Classifier uses the probability of each input attribute for the prediction with the probability that an event has occurred. This classifier identifies the most important cause or feature of the disease among the attributes of the dataset.

It applies Bayes' Theorem in the following manner:

$$P(y|X) = \frac{P(X|y)P(y)}{P(X)} \tag{2}$$

y: class label
X: an observation set of 'n' features
P(y): probability of event y
P(X): probability of event X
P(y—X): probability of event y such that event X has already occurred
KNN The prediction is made by searching through the k-most similar or nearest observations in the entire training dataset. The result is the collective contribution of the most similar samples.

The criterion for finding similarity depends on the dataset type. For categorical data Hamming data can be used and for other type Euclidean distance. This classifier uses competitive learning scheme, that is the new instance tries to triumph or at least be similar to most data observations.

Decision Trees. This Classifier is a tree like structure with the most important attribute closest to the root. The significance of attribute for splitting of the data is based on the following approach:

The dataset is categorical with two classes, let A(1) and B(2). Entropy is calculated for all the attributes before split and after the split. The attribute(X) to be split has n labeled values in the dataset.

E(X): $p(A)log(p(A)) - p(B)log(p(B))$
W(X): $(\sum_{i=1}^{n}(p(X_i) * E(X_i)))$
I(X): $E(X) - W(X)$
E(X): Entropy of X before split
W(X): Weighted Entropy of X after split
I(X): Information Gain of X

p(A): probability of class A

p(B): probability of class B

The attribute having the highest Information Gain is selected as the node for splitting. *Random Forest Classifier.* This particular Classifier algorithm shapes a forest with loads of decision trees. It guarantees more vigorousness because of the augmented number of decision trees. Every decision tree has its count in the output. Unlike Feature Selection it searches for the best features amongst a random subset of features taken at a time. *SVM.* This Classifier uses a separator and is mostly used for non-linearly separable data. It approaches to find an optimal hyperplane to separate all the classes of the dataset. The data is transferred to a high-dimensional space. The best separation is achieved by the plane that has the largest distance to the nearest training observation of any class. *Artificial Neural Network(ANN).* In the procedure to diagnose Liver Disease correctly, the designed Back Propagation ANN needs to have proper Learning Rate, Momentum Rate and hidden layer number. All these factors contribute to the optimum result as they tender the learning speed and power of the network used. In the designed network, Number of inputs initially used for training were 10 and then again ANN was trained with varying inputs due to the different feature selection technique used each time. Activation function used in Input Layer is Rectified Linear Unit Activation and for other layers Sigmoid Activation function. This Neural Network is implemented using keras package in python.

Table 1. Structure of designed Neural Network

Number of inputs	7, 8, 9
Number of hidden layers	2
Number of neurons in both hidden layer	500
Number of output	1
Epoch	100

5   Proposed Methodology

5.1   Data Visualization

This part helps to convey the complete information of the dataset distinctly and efficiently by using statistical graphics, plots, other tools. Numerical data may be concealed using lines, or bars, to visually decipher a quantitative message to the user. Following information can be drawn:

Visualizing Skewed Continuous Features. The skewed data represents one class more vastly than the other and hence leads to inefficient training. Blessedness towards majority label is avoided by normalizing skewed data (Figs. 1 and 2).

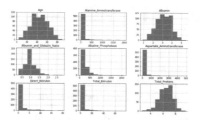

Fig. 1. Skewed data

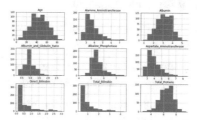

Fig. 2. Unskewed data

Attributes in Unhealthy People. An individual is more prone to Liver Disease at an age range of 40–65 years (Fig. 3 and Table 2).

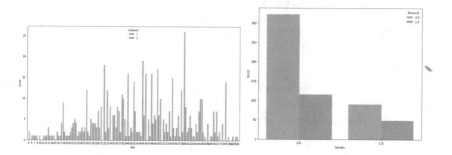

Fig. 3. a) Age based categorization b) Gender based categorization

Table 2. Percentage of attributes in unhealthy people

Attribute	Percentage
Alkaline Phosphotase	55.7
Aspartate aminotransferase	24.0
Total proteins	1.1
Alamine Aminotransferase	17.4
Total Bilirubin	0.7
Albumin	0.5
Direct Bilirubin	0.3
Albumin and Globulin ratio	0.2

Gender. Of all the total Males and Females dataset available, Males are more prone to Liver disease.

5.2   Data Preprocessing

In this phase following tasks have been performed:

Data Cleaning. It is done to detect and correct the inaccurate records in the dataset. 4 rows consisted of the Nan Values and hence they have been removed leaving the dataset to 579 rows and 10 columns.

Data Normalization. Values have been scaled to within 0–1 range with the min-max normalization approach.

Oversampling. Adjusting the distribution of classes for a balanced dataset is a prerequisite. The minority class have been upsampled.

Binary Encodings. The representation of Gender in the form of string in the dataset is changed to binary digit, Male is mapped to 0 and Female to 1.

Feature Selection. The Feature Selection techniques mentioned in Section II are enforced one at a time on the upsampled data and selected feature values are stored in variant sets. The cardinality of features discarded in a particular set varies from one to three.

Trained and Tested on Various Classifiers. Each feature set is then trained and tested singly on all the Classifiers described in Sect. 2. For each classifier the best feature selection technique is acclaimed on the ground of accuracy value and finally the predicted values of only that feature set is noted for all the classifiers.

Prediction of the Target Value. For each test case, the result is concluded on the basis of frequency of predicted target values by all the classifiers. The highest voted class is the corollary.

6   Results

The above methodology provided us the feature selection technique to be used with the various classifiers (Tables 3, 4, 5, 6, 7, 8, 9 and 10).

Table 3. Logistic regression classifier

Feature selection	Accuracy
Univariate	0.71
RFE	0.71
ExtraTree	0.72

Table 4. Naive Bayes classifier

Feature selection	Accuracy
Univariate	0.65
RFE	0.67
ExtraTree	0.68

Table 5. KNN classifier

Feature selection	Accuracy
Univariate	0.84
RFE	0.83
ExtraTree	0.83

Table 6. Decision tree classifier

Feature selection	Accuracy
Univariate	0.84
RFE	0.83
ExtraTree	.83

Table 7. Random forest classifier

Feature selection	Accuracy
Univariate	0.85
RFE	0.82
ExtraTree	0.83

Table 8. SVM classifier

Feature Selection	Accuracy
Univariate	0.70
RFE	0.69
ExtraTree	0.69

Table 9. ANN

Feature selection	Accuracy
Univariate	0.70
RFE	0.73
ExtraTree	0.74

Table 10. Selected classifiers

Classifier	Feature selection
Logistic Regression	ExtraTree
Naive Bayes	ExtraTree
KNN	Univariate
Decision Tree	Univariate
Random Forest	Univariate
SVM	Univariate
ANN	ExtraTree

The sample for testing is passed through all the above classifiers with the respective Feature selection techniques and result of each is noted. The same class predicted by more than 3 classifiers is the final outcome. This technique proves to be more accurate with the accuracy of 0.92 than predicting value through a single classifier. Though the computation time is more but results are more accurate.

7   Conclusion and Future Scope

This paper presents the correlative utilization of the functioning of the various classifiers. The work done propose an approach for correct prediction of Liver Disease at preparatory stage using various Classifiers with the adoption of Feature Selection techniques. Future work includes setting more diversity in the model. Similar approach can benefit diagnosis of other diseases too obligatory for improving ailing condition of patients.

References

1. Hsiao, H.C.W., Chen, S.H.F., Tsai, J.J.P.: Deep learning for risk analysis of specific cardiovascular diseases using environmental data and outpatient records. In: IEEE 16th International Conference on Bioinformatics and Bioengineering (2016)
2. Dinesh, K.G., Arumugaraj, K., Santhosh, K.D., Mareeswari, V.: Prediction of cardiovascular disease using machine learning algorithms. In: International Conference on Current Trends Towards Converging Technologies (2018)
3. Auxilia, L.A.: Accuracy prediction using machine learning techniques for indian patient liver disease. In: 2nd International Conference on Trends in Electronics and Informatics (2018)
4. Kononenko, I.: Machine learning for medical diagnosis: history, state of the art and perspective. Artif. Intell. Med. **23**, 89–109 (2001)
5. Bharti, S., Singh, S.N.: Analytical study of heart disease comparing with different algorithms. In: International Conference on Computing, Communication and Automation (2015)

6. Kanchan, B.D., Kishore, M.M.: Study of Machine learning algorithms for special disease prediction using principal of component analysis. In: International Conference on Global Trends in Signal Processing, Information Computing and Communication (2016)
7. Kukar, K., Grošelj, K., Fettich, J.: Analysing and improving the diagnosis of ischaemic heart disease with machine learning. Artif. Intell. Med. (1999)
8. Singh, J., Kamra, A., Singh, H.: Prediction of heart diseases using associative classification. In: 5th International Conference on Wireless Networks and Embedded Systems (2016)
9. Jabbar, M.A., Deekshatulu, B.L., Chandra, P.: Alternating decision trees for early diagnosis of heart diseases. In: International Conference on Circuits, Communication, Control and Computing (2014)
10. Wu, Y., Hsu, I., Nguyen, A.(A.), Poly, T.N., et al.: Prediction of fatty liver disease using machine learning algorithms. Comput. Methods Programs Biomed. (2018)
11. Drotar, P., Gazda, J.: Two-step feature selection methods for selection of very few features. In: 3rd International Conference on Soft Computing and Machine Intelligence (2016)
12. Akın, M., Ceylan, M.: Comparison of artificial neural network and extreme learning machine in benign liver lesions classification. In: Medical Technologies National Conference (TIPTEKNO) (2015)
13. Anisha, P.R., Reddy, C., Prasad, L.V.: A pragmatic approach for detecting liver cancer using image processing and data mining techniques. In: Signal Processing and Communication Engineering Systems (SPACES). IEEE (2015)
14. Abdar, M.: A survey and compare the performance of IBM SPSS modeler and rapid miner software for predicting liver disease by using various data mining algorithms. Cumhuriyet Sci. J. **36**, 3230-3241 (2015)
15. Weng, C.H., Huang, T.C.K., Han, R.P.: Disease prediction with different types of neural network classifiers. Telematics Inform. **33**, 277–292 (2016)

Travelling Salesman Problem Optimization Using Hybrid Genetic Algorithm

Rahul Jain[1(✉)], M. L. Meena[2], and Kushal Pal Singh[3]

[1] Department of Mechanical Engineering, University Departments, Rajasthan Technical University, Kota, Rajasthan, India
rjain.npiu.me@rtu.ac.in, rjmahesh207@gmail.com
[2] Department of Mechanical Engineering, Malaviya National Institute of Technology Jaipur, Jaipur, Rajasthan, India
mlmeena.mech@mnit.ac.in
[3] Department of Mechanical Engineering, G L Bajaj Institute of Technology and Management, Greater Noida, Uttar Pradesh, India
kushal.singh@glbitm.ac.in

Abstract. In present, travelling salesman problem (TSP) has given more priority due to its analogy real applications like transportation of goods, distribution of the supplies, power cable distribution, tourism, routing of public transport, laser printing, drilling operation. Genetic algorithm (GA) class is used to find the optimal solution but as problem increases, robustness decreases due to very slow convergence rate. In this study, a hybrid GA is proposed in order to optimize the TSP efficiently. The robustness was increased at higher distance by the developed hybrid GA.

Keywords: Travelling salesman problem · Genetic algorithm · Active genetic algorithm · Hybrid GA

1 Introduction

TSP is a mathematical problem which is treated by the Irish mathematician Sir William rowan hamilton in a Hamilton's icosian game where players complete the tour by passing through the 20 points having specified connections in 18th century. Though, the general form of the TSP has been studied by mathematicians in the 1930s by Karl menger. TSP defines the problem such that the distance between the 'n' cities should be minimized and they are connected with the Hamilton path. Though TSP [1] is for salesman, many applications are related like distributor related problems, Collector related problems, airlines crew problem, drilling operation on circuit board, laser printing applications, tourism, routing public transport. TSP concept is applicable for movement of people, machine and material.

2 Problem Formulation

TSP [2–7] defines the connectivity among the 'n' cities and objective is to minimize the distance of the tour such that all cities visited by person only once. TSP is considers

© The Author(s), under exclusive license to Springer Nature Switzerland AG 2021
M. Tripathi and S. Upadhyaya (Eds.): ICDLAIR 2019, LNNS 175, pp. 350–356, 2021.
https://doi.org/10.1007/978-3-030-67187-7_36

as Non Polynomial Hard problem, as complexity increases, work space of the problem increases exponentially.

Let Xij be the feasible path from city 'i' and city 'j' such that if exist then Xij $= 1$ otherwise Xij $= 0$, Dij be the distance between the city 'i' and city 'j', Ci and Cj are variables of the node 'i' and 'j' and 'n' be the number of cities in the problem.

- Objective function:

$$\mathbf{Minimize} = \sum_{i=1}^{n} \sum_{j=1}^{n} D_{ij} X_{ij}$$

- Constraints:

First, 'i' city arrived at from exactly one other city.

$$\sum_{i=1, i \neq j}^{n} X_{ij} = 1 \quad \text{where, } j = 1, 2, \ldots, n.$$

Second, from each city there is a departure to exactly one 'j' city.

$$\sum_{j=1, j \neq i}^{n} X_{ji} = 1 \quad \text{where, } i = 1, 2, \ldots, n$$

Third, Sub-tour elimination equality.

$$\mathbf{C_i - C_j + nX_{ij} \leq n-1} \quad \text{where, } i = 1, 2, \ldots, (n-1) j = 2, 3, \ldots, n$$

MATLAB: It is an advanced mathematical software which acts as the platform for the execution of the program codes [8] of GA, MGA and proposed Hybrid GA in order to optimize the given TSP [9].

3   Problem Statement

TSP consists of 28 capitals of states of India (including Delhi), find the route such that person travels all capitals with minimum distance as shown in Fig. 1.

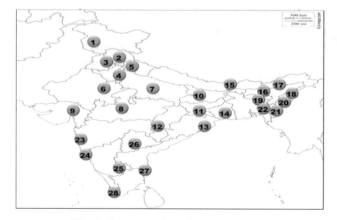

Fig. 1. Representation of all cities of TSP

4 Genetic Algorithm

GA was first proposed by J. H. Holland which is based on the evolution theory given by Sir Darwin. Algorithm was designed on the principle – "Survival for the fittest". It is known for handling large work space where finding the suitable solution is difficult [10–15]. It is designed search algorithm in the well-organized manner such that certain steps are followed in order to optimize the solution which includes five parts such as encoding, evaluation, cross-over, mutation and decoding. Above steps are followed in order to reach a definite solution for the given problem.

4.1 Encoding

It is the initial phase of the GA where language of the solution is selected to represents the output of the given problem. Encoding can be done in alphabetic form, Binary form, Numerical form, Symbolic form or other forms in which getting a solution and understanding it becomes easy. For the given problem, numerical form is selected. Initial random population is drawn as per the Encoding form of the solutions.

4.2 Evaluation

Termination criterion and selection of the chromosomes (solutions) are determined in this second phase. Selection [7] is based on the fitness function. Fitness function is defined with respect to objective function i.e., whether it is for maximum or minimum case.

For Maximum, $\mathbf{F(x) = f(x)}$.
For Minimum, $\mathbf{F(x) = k/(k + f(x))}$.
Where, $F(x) =$ Fitness function of the GA.
$f(x) =$ Objective function of the GA.
$k =$ Constant value.

4.3 Cross-Over

Selected chromosomes are shifted in the mating pool for interaction in order to produce offspring. It is the first genetic operator of the algorithm which acts as motivator for the progress of the solution. Cross-over [16] governs the feature of exploitation of the work space. Different techniques are used for mating process like single point cross-over, double cross-over, uniform cross-over.

4.4 Mutation

After cross-over phase, it is the second motivator which is used to regain the loss information in the previous phase as well as to jump from the local optimum solution. It governs the feature of the exploration of the work space.

4.5 Decoding

It is the last phase of GA, new solutions which are generated from genetic operators, are decoded and evaluated to check the required conditions for the optimality of the solutions. If the termination criteria are matched, it leads to the termination of the GA, otherwise new iteration (Generation) will continue from the evaluation phase to the last phase i.e., decoding till the termination criteria is matched. Figure 2 shows the flow chart of the GA of all five phases in the organized manner.

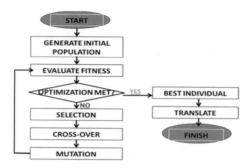

Fig. 2. Flowchart of GA

5   ACTIVE Genetic Algorithm (AGA)

It is the proposed Hybrid GA for the TSP. Algorithm is based on GA and Modularization technique with the concept of active and in-active region in order to boost the optimality to great extent. It consists of seven phases which includes (in order wise) – encoding, sub division, evaluation, cross-over, mutation, sorting & grouping and decoding. Function of the five phases of GA is same but other two phases i.e., sub division and sorting & grouping, makes difference from GA. Both the phases are discussed below:

5.1 Sub Division

It is the second phase related to the AGA, function of this phase is to divide the TSP into three sub-TSPs and further divide into active and in-active zone. Active zone means the only front part of the Sub-TSP interact with other front part of sub-TSP to form the solution for the given problem. In-active zone means the rest part of sub-TSP which does not take part in the interaction with other sub-TSP but it interacts with the active zone of its sub-TSP respectively to complete the solution for the sub-TSP. War is an example to understand the sub-division. In war, Front line interacts with the front line of other and rest is to support the front line which is similar to active and inactive zone in which only active zone will interact with other active zone.

5.2   Sorting and Grouping

It is the second last phase of AGA which takes population from mutation phase. It split each chromosome on the basis of the sub division phase and group them as per the active and inactive criteria to form the new knowledgeable chromosome for the given problem then it passes all new chromosomes to the decoding phase for re-evaluation. Grouping phase is used to re-group the sub-tours in one solution which is based on hybrid criteria i.e., learning. It helps the new chromosomes to improve the quality in order to counter the problem. Figure 3 shows the flowchart of AGA and Fig. 4 show the graphical representation of the AGA.

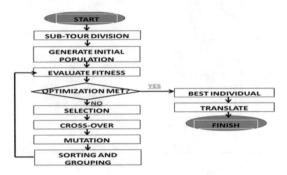

Fig. 3. Flowchart of AGA

6   Implementation of AGA

As per the GA, first Encoding phase introduces the chromosomes in the required form so that the solution is easier to understand. Then random population is drawn for the

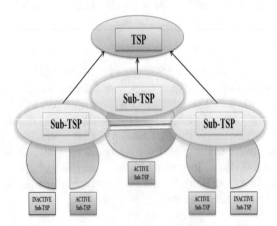

Fig. 4. Graphical overview of AGA

Evaluation phase for the GA but for AGA, first sub Division phase is introduced in order to distinguish for sub-TSPs and further for active and inactive zones in the respective sub-TSPs (Fig. 4).

7 Results

The program codes for the GA and AGA are designed on the MATLAB 2014 on i5 Intel processor. The results are discussed in the scatter plot of the respective algorithm for 200 iterations. Graphs show the minimum solution of each generation till it reaches the end criteria. Figure 5 shows the scatter plot for GA and Fig. 6 shows the scatter plot for AGA. These figures shows that at higher amount of iterations AGA will help to achieve more accurate results as compared to GA with greater robustness.

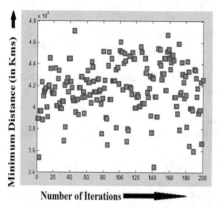

Fig. 5. Scatter plot of GA

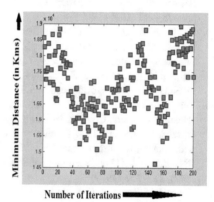

Fig. 6. Scatter plot of AGA

Table 1 shows the summary of the results.

Table 1. Summary of results

	Distance (Kms)	Difference (Kms)	% error
Optimum solution of TSP	14436	-	-
Solution by GA	34463	20027	138.73
Solution by hybrid GA (AGA)	14612	176	1.22

8 Concluding Remarks and Future Work

As per the Table 1, AGA is superior to the rest of the algorithms and overall performance is much better than GA. The error percentage of the AGA with respect to optimum solution is nearly 2%, this show the convergence of AGA. The concept of active and inactive zones of sub-TSP makes it possible for AGA to confine range for the better solutions as shown in the plots.

Since the solution has not reached the optimum value by AGA this means we have to improve the AGA by making changes in different parameters so that it converges to the optimum solution. Implementation of AGA in the real application is the future work which shows the strength as well as the weakness.

References

1. Filip, E., Otakar, M.: The travelling salesman problem and its application in logistic practice. WSEAS Trans. Bus. Econ. **8**(4), 163–173 (2011)
2. Abdulkarim, H.A., Alshammari, I.F.: Comparison of algorithms for solving traveling salesman problem. Int. J. Eng. Adv. Technol. **4**(6), 76–79 (2015)
3. Nilsson, C., Tewfik: Heuristics for the Traveling Salesman Problem. Linkoping University, pp. 473–480 (1996)
4. Hu, B., Raidl, G.R.: Solving the railway traveling salesman problem via a transformation into the classical traveling salesman problem. In: Hybrid Intelligent Systems, HIS 2008. Eighth International Conference, pp. 73–77. IEEE (2008)
5. Liu, S.B., Ng, K.M., Ong, H.L.: A new heuristic algorithm for the classical symmetric traveling salesman problem. World Acad. Sci. Eng. Technol. **1**(1), 34–38 (2007)
6. Wang, Y.: The hybrid genetic algorithm with two local optimization strategies for traveling salesman problem. Comput. Ind. Eng. **70**, 124–133 (2014)
7. Ghaziri, H., Osman, I.H.: A neural network algorithm for the traveling salesman problem with backhauls. Comput. Ind. Eng. **44**(2), 267–281 (2003)
8. Cao, Y.J., Wu, Q.H.: Teaching genetic algorithm using MATLAB. Int. J. Electr. Eng. Educ. **36**(2), 139–153 (1999)
9. Renaud, J., Boctor, F.F., Laporte, G.: A fast composite heuristic for the symmetric traveling salesman problem. INFORMS J. Comput. **8**(2), 134–143 (1996)
10. Bryant, K., Benjamin, A.: Genetic algorithms and the traveling salesman problem, pp. 10–12. Department of Mathematics, Harvey Mudd College (2000)
11. Holland, J.: Adaptation in Natural and Artificial Systems. University of Michigan Press, Ann Arbor (1975)
12. Király, A., Abonyi, J.: A novel approach to solve multiple traveling salesmen problem by genetic algorithm. In: Computational Intelligence in Engineering, pp. 141–151. Springer, Heidelberg (2010)
13. Man, K.F., Tang, K.S., Kwong, S.: Genetic algorithms: concepts and applications. IEEE Trans. Industr. Electron. **43**(5), 519–534 (1996)
14. Pullan, W.: Adapting the genetic algorithm to the travelling salesman problem. In: The 2003 Congress on Evolutionary Computation. CEC 2003, vol. 2, pp. 1029–1035. IEEE (2003)
15. Zhu, J., Zhou, D., Li, F., Fu, T.: Improved real coded genetic algorithm and its simulation. J. Softw. **9**(2), 389–397 (2014)
16. Abdoun, O., Abouchabaka, J.: A comparative study of adaptive crossover operators for genetic algorithms to resolve the traveling salesman problem. Int. J. Comput. Appl. **31**(11), 49–57 (2012)

VBNC: Voting Based Noise Classification Framework Using Deep CNN

Sandeep Chand Kumain[1,2(✉)] and Kamal Kumar[2]

[1] Department of Computer Science and Engineering, Tula's Institute, Dehradun, India
skumain@tulas.edu.in
[2] Department of Computer Science and Engineering, National Institute of Technology, Uttarakhand, Srinagar, India
kamalkumar@nituk.ac.in

Abstract. In this revolutionary digital world, the role of images and videos for transmitting the information is much more than the text. Now, people prefer short, concise and informative data over the large volume of texts like the information in the form of images is preferable. Noise availability in an image is a major challenge in digital image processing and this presence of noise misleads the results during various computer vision operations. The image denoise step is must require in this case. Multiple noise reduction algorithms are developed and used till now, but mostly the work is done with the known information of noise. In this research work, the author(s) focused on noise type identification in an image that will helpful for the images having unknown information about the type of noise is present. Further, it will help to select the appropriate algorithm for noise reduction purposes. For this purpose, the author(s) proposed a voting based noise classification framework that utilized deep learning technology.

Keywords: Classification · Noise · Digital image processing · Deep learning · Convolutional Neural Network

1 Introduction

In this modern progressive world, images and videos are the primary sources of transferring the information and most preferable in the present days. Images undergo several steps such as Image capturing, storage, compression and transmission before it displayed to the users and during these steps there is a chance of noise incorporation in an image [1]. The presence of noise misleads the results in several image processing and computer vision operations. Gaussian noise, Salt and pepper (impulse), Speckle noise, Poison noise are the most common noises which degrade the quality of an image [2,3].

For better understanding of an image, the de-noising is very important step. To recover the original image from noisy images, one should not only reduce the noise but also keeping the original details present in an image. Recently researches has proposed several de-noising algorithms of which majority one rely upon the noise information available before hand. Lack of such information renders selection of appropriate de-noising algorithm an uphill task [3,4].

M. Tripathi and S. Upadhyaya (Eds.): ICDLAIR 2019, LNNS 175, pp. 357–363, 2021.
https://doi.org/10.1007/978-3-030-67187-7_37

In this study, the author(s) proposed a Voting Based Noise Classification (VBNC) model, which is based on deep learning concept and predict the type of noise present in an image. For, the experimental results by mixing artificial noise at a different level the noisy data-set is prepared and the same is discussed in the related work section. The further section of the paper is as Sect. 2 is about Related work. Section 3 is about the proposed model. Section 4 is about the experimental results and discussions and Sect. 5 concludes the work and states the future scope of the work.

2   Related Work

Various noise reduction algorithms are developed till now. The effect of noise is seen in Synthetic Aperture Radar (SAR) images, Radar images and Medical images. The various filters developed for noise reduction purposes follow the linear or non-linear methodology. During the process of noise reduction, important details of an image must need to be preserved. For identification of the amount of Gaussian noise present in an image a CNN based noise level detection model is proposed by chauh et al. [5]. By utilizing the stochastic gradient descent optimization technique a noise classifier which is based on CNN is proposed by Hui Ying Khaw et al. [4]. In Convolutional Neural Network (CNN) a digital image is produced as input and based on the extraction of distinctive features by the sequence of convolutional and sampling layers the classification task is done [6, 13]. In several applications like handwritten recognition or character classification [8], face classification [9], series identification in banknotes [10], the CNN classification techniques are widely used and provide excellent results. By utilizing the stochastic online gradient learning method Jain and Seung [11] minimize the reconstruction error of the noisy images, this proposed method of Seung et al., outperform the Markov random field method of denoising natural images with less computational difficulties. Through this research work the LeNet and AlexNet model functionality [12] is utilized in the final decision process with the proposed CNN model.

Further Subsect. 2.1 describes the prepared data-set, Subsect. 2.2 discusses the research challenge.

2.1   Data-Set Description

For the model training and testing purpose first the noisy data set is prepared by mixing the artificial noise as per Table 1 specification. This prepared data-set is utilized in our experimental result analysis.

Table 1. Description of noisy data-set

Class	Type	Level	Image	Description
0	Gaussian & Impulse	10	8000	Both the noises are mixed in same ratio for these type of images
1	Gaussian	10	8000	0 mean and variance level 1,2,3,4,5,10,20,30,40,50 %
2	Non-Noisy	1	8000	Randomly select 8000 non-noisy images
3	Salt & Pepper	10	8000	Noise density 1,2,3,4,5,10,20,30,40,50 %
4	Speckle	10	8000	Speckle noise with variance level 1,2,3,4,5,10,20,30,40,50 %

2.2 Research Challenge

As the noise reduction step is necessary for image processing and computer vision application, the unknown information about which type of noise present in an image makes it more challenging to address. The classification method is suitable for providing a solution to this type of problem. This classification result is based on the prediction and based on that predicted value the suitable algorithm can be applied for noise reduction purposes. Through this research, the voting-based system for taking the final decision about which type of noise present in an image is proposed. For this purpose along with proposed architecture, the well known LeNet and AlexNet architecture is also trained with the developed data-set and the final decision of the presence of noise is decided as per the steps discussed in the next section.

3 Proposed Model

For the identification of noise in an image, the classification strategy is very effective and helpful for the appropriate denoised algorithm selection based on noise type identification. Instead of applying one model for classification of noise the author(s) proposed a voting based framework. For this, the proposed CNN architecture, LeNet architecture and AlexNet architecture model is trained with the data-set and utilized during the decision making step. The Proposed Model is shown in Fig. 1 which worked on a certain set of rules for final prediction. The input noisy image is tested with the LeNet trained

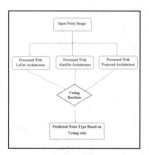

Fig. 1. Proposed model flow diagram

model, AlexNet trained model and proposed architecture trained model and the final decision is made as per the following rule.

1. For classification purposes, three models are used so the final decision for the type of noise availability in an image is based on the majority.
2. If there is a tie (means all of the models show different results for prediction), The type of noise is predicted based on author(s) proposed architecture because it has the highest accuracy prediction among the three.

Table 2. Model description

S. NO.	Operation	Kernal size	Stride	Parameters
1	Input Image (300, 300)	-	-	-
2	Convolution + ReLU	3*3 @ 8	1	80
3	Convolution + ReLU	3*3 @ 16	1	1168
4	Max - Pooling	2*2	2	0
5	Convolution + ReLU	3*3 @ 16	1	2320
6	Convolution + ReLU	3*3 @ 32	1	4640
7	Max - Pooling	2*2	2	0
8	Convolution + ReLU	3*3 @ 32	1	9248
9	Convolution + ReLU	3*3 @ 64	1	18496
10	Max - Pooling	2*2	2	0
11	Convolution + ReLU	3*3 @ 64	1	36928
12	Convolution + ReLU	3*3 @ 128	1	73856
13	Max - Pooling	2*2	2	0
14	Convolution + ReLU	3*3 @ 128	1	147584
15	Convolution + ReLU	3*3 @ 256	1	295168
16	Max - Pooling	2*2	2	0
17	Dropout Layer (.20)			0
18	Dense Layer 1 (2048) + ReLU			52430848
19	Dense Layer 2 (1024) + ReLU			2098176
20	Dense Layer 3 (5) + Softmax			5126

Total Trainable Parameter: 55,123,637

3.1   Proposed CNN Architecture

The proposed model is designed to identify the type of noise present in an image. The proposed model classifies the noise namely in five categories like noise free, Gaussian, Salt and pepper, Speckle and combination of Gaussian and Impulse noise (commonly found in medical images). With this developed architecture shown in Table 2 the training process is done with the data set discussed in Sect. 2. During the development of this model architecture the Convolutional Layer, Sampling Layer functionality is utilized. For reducing the problem of overfitting the Dropout Layer functionality is utilized, along with this for providing the non-linear feature for the developed classifier the ReLU (Rectified Linear Unit) function is used. The other two models are trained with the same data-set by utilizing their default architecture. The result is discussed in the next section.

4   Experimental Results and Analysis

For the overall experiment, all the discussed models are trained with 28000 images (from each class 5600 images are taken), Validated with 6000 images and finally, all these models are tested with 6000 images. For evaluation and comparision purpose the following perfromance parameters are used.

```
Confusion Matrix
[[799  82  13 221  85]
 [ 81 794  70 116 139]
 [ 19  63 889  87 142]
 [317 135 167 409 172]
 [154 116 274 127 529]]
```

Classification Report

	precision	recall	f1-score	support
Class 0	0.58	0.67	0.62	1200
Class 1	0.67	0.66	0.66	1200
Class 2	0.63	0.74	0.68	1200
Class 3	0.43	0.34	0.38	1200
Class 4	0.50	0.44	0.47	1200
avg	0.56	0.57	0.56	6000

(a) LeNet model confusion matrix   (b) LeNet model classification report

Fig. 2. Illustration of LeNet model results

4.1   Performance Evaluation Parameter

In this research work for the overall evaluation, classification report and confusion matrix is used as a performance parameters. The key terms for this is as follows [13].

A) Precision: Out of total positive predicted example how many sample are correctly classified is basically used to calculate the precision value. Precision value is calculated as per the following equation

$$precision = \frac{TruePositive}{TruePositive + FalsePositive} \qquad (1)$$

B) Recall: For a perticular class how many samples are correctly classified out of total number of samples is indicate the recall value. Recall value is calculated as per the following equation

$$recall = \frac{TruePositive}{TruePositive + FalseNegative} \qquad (2)$$

C) F1-Score: It is basically a weighted average of Precision and Recall. In this case the worst value is represented by 0 and the best value is represented by 1.

D) Confusion Matrix: This is the method of representing the result in tabular form. It is useful for the analysis of the behaviour of trained model with the test data set. The confusion matrix which is pure diagonal represent 100% accuracy of that trainned model.

Confusion Matrix

```
[[1147    25     0    25     3]
 [  12  1144    16    13    15]
 [   0     3  1189     0     8]
 [  68     4    35  1091     2]
 [  11    29   213     6   941]]
```

(a) AlexNet model confusion matrix

Classification Report

	precision	recall	f1-score	support
Class 0	0.93	0.96	0.94	1200
Class 1	0.95	0.95	0.95	1200
Class 2	0.82	0.99	0.90	1200
Class 3	0.96	0.91	0.93	1200
Class 4	0.97	0.78	0.87	1200
avg	0.93	0.92	0.92	6000

(b) AlexNet model classification report

Confusion Matrix

```
[[1171     0     0    28     1]
 [   0  1198     0     0     2]
 [   0     1  1190     1     8]
 [   5     0     2  1193     0]
 [   1    16    63     0  1120]]
```

(c) proposed model confusion matrix

Classification Report

	precision	recall	f1-score	support
Class 0	0.99	0.98	0.99	1200
Class 1	0.99	1.00	0.99	1200
Class 2	0.95	0.99	0.97	1200
Class 3	0.98	0.99	0.99	1200
Class 4	0.99	0.93	0.96	1200
avg	0.98	0.98	0.98	6000

(d) Proposed model classification report

Fig. 3. Illustration of AlexNet and Proposed Model results

Table 3. Prediction with real life data

Image	Source	LeNet	AlexNet	Proposed	Prediction
	Mobile	Gaussian	Non-noisy	Non-noisy	**Non-noisy**
	Internet	Salt & Pepper	Speckle	Speckle	**Speckle**
	Internet	Salt & Pepper	Gaussian & Impulse	Gaussian & Impulse	**Gaussian & Impulse**
	Camera	Gaussian	Speckle	Noise-noisy	**Non-Noisy**

4.2   Results

The experimental results as per the performace parameters discussed above is shown as per Fig. 2 and Fig. 3. Classes indicated in confusion matrix represent the type of noise as per Table 1.

Based on the result obtained from the above discussed trainned model the final decison about the presence of noise in an image is predicted by the framed rules discussed earlier. The result with some real world images is shown in Table 3.

5   Conclusion and Future Work

The classification technique is very useful for the selection of an appropriate algorithm for image denoising. There might be a possibility of the wrong prediction because achieving a 100% trained model is a challenging task. To overcome the limitation a hybrid model is proposed in which instead of using the prediction of one model the three models are trained for the same purpose and a final decision is taken based on the voting. Future work will be focused on improving the performance of proposed architecture and improve the result for noise type prediction.

References

1. Sampat, M.P., Markey, M.K., Bovik, A.C.: Computer-aided detection and diagnosis in mammography. Handb. Image Video Process. **2**(1), 1195–1217 (2005)
2. Gonzalez, R.C., Woods, R.E.: Digital Image Processing, 3rd edn. Pearson Education (1992, 2008)
3. Beaton, R.J.: Quantitative models of image quality. In: Proceedings of the Human Factors Society Annual Meeting, vol. 27, no. 1, pp. 41–45. Sage CA, Los Angeles, October 1983
4. Khaw, H.Y., Soon, F.C., Chuah, J.H., Chow, C.O.: Image noise types recognition using convolutional neural network with principal components analysis. IET Image Process. **11**(12), 1238–1245 (2017)
5. Chuah, J.H., Khaw, H.Y., Soon, F.C., Chow, C.O.: Detection of Gaussian noise and its level using deep convolutional neural network. In: TENCON 2017-2017 IEEE Region 10 Conference, pp. 2447–2450. IEEE, November 2017
6. Fang, J., Zhou, Y., Yu, Y., Du, S.: Fine-grained vehicle model recognition using a coarse-to-fine convolutional neural network architecture. IEEE Trans. Intell. Transp. Syst. **18**(7), 1782–1792 (2016)
7. Huang, Y., Wu, R., Sun, Y., Wang, W., Ding, X.: Vehicle logo recognition system based on convolutional neural networks with a pretraining strategy. IEEE Trans. Intell. Transp. Syst. **16**(4), 1951–1960 (2015)
8. LeCun, Y., Boser, B.E., Denker, J.S., Henderson, D., Howard, R.E., Hubbard, W.E., Jackel, L.D.: Handwritten digit recognition with a back-propagation network. In: Advances in Neural Information Processing Systems, pp. 396–404 (1990)
9. Lawrence, S., Giles, C.L., Tsoi, A.C., Back, A.D.: Face recognition: a convolutional neural-network approach. IEEE Trans. Neural Netw. **8**(1), 98–113 (1997)
10. Feng, B.Y., Ren, M., Zhang, X.Y., Suen, C.Y.: Automatic recognition of serial numbers in bank notes. Pattern Recogn. **47**(8), 2621–2634 (2014)
11. Jain, V., Seung, S.: Natural image denoising with convolutional networks. In: Advances in Neural Information Processing Systems, pp. 769–776 (2009)
12. Deep learning architecture and popular deep learning models. https://medium.com/@sidereal/cnns-architectures-lenet-alexnet-vgg-googlenet-resnet-and-more-666091488df5. Accessed 18 Feb 2019
13. Performance parameter for classification problem. https://www.geeksforgeeks.org/confusion-matrix-machine-learning/. Accessed 18 Feb 2019

Author Index

M. Tripathi and S. Upadhyaya (Eds.): ICDLAIR 2019, LNNS 175, pp. 365–366, 2021.
https://doi.org/10.1007/978-3-030-67187-7

Printed in the United States
By Bookmasters